Springer Tracts in Natural Philosophy

Volume 10

Edited by C. Truesdell

Co-Editors: R. Aris · L. Collatz · G. Fichera · P. Germain

J. Keller · M. M. Schiffer · A. Seeger

Mario Bunge

Foundations of Physics

With 5 Figures

Springer-Verlag New York Inc. 1967

Dr. Mario Bunge
Professor of Philosophy
McGill University, Montreal

ISBN 978-3-642-49289-1   ISBN 978-3-642-49287-7 (eBook)
DOI 10.1007/978-3-642-49287-7

Title-No. 6738

Preface

This is not an introduction to physics but an analysis of its foundations. Indeed, the aims of this book are: (1) to analyze the form and content of some of the key ideas of physics; (2) to formulate several basic physical theories in an explicit and orderly (i.e., axiomatic) fashion; (3) to exhibit their presuppositions and discuss some of their philosophical implications; (4) to discuss some of the controversial issues, and (5) to debunk certain dusty philosophical tenets that obscure the understanding of physics and hinder its progress. To the extent to which these goals are attained, the volume can serve as a companion to studies in theoretical physics aiming at deepening the understanding of the logical structure and the physical meaning of our science.

In order to keep the book slender, whole fields of basic physical research had to be excluded — chiefly many-body physics, quantum field theories, and elementary particle theories. A large coverage was believed to be less important than a comparatively detailed analysis and reconstruction of three representative monuments: classical mechanics, general relativity, and quantum mechanics, as well as their usually unrecognized presuppositions. The reader is invited to join the project and supply some of the many missing chapters — or to rewrite the present ones entirely.

Foundations research is not popular: most philosophers shrink at its technicalities and most scientists feel the urge to go forward rather than retrace their steps, understand what they have gone through, analyze it critically and thereby be in a better position to plan what to do next. Only a few physicists have been allowed to work in the foundations of physics: GALILEI, KEPLER, LEIBNIZ, HUYGHENS, NEWTON, EULER, D'ALEMBERT, AMPÈRE, HELMHOLTZ, HERTZ, MACH, DUHEM, BOLTZMANN, OSTWALD, POINCARÉ, PLANCK, EINSTEIN, BOHR, BORN, SCHRÖDINGER, DE BROGLIE, HEISENBERG, and not many more. They could afford it because they had done "hard" work. The rest must pretend that there are no foundation problems. Yet no one can help asking and answering questions belonging to foundations research — only, this is usually done with some shame and therefore hurriedly and sloppily. Even those eminent scientists were amateurs in the field of foundations research if judged by the current standards prevailing in the discipline: when it came to dissect their own creatures they did not avail themselves of the necessary tools — mathematical logic, metamathematics, scientific semantics, methodology, and other disciplines which, after all, have attained maturity only in recent times.

Foundations research in physics compares unfavorably with its partner the foundations of mathematics. While practically no physicist knows what 'foundations research' means, mathematicians have learned to esteem it because it contributes to clarifying and organizing the body of acquired knowledge and, by showing its presuppositions and limitations, it suggests new research lines: in short it contributes to the maturation of their science. Why not learn from mathematicians, pushing foundations research ahead and advancing it far enough that it may play a creative role and contribute to the maturation, not just the swelling, of physical science? There is little excuse for failing to attempt it, as all physical theories teem with logical and semantical difficulties, and the great majority of them are in their infancy as regards logical organization and physical interpretation. The prime matter — supplied by the physicist — and the tools — wrought by the mathematician, the logician and the philosopher of science — are there. The problems are exciting and many of them brand new: why should they not attract a good many thoughtful physicists, whether young or experienced, and become an established branch of physics?

I am grateful to

Professors PETER G. BERGMANN, PETER HAVAS and WALTER NOLL, and Doctors BERNARD COLEMAN, JOHN L. MARTIN and E. J. POST, each of which criticized some fragments of the manuscript;

my wife Dr. MARTA C. BUNGE and my son Dr. CARLOS F. BUNGE, Professor MILLARD BEATTY and several former students, particularly Professor ANDRÉS J. KÁLNAY and Mr. WILLIAM G. SUTCLIFFE, all of whom made me aware of several difficulties;

the Physics Departments of the Universities of Buenos Aires, La Plata, Delaware and Freiburg, which gave me the opportunity to teach seminars on the subject, to students who had not surrendered their right to understand and criticize;

the far-sighted and liberal Alexander-von-Humboldt-Stiftung, for a research fellowship at the Albert-Ludwigs-Universität in Freiburg i. Br. during the last stage of this research;

Professors S. FLÜGGE and H. HÖNL for their cordial hospitality at the Freiburg University;

and above all Professor CLIFFORD TRUESDELL for urging me to undertake this project.

Physikalisches Institut der Universität
Freiburg i. Br., November, 1966   MARIO BUNGE

Table of Contents

Chapter 2
Protophysics

Chapter 3
Classical Mechanics

Chapter 4
Classical Field Theories

Chapter 5

Quantum Mechanics

Special Symbols

Names of subjects

CED	Classical electrodynamics
CEM	Classical electromagnetism = MAXWELL's theory
CM	Continuum mechanics = Classical mechanics
CP	Calculus of Probability
EG	Euclidean Geometry
e.m.	electromagnetic
FR	Foundations research
G	General "dynamics"
GO	Geometrical optics
GR	General relativity = EINSTEIN's gravitation theory
H	HAMILTON's "dynamics"
HJ	HAMILTON-JACOBI's "dynamics"
L	LAGRANGE's "dynamics"
LT	Local time theory
NP	Newton-Poisson theory of gravitation
n.r.	nonrelativistic
PC=	Predicate calculus with identity = ordinary logic
PEG	Physical Euclidean geometry
PM	Particle mechanics
QED	Quantum electrodynamics
QFT	Quantum field theory
q.m.	quantum mechanical
QM	Quantum mechanics
QMM	Quantum mechanics of measurement
S	General theory of systems
SM	Statistical mechanics
SR	Special relativity
SRK	Special relativistic kinematics
UT	Universal time theory

Logical symbols

¬	not
∧	and
∨	or
⇒	implies = if ... then ... = sufficient condition
⇔	equivalence = if and only if (iff) = necessary and sufficient condition
∃x	there is at least one x
∃!	there is exactly one
∀x	for all x

Metatheoretical symbols

w.f.f.	well formed formula
A	axiom, postulate, assumption, hypothesis, premise
t, Thm.	theorem, logical consequence

$\vdash$ entails
$Cn(A)$ set of consequences of the assumption(s) A
$\overset{\mathrm{df}}{=}$ equal by definition
$Df.$ definition
$\mathsf{T} = \langle A, \vdash \rangle$ theory based on the axiom set A

Semantical symbols

$\mathscr{D}$ designates, names
$\mathscr{R}$ refers to, represents
$\triangleq$ mirrors, models
$\mathscr{E}$ extension
$\mathscr{I}$ intension
$\mathscr{M}ean$ meaning
$\overset{\mathrm{rt}}{=}$ equal by referition

Mathematical symbols

$\cup$ logical sum (union)
$\cap$ logical product (intersection)
$\subset$ is included in
$\in$ is a member of
$|$ such that, given that
$\{x \mid Px\}$ the set of all x such that Px
$\langle x, y \rangle$ ordered pair of individuals x and y
$X \times Y$ Cartesian product of the sets X and $Y = \{\langle x, y \rangle \mid x \in X \wedge y \in Y\}$
$f: X \to Y$ f maps X into Y
$\emptyset$ the empty set
C the set of complex numbers
I the set of integers
N the set of natural numbers
R the set of real numbers
R^+ the set of nonnegative reals

Protophysical symbols

Σ set of physical systems σ
$\dot{\in}$ is a proper part of
$\dot{+}$ physical addition
$\dot{\times}$ physical product
$\dot{\to}$ gives rise to
$\sigma^N = \dot{\Sigma} \sigma_i$ system of N physical parts $\sigma_i \in \Sigma$
σ_0 the null individual of the kind Σ

Introduction

Foundations Research

1. Object: Fundamental Theories

Pure scientific research is either basic or based. Basic research is concerned with source theories, such as classical mechanics, and with basic empirical procedures, such as time measurement. The conceptual and empirical tools wrought by basic research are then worked out in based research, e. g., solid state physics, which in turn poses problems requiring further basic research, and in addition keeps applied research alive.

Every basic theory and every basic empirical procedure has in turn some foundation or other. The foundation of a theory is the set of its assumptions, both tacit and explicit. And the foundation of an empirical procedure recognized in science is the set of theories by means of which the procedure is designed, carried out, and interpreted. In every case, then, the foundations of a science are ultimately theoretical — yet neither self-sufficient nor eternal. The branch of scientific research concerned with setting up and scrutinizing such foundations is called *foundations research* — FR for short. In other words, the object of FR is the set of source theories — i. e. of theories that help build other theories and make experiments. It is a second degree activity in the sense that while most physicists burrow into nature, the foundational worker burrows into physics. For this reason FR is apt to be regarded as a second rate occupation by all those who assign ideas a second place.

2. Aims: Analysis and Synthesis

The goal of FR is twofold: to perform a critical analysis of the existing theoretical foundations, and to reconstruct them in a more explicit and cogent way — thereby solving Hilbert's *6th* problem (HILBERT, 1901, 1918).

On the critical side, the foundations worker may do the following:

(a) To examine the philosophical presuppositions of physics: to find out wich ones are involved in today's physics, to establish whether they facilitate research (e.g., "Nature is lawful") or tend to block it (e.g., "There is no point in burrowing into the elementary particles"). Ex.: to inquire whether measurement, as handled by quantum mechanics, involves the thesis that microobjects exist only in so far as it pleases the operator.

(b) To discuss the status of key concepts, formulas, and procedures of physics: to examine the logical forms of physical ideas, their possible reference, the ways theories are tested, and the theoretical presuppositions of empirical procedures. Ex.: to investigate whether the quasi-particles — e.g., phonons — are real entities, or rather processes.

(c) To shrink or even eliminate vagueness, inconsistency, and other blemishes: to clean up the foundations with the help of the best available tools. Ex.: to elucidate the meaning of 'reference frame' wherever it occurs.

As to the constructive tasks of FR, here go some of the most important:

(a) To bring order to the various fields of physics by axiomatizing their cores: to recognize their basic concepts and to relate these into basic statements (axioms), thereby building theories proper out of bodies of loosely or even incorrectly related formulas. Ex.: to propose a mathematically correct and physically adequate axiom system for statistical mechanics.

(b) To examine the various proposed axiomatic foundations: to check whether they specify the mathematical structure and the physical meaning of their basic (undefined) concepts, whether these are mutually independent, and so on. Ex.: to investigate whether the usual formulations of quantum mechanics are consistent.

(c) To discover the relations among the various physical theories: to find out what presupposes what and what entails what. Ex.: to investigate whether thermodynamics is a subtheory of statistical mechanics.

In short, the foundations worker can assume two roles: that of the gadfly which keeps the square physicist awake; and that of the ovenbird, which builds his house out of mud — only to see it demolished by a colleague. Clearly, the analytic work will be most accurate and efficient when performed within the closed context of an axiomatic theory: this is why our analyses of specific physical theories will appear as remarks on their axioms.

3. Why Foundations?

But does physics have any foundations at all? If it does, why should we want to inquire into them? And with what means?

The first question is rather embarrassing: do not we know that the bases of physics are shaken and even replaced from time to time, and do not we suspect that physics can be likened to a system of mutually dependent parts rather than to a multiple storey building? True: the foundations of physics are shaky rather than firm and there is no reason to suppose that FR can or should give them any rigidity for if it did it would turn our science into dogma. But the provisional nature of foundations does not render them nonexistent: insecure and therefore subject to change they are, but this provisional character of theirs does not make them any the less fundamental at any given moment. As to the organic character of physics, the fact that it must be pictured as a net of mutually interacting parts rather than as a layer system must be granted as well. But it is only when we focus on first principles that we begin to see how they intertwine. Thus if we are busy applying quantum mechanics to chemistry we may not realize that it has some fragments of classical physics built into it, which we may on the other hand recognize if we uncover the basic assumptions. In short, the foundations of physics are not to be likened to those of a building and perhaps the very word 'foundation' for our discipline is a misnomer. Yet the problems are there, many of them are deep and difficult, and we need not quarrel about a word. Let this be a preliminary answer to the first question; as we proceed the answer will emerge in a more complete and clear way.

As to the second question, namely whether it is worth-while to bother about foundations, this is a perfectly legitimate question as well. Do not so few people worry about such problems and are not so many of the worriers amateurs or even cranks? True but irrelevant: also the foundations of mathematics have been until recently the concern of a handful of mathematicians — such as GAUSS, RIEMANN, WEIERSTRASS, DEDEKIND, FREGE, CANTOR, WHITEHEAD, HILBERT, RUSSELL, TARSKI, and the BOURBAKI. Many more did engage in the same kind of research but failed to produce any results, let alone the revolutionary ideas those foundation workers invented. Yet today the situation has altered radically: the foundations of mathematics are an established, flourishing, and fertile branch of mathematics. It is up to the physicists to see that something similar happens with their science. Why should it be desirable? Firstly, because foundations research supplies a better understanding of what has been achieved; secondly, because a clear understanding of what we have got involves an understanding of its limitations and is thereby apt to suggest ways of improving it and, occasionally, wholly new avenues of approach; thirdly, because keeping FR alive helps pure research as a whole, which is particularly important at a time when practical interests tend to curtail and distort it. The desirability of FR does not render it compulsory. Most research is foundations-free, in the sense that it is covariant, or nearly so, under changes in foundations. But some tricky physical problems do lead to foundation questions: the question then is not whether these can be dodged but how they can best be tackled. Which leads us to the third question, concerning the means of FR.

1*

4. How FR?

To achieve some of the goals of FR we shall have to bear in mind the fundamental theories of physics: mechanics, electrodynamics, relativity, and quantum mechanics. The need for physics is self-explanatory: we shall be dealing with it all the time. The nonphysicist is too far from physical research to get a close view of it. Even the applied physicist is usually much too engulfed in his specialized research to notice that the theories he is using are in need of critical scrutiny and reconstruction. We shall then keep in mind the fundamental theories of physics — essentially those taught at any university worth its name. But physics, though necessary for our work, is insufficient: we shall also need some dose of philosophy if we are to recognize our problems at all and if we are to state them clearly.

The philosophy we need consists, first of all, in a critical attitude: one of critical alertness and taste for fundamental problems. In the second place we shall need some of the most elementary tools of analysis and reconstruction built by logic, metatheory (started by metamathematics), philosophical semantics (the theory of reference), and the general philosophy of science. Logic is now part of the mathematical equipment of most physics apprentices — except when they have the bad luck of being taught computation techniques rather than mathematical theories. Therefore we shall cast only a quick glance at it. We shall also expound some metatheoretical ideas and work out the semantics we need to clarify the meaning of physical theories. Finally, we shall discuss and apply a number of key ideas of the contemporary philosophy of science. These tools will be assembled and displayed in Ch. 1.

What about "the" theory of knowledge (epistemology) and "the" theory of the most basic and pervasive traits of reality (ontology)? We cannot include them in the same category as the previously mentioned disciplines because they are still undisciplined. To begin with they are not theories proper (see Ch. 1, 4) but sets of views. Moreover they are too often far removed from science: they deal mainly with ordinary knowledge, which is characterized by particular ideas and superficial inductive generalizations rather than by comprehensive and deep theories. This does not entail that we are going to take epistemological and ontological problems lightly: on the contrary, it will be our task to uncover some of the epistemological and ontological hypotheses presupposed by physical research and suggested by its outcomes. We shall face them all along the book and particularly in Ch. 2. But a formal training in epistemology and ontology is unnecessary and even apt to becloud our search because of the backward state of those fields.

In short, of the two batches of tools required by FR — physics and philosophy — physics will be by far the most important and the one which we shall take for granted. The rest can be picked up along the way.

5. Present State of FR

FR is, though underdeveloped, very old — nearly as old as physics itself. Indeed, ARCHIMEDES axiomatized his fluid statics and PTOLEMY discussed the relations of theory to reality (a semantical problem) and to experience (a methodological problem). GALILEI embarked on the same discussion and added problems and new ideas. NEWTON reintroduced the axiomatic method and discussed the meaning of his basic concepts. MAXWELL, BOLTZMANN, EINSTEIN, BOHR and some other creative physicists have discussed foundation problems as well as methodological and philosophical issues. But apart from a few isolated cases the positive achievements of FR have been meager. In particular, the axiomatization of physical theories and the metatheoretical study of such systems and of their mutual relations is still a task before us, almost as much as it was when HILBERT put it forth in 1900. In one respect, though, we are better off than HILBERT and his generation: we can make use of the logical, mathematical, metamathematical and semantical instruments wrought by HILBERT, TARSKI, and a hundred other FR workers — most of them our contemporaries. So much so, that we can hope to get an adequate axiomatization of quantum mechanics (wrongly assumed to have been given by von NEUMANN) and even of classical mechanics. Because, incidentally, there is no universally accepted organization of classical mechanics, so much so that millions of students are still taught MACH's unsuccessful demolition of it.

We have fared somewhat better as regards the critical tasks of FR: a number of physics and philosophy journals are bristling with critical analyses. Notwithstanding, the criticisms of foundations is, with a few exceptions, far from having attained a scientific level. Three main obstacles are blocking the progress of this aspect of FR. In the first place, more often than not the critic is an amateur: he is either a philosopher who has never done a piece of scientific research and is therefore unable to "see" the problems that will catch the eye of the physicist, or he is a physicist with no philosophical background, in which case he may entertain philosophical ideas so unbaked or anachronistic that the philosopher will either laugh them away or, worse, accept them as articles of faith on the strength of the propounder's scientific prestige.

A second obstacle is constituted by a batch of die-hard prejudices concerning scientific research, tenets that are often shared by both philosophers and scientists. Chief among these hindrances to the analysis and renewal of ideas is the very distrust of ideas, and the associated deification of pure experience — even among theoreticians. Some conspicuous manifestations of this diffidence are: (a) the belief that experience can and must by itself define every physical concept; (b) the doctrine that every theory is just an a posteriori razionalization of experience, and (c) the opinion that theories are not invented but rather inferred from data by induction. If this prevailing philosophy of physics were true there would be no need for FR: there would never remain place for alternative theories, much less for alternative programmes for theory building, and no place for criticism either: experience would be self-sufficient (but then a blind manipulation) and the FR investigator would be one more parasite, and not even a useful parasite like the theoretician.

This brings us to a third chief obstacle to the growth of FR, namely the dislike of criticism for seeming to be sterile and even unkind. It is sometimes argued that scientific research is a cooperative enterprise — true — and that new facts, which alone matter — false — are not discovered by criticism but by hard, constructive and cooperative work — a work consisting in attentively staring at things and busily manipulating them. This, the nice (and dull) view of research, is false. Firstly because the investigator himself must adopt a critical attitude if he wants to avoid both sheer speculation and dogma; secondly because scientific research in an advanced science like physics is not piling up data but conceiving ideas, working them out and checking them both conceptually and empirically — and conceiving and evolving a new idea involves criticizing some old ones, and checking the fresh idea is subjecting it to critical scrutiny in the light of further ideas and of experience. Unless we get rid of the antitheoretical tenets we will be unable to even set ourselves the task of criticizing and improving on the foundations of our science. Which goes to show that the critical and the constructive tasks of FR are inseparable from one another. A certain ALBERT EINSTEIN put it better: "the physicist cannot simply surrender to the philosopher the critical contemplation of the theoretical foundations; for he himself knows best and feels more surely where the shoe pinches" (EINSTEIN, 1950).

6. Outlook

As suggested before, FR is in physics still embryonic: much of it is at a protoscientific level and most of it remains to be done. There has been little steady work in FR: most of it has been prompted by some deep crises. This is understandable: only under the pressure of failure are scientists willing to reexamine their principles. Yet crises can be eased by preventive work in FR: by periodically taking apart, cleansing, streamlining and reassembling the basic ideas. An FR pill a day keeps the undertaker away.

What can we expect from FR in the next few decades? This question calls for guessing rather than scientific prediction since, although we may pretend to have all the relevant information, we have not a single law concerning the development of physics. The author's guess is reasonably optimistic: he wishfully thinks FR will mature rather quickly and attain a scientific level throughout within the century. There are various grounds for this hope. Firstly, the need for a critical examination and reconstruction of foundations is being felt by an increasing number of scientists, as witnessed by the increasing bulk of papers on those subjects. More and more physicists are realizing that the present rather chaotic accumulation of undigested data, and of theories that are fragmentary and/or inconsistent, is unhealthy. Looking back to the glorious 1920's many feel that physics is now growing without maturing, and that only a cleansing of its present foundations can uncover its basic defects and perhaps suggest the deeper approach needed to cope with the proliferation of empirical information and of fragmentary theories, notably in the field of "elementary" "particles".

As every other intellectual crisis in the past, the current crisis in physics stimulates interest in FR. Just as GALILEI analyzed the foundations of Aristotelian dynamics, NEWTON those of Cartesian physics, and EINSTEIN those of Newtonian mechanics, so some thoughtful contemporaries are worried about the foundations of our present theories. Of course this is not the dominant attitude: even a Nobel prize winner can refuse to look down at the unclear foundations of "elementary" "particle" physics, arguing that, being far from a final synthesis, we cannot discover what the foundation problems are. But there are no final syntheses in science anyway and there is no need to wait until a nice quiet corpse is available: the discussion of the very kinds of approaches to the construction of new theories may be illuminating. FR is as useful during periods of growth as it is during stagnation stages: it can always help realizing mistakes and seeing new ways out.

A second ground for optimism is that the logical and metatheoretical tools for FR are there: they were invented chiefly for the purpose of cleansing the foundations of mathematics and they are in store waiting to be applied methodically to the analysis and reconstruction of fundamental physical theories. Never before has FR in physics been in such a favorable position.

A third ground is that a number of philosophical tenets that stood in the way of constructive work in FR are in the process of being discarded. Thus operationalism and inductivism are currently less prestigious among philosophers than among physicists, who in this respect lag behind the more enlightened philosophers. While many physicists are still trying to fit theories into antitheoretical philosophies like operationalism, these philosophies are *de jure* dead as a result of philosophical criticism. Indeed, every philosophical school is in a state of deep crisis and disintegration and, at the same time, a freer style of philosophizing is emerging from those ruins, one characterized by logical rigor and by a taste for some of the pangs of scientific research. Simultaneously, FR has become so technical that the nonspecialized philosopher leaves it alone. The only considerable enemy FR faces is its own fifth column — operationalism. This again constitutes an unprecedent favorable condition for the development of FR.

In short, the time seems to have come for a quick maturation of FR in physics: physicists are beginning to take it seriously; mathematicians, logicians and philosophers have wrought many of the necessary tools; and philosophers are either retiring from the field or becoming acquainted with it to the point of making useful contributions to FR. Because, incidentally, the research into the foundations of physics is as much a part of physics as mathematical FR is a part of mathematics. The fact that curricula designers are slow in recognizing this should not discourage it: after all, in the beginnings relativity and wave mechanics were often taken for philosophical speculation, and the dynamics of continuous media is nowadays regarded as mathematics applied to engineering. The important thing is to get some work done: status recognition will come sooner or later if the discipline itself acquires stature.

Chapter 1

Toolbox

Workers in the foundations of mathematics and physics are concerned with finding out the status of concepts, statements (heuristic clues, postulates, theorems, definitions, data) and systems of hypotheses (theories) as well as with clarifying and rearranging them in a logical order. The physicist faces in addition two sets of problems that do not come up in mathematics: the relation of physical ideas to reality — a semantical problem — and the procedures, only partly empirical, whereby the assumed idea-fact relationship is put to the test — a methodological problem. The performance of any such task requires handling certain tools that will presently be introduced in a painless way. Those wishing to get hold of further and more delicate tools of analysis are advised to study a good logic book (e.g., HERMES, 1963; QUINE, 1951; STOLL, 1963; SUPPES, 1957; or TARSKI, 1941), something on meta-mathematics and semantics (e.g., BETH, 1959; ROBINSON, 1963; TARSKI, 1956), and the author's methodology and philosophy of science treatise (BUNGE, 1967b).

1. Form and Content

Every physical idea is expressed in some language and has a logical structure and a content or meaning. Therefore the analysis of physical discourse is threefold: linguistic, logical, and semantical. A linguistic analysis is prior to any other one because we come across scientific ideas mostly through linguistic envelopes. But since external wrappings are thin and largely conventional, we shall take only a hurried glance at them: after all, although linguistic analysis dissolves some muddles it leads to no new ideas.

1.1. Language

A language is a system of signs — not just a set of signs but a fairly organized body of such. A *sign* is a physical thing, such as a mark on a piece of paper, and so is an *expression*, i.e. a finite string of signs. But the linguistic relations among signs is not physical, nor is it the idea a sign is supposed to stand for. A *physical language* is a system of signs, or symbolism, devised to convey ideas about physical objects, such as light waves, and physical procedures, such as intensity comparisons. (On this description, phrases like 'Δx represents the observer's uncertainty concerning the location of the particle' do not belong to a physical language.) Every branch of physics has its own language: its peculiar vocabulary and grammar. The language of physics is the union of all physical languages.

A *vocabulary* is a set of signs or terms — the linguistic units. And a *grammar* is a body of rules by which the corresponding signs are put together and transformed into one another. Take, for instance, the language of solid state physics: its vocabulary consists of ordinary language words (e.g., 'solid'), mathematical symbols, signs belonging to basic physicalese (e.g., 'electric conductivity'), and terms of its own (e.g., 'lattice'). In addition to this vocabulary, which is partly ordinary and partly technical, we find all the morphological rules of ordinary language plus those of the fragments of mathematics employed heretofore in that branch of physics. Every physical language obeys two masters: ordinary language and mathematics — the language of science.

The morphological rules of a language endow its vocabulary with a structure. These rules are of two kinds: formation rules and transformation rules. The *formation rules* of a language are the prescriptions that regulate the construction of complex signs out of simpler ones and, conversely, the analysis of expressions. Ex.: "The sign '$>$' shall occur between two signs designating objects of a kind." We might as well adopt another rule, exemplified by '$> xy$'. This suggests that formation rules are conventional (arbitrary). But conventions are not all equivalent. Thus '$> xy$' is more reasonable than '$x > y$' because only binary (two-place) relations can be written in the form 'Rxy': for higher degree relations we should adopt the uniform notation '$Rxyz...$'. As to the *transformation rules* of a language, they govern the conversion of formulas into one another. Ex. 1: "'$> xy$' is exchangeable with '$< yx$'" is a rule of transformation in the language of arithmetic. Likewise, "'x is hotter than y' is exchangeable with 'y is colder than x'", is a transformation rule in thermology. Ex. 2: "The inequality '$> xy$' is equivalent to the formula 'There is a positive z such that $x = y + z$'" is a transformation rule in the language of arithmetic but, unlike the rule of Ex. 1, it has no thermological analog. Every set of morphological rules is peculiar to a language.

By combining terms in accordance with the formation rules of a given language we can build self-contained expressions of various kinds. In scientific research one is particularly interested in the following species: (a) *questions*, e.g., 'Is Cd conducting?'; *proposals*, e.g., 'Let us assume that the Cd crystal has no dislocations'; (c) *rules*, e.g., 'In order to compute the electric conductivity of Cd take the electron-lattice interaction into account'; (d) *sentences*, e.g., 'The electric conductivity of Cd at normal temperature and pressure in (ohm-cm)$^{-1}$ is $5 \cdot 10^{-4}$'. We might add commands, threats, promises, pleas, whinings and other equally self-contained expressions for they all occur in the verbalizations of actual pieces of research — particularly of those under contract. But since they do not occur in the verbalizations of pieces of finished research — our main concern — and since their logic is unknown, we shall leave them aside.

Every one of the above illustrations has a definite meaning and, in addition, a definite value of some sort: the given question is a fruitful one, the proposal is reasonable within limits, the rule is grounded on theoretical knowledge, and the sentence expresses a nearly true statement or proposition. All those expressions would lose their meaning if an indefinite symbol, say 'x', were substituted for the definite symbol (constant) 'Cd'. Yet it is by replacing definite by indefinite terms that we get the generality peculiar to science. Thus as long as we heap sentences like '$1 > 0$' and '$1 + 1 = 2$' we do not get an arithmetic, just as collecting particular physical information does not yield a physical science. Science is general and involves abstraction of some kind or other. In the case of the concept "greater than" we get generality by replacing the individual constants $0, 1, 2$, etc. by variables. Now a formula like '$> xy$' is not a sentence, for it states nothing definite: it is a *sentential function* — and so are 'x is charged', 'Let f be a function', and 'What are the x such that $J_n(x) = 0$?' A sentential function is a self-contained expression in which at least one (free) variable sign occurs. And a *variable* symbol is in turn a sign that names an arbitrary member of a given class called the *range* of the variable. Thus, in 'x is conducting', x is a variable that can take up a number of (nonnumerical) values — e. g., Cu, Fe, and Co. Every such individual term is called a *constant*.

A glimpse at language has brought us to its border with logic: in fact the linguistic categories we have reviewed are as many linguistic wrappings of logical categories. To these we now turn.

1.2. Logic

1.2.1. Signs and Ideas. Every language has its own grammar, and the grammar of any ordinary or historical language is partly conventional, rather imprecise, and never quite exceptionless. If we want a body of exceptionless rules and laws determining a definite structure we must turn from signs to ideas and from natural languages to the artificial languages expressing organic bodies of ideas. If we gather all that is common to, or presupposed by, the various scientific disciplines, we come up with a general syntax of scientific ideas or basic grammar of science: formal (or symbolic, or mathematical) logic. Since the middle of the 19*th* century mathematical logic has grown into an exact science by virtue of having adopted the mathematical mode of thinking. Unlike the rules and laws of the ordinary languages, those of logic are not accepted by tradition or on aesthetic grounds but to the extent to which they facilitate rational argument and the accurate analysis thereof.

Mathematical logic does not apply exactly to natural language but rather to the bodies of ideas expressed in some artificial language. In fact (*a*) logic is concerned with concepts, statements, arguments and other ideal objects rather than with the more or less conventional and unruly signs by which such ideas are expressed; and (*b*) logic applies to clear cut ideas rather than to the comparatively vague ideas and muddled arguments of ordinary discourse. Take, for instance, the concept "long": logic will be unconcerned with the various linguistic expressions of this concept, such as *longus* or *largo*. Moreover it will demand that the length concept be somewhat refined before entering the ideal relations of logic: otherwise paradoxes and contradictions might arise. Thus in daily life we might accept a contradiction such as "Interplanetary distances are both long and short" if the context makes it clear that the term 'long' in the sentence is short for "long compared with atomic distances" while 'short' is tacitly understood to mean "short compared with galactic distances". And in talking we may inadvertently utter the paradox 'Long is short' while trying to say that the word 'long' is short. Just as contradictions due to vagueness can be removed by adding suitable qualifiers, so paradoxes due to careless talk can often be removed by purely linguistic devices — in our case by enclosing the troublesome word in simple quotes, namely: "'Long' is short." This makes it clear we are not using the length concept but are referring to one of its names. In general, to indicate signs we shall use simple quotes and to mention concepts double quotes. Ex.: '*p*' names (designates) "pressure".

1.2.2. Concepts. Ideas are either concepts or bodies of such. The concept is the unit handled by logic. We are not defining "concept" because we need it to characterize more complex conceptual systems. (The same holds for every other basic concept: it must be taken as undefined or primitive.) Instead, we shall characterize it by way of examples: "body", "is a part of" and "is between" are concepts and, in particular, predicates.

As regards their form, concepts may be classed into *individuals* and *predicates*. In the statement "2 is even" the individual is "2" — or simply 2, as in formal science there should be little danger of taking a concept for a thing. And the predicate is here "is even". In compact symbolization: "$E\,2$", not to be mistaken for "$E\,'2'$", which is non-sensical as '2' is a sign (a numeral) with no mathematical properties. The circumstance that 2 can be reconstructed or analyzed as the set of all pairs (FREGE), or as the set whose sole members are 0 and 1 (v. NEUMANN), is beside the point: what is an individual on a certain level of analysis may turn out to be a class on a different level, just as what is an elementary particle in low energy physics may turn out to be a complex system in high energy physics.

1.2.3. Statements. Ordinary logic restricts itself to formulas of a single kind: those expressing actual or potential statements. Problems, rules, proposals and other kinds of ideas are left aside. An actual statement or *proposition* states something definite — e.g., "$3^2 + 4^2 = 5^2$". A potential statement or *propositional function* contains blanks (variables) which render it indefinite in one or more respects — e.g., "$x^2 + y^2 = z^2$". Whereas propositions are expressed by sentences, propositional functions are designated by sentential functions (see 1.1).

A propositional function is a potential statement in the sense that it can be converted into one or more propositions. There are two main ways in which this transformation can be effected, and both consist in tampering with the variables. One is to *specify* or fill in the blanks, i.e. to assign definite value(s) to the variable(s) occurring in the function. Ex.: upon setting $x = 3$ in "x is even" we get a false proposition. A second way of generating propositions out of propositional functions is to *quantify*. Ex.: from "x is thinkable" we can form "For some x, x is thinkable" or "For every x, x is thinkable". The former is an analysis of the proposition "Some things are thinkable", the latter of "Anything is thinkable". The prefix "for some x" is called an *existential quantifier*, symbolized '$(\exists x)$'. The prefix "for any x", symbolized '$(\forall x)$', is called the *universal quantifier*. Other usual quantifiers are "$\exists!$" (there is exactly one), "$(\exists x)_n$" (there are n individuals x), "$(\forall x)_S$" (every x in the set S), and "$(\exists x)_S$" (some x in S).

Note that upon specification or assigning definite values we get rid of variables whereas when we prefix quantifiers to propositional functions we fix the range of the variables. In the former case variables disappear altogether for the benefit of particulars whereas in the latter case variables lose their freedom: they become bound. When no quantifier is prefixed to a propositional function, the latter is, for deductive purposes, as good as a universal proposition. Ex.: if the domain of the independent variable in an ordinary differential equation is not specified, the equation is taken to hold at all points.

Propositional functions should not be mistaken for descriptive or *designating functions* such as "the preceding observation", "the length of b", "the square of x", "the laplacian of f", or "the integral of $f(x)$". None of these is a statement or can become a statement. Not even binding x in the last example do we get a proposition: in fact by adding the integration limits all we get is a *definite description* instead of an indefinite one. Notice incidentally that unless f is specified it is a predicate variable not a predicate constant. Similarly, the f in the Newton-Euler equation of motion stands for a predicate variable; and since the law is supposed to hold for any force function, we infer (by virtue of the any-all equivalence) that it holds for all force functions.

1.2.4. Connectives. Let us carry our analysis a step further by introducing the notion of compound or molecular proposition. If p is a proposition, then its negate $\neg p$ is another proposition — this stipulation being a formation rule of the propositional calculus. The law of *contradiction* can be written: $\neg (p \wedge \neg p)$, where ' $\wedge$ ' stands for the conjunction "and". The preceding formula, a key theorem of the propositional calculus, can be read "It is not the case that p and not-p". And the *excluded middle* law can be written: $p \vee \neg p$, where ' $\vee$ ' symbolizes the disjunction "or". The operators $\neg$, $\wedge$ and $\vee$ are among the *connectives* enabling us to build compound propositions out of simpler ones. Thus if p and q are propositions, then $p \wedge q$ is another proposition called the *conjunction* of p and q, and $p \vee q$ is yet another statement called the *disjunction* of p and q. (The preceding sentence embodies two formation rules and two nominal definitions.) Ex. 1: "Au and Ag are metals" is the conjunction: "Au is a metal $\wedge$ Ag is a metal". Ex. 2: "$n = 1, 2, 3$" is short for the disjunction: "$n = 1 \vee n = 2 \vee n = 3$".

Further connectives are the *implication* $\Rightarrow$, read 'if... then —' and the *equivalence* $\Leftrightarrow$, read '...if and only if —' or 'iff' for short. Ex. 1: "If a statement is testable then it is meaningful", short for "For every x in the set S of statements, if x is testable then x is meaningful" — briefly $(\forall x)_S (T x \Rightarrow M x)$. Ex. 2: "A statement is physically meaningful if and only if it refers to a physical object", short for "For every x in the set S of statements, x is physically meaningful if and only if there is at least one physical object y such that x refers to y" — briefly $(\forall x)_S [M x \Leftrightarrow (\exists y)(P y \wedge \mathscr{R} x y)]$. Notice that the scope or reach of the universal quantifier is here the whole propositional function to its right.

In ordinary logic a proposition is either true (T) or false (F), and the truth value of any compound (molecular) proposition can be computed from the truth values of its components with the help of the following *truth tables*:

p	q	$p \wedge q$	$p \vee q$	$p \Rightarrow q$	$p \Leftrightarrow q$
T	T	T	T	T	T
T	F	F	T	F	F
F	T	F	T	T	F
F	F	F	F	T	T

By choosing one or more connectives as *primitive* or undefined concepts the remaining connectives can be built in terms of the former. Thus since "Not either p or q" amounts to "Neither p nor q", we write

$$\neg (p \vee q) \Leftrightarrow \neg p \wedge \neg q. \tag{1.1}$$

By replacing p and q by their contradictories or negates, and recalling that in ordinary logic (though not in alternative systems of logic) $\neg \neg p$ amounts to p, we can define conjunction in terms of disjunction and negation:

$$p \wedge q \overset{\text{df}}{=} \neg (\neg p \vee \neg q). \tag{1.2}$$

Likewise we may introduce an implicit definition of "or" in terms of "and" and "not":

$$p \vee q \overset{\text{df}}{=} \neg (\neg p \wedge \neg q). \tag{1.3}$$

In the same vein we may define the implication:

$$p \Rightarrow q \overset{\text{df}}{=} \neg p \vee q, \qquad p \Rightarrow q \overset{\text{df}}{=} \neg (p \wedge \neg q) \tag{1.4}$$

and the equivalence:

$$p \Leftrightarrow q \overset{\text{df}}{=} (p \Rightarrow q) \wedge (q \Rightarrow p). \tag{1.5}$$

The conditional "$p \Rightarrow q$" amounts to "p is sufficient for q" or, equivalently, "q is necessary for p". Consequently according to (1.5) the biconditional "$p \Leftrightarrow q$" amounts to "p is necessary and sufficient for q". Notice that the expressions standing on the two sides of a sign '$\overset{\text{df}}{=}$' can be exchanged, for $\overset{\text{df}}{=}$ is a kind of identity — *identity by definition*. On the other hand equivalences authorize no similar exchange or elimination: indeed even when $p \Leftrightarrow q$ holds, p and q may differ in meaning, whence both constituents may have to be kept. For example, "Slippery iff wet" does not warrant identifying "slippery" and "wet". Definitions can be rewritten as equivalences but not conversely.

One of the aims of logic is to establish (prove) universal formal equivalences such as, e.g., the *law of contraposition*

$$(p \Rightarrow q) \Leftrightarrow (\neg q \Rightarrow \neg p). \tag{1.6}$$

This logical truth can be derived, e.g., by rewriting either of the formulas (1.4) as equivalences and adding the law of the commutativity of the disjunction and the conjunction respectively, i.e.

$$(p \vee q) \Leftrightarrow (q \vee p), \qquad (p \wedge q) \Leftrightarrow (q \wedge p). \tag{1.7}$$

All this can be done rigorously by starting from first principles (axioms and inference rules). We shall not carry this any further because we shall avail ourselves of logic as a tool described in logic books. A simple example will suggest how to use this tool.

1.2.5. Analysis. Formal logic is the tool for analyzing statements and arguments: it is the theory of logical form and deduction. Take the proposition "Materials are deformable", which at first blush is logically simple; actually it contains a variable and a connective. In fact, the statement can be analyzed into "For every x, if x is a material then x is deformable". In obvious symbols,

$$(\forall x)(M x \Rightarrow D x) \tag{1.8}$$

The equivalences established in the propositional calculus and in the predicate calculus (calculus of functions) enable us to rewrite (1.8) in a number of ways. For example, using (1.4) the arrow can be eliminated in favor of the negation and the conjunction:

$$(\forall x) \neg (M x \wedge \neg D x). \tag{1.9}$$

Further, since "not every" amounts to "some not", it will be possible to replace "every" by "not some not", transforming (1.9) into

$$\neg (\exists x) \neg\neg (M x \wedge \neg D x) \tag{1.10}$$

which, by virtue of the double negation theorem, boils down to

$$\neg (\exists x)(M x \wedge \neg D x). \tag{1.11}$$

All these have been purely logical transformations. Recalling the meanings assigned to 'M' and 'D', we may return from abstraction and read the last formula as 'There are no materials that are not deformable', or just 'There are no undeformable materials'. But formal logic is concerned with form not with physical objects. In other words, the laws of logic are independent of matters of fact, and this simply because logic is not about facts but about ideas and, indeed, about one aspect of ideas, to wit their logical form or structure. Therefore, the belief that revolutions in experimental science may shatter logic is unfounded: logic is immune to fact — just as the physicists who mistake $\Rightarrow$ for the causal relation and talk about a logic of facts, are immune to logic.

1.2.6. Logical Truth and Entailment. The definitions and equivalences (1.1)—(1.7) are a sample of an infinite set of *logical truths* or *tautologies*. These are rough weather propositions, statements that hold no matter what the content (if any) of their components may be nor how adequate their external reference (if any) may be. Thus $p \Rightarrow p \vee q$ is logically true: it holds for every p and every q irrespective of their meaning and truth: if the antecedent p is true, the consequent $p \vee q$ cannot be false because the disjunct p makes it automatically true in such case; and if the antecedent is false then the whole conditional is true no matter what the value of the consequent may be, as shown by (1.4).

So far we have not dealt with the (metalogical) relation of entailment or deducibility. When writing '$p \Rightarrow q$' we do not mean that p is taken as a premise from which the conclusion q is deduced: the conditional is just a peculiar combination of negation and disjunction and it was introduced by (1.4) in a purely syntactical (formal) way without employing the truth concept. On the other hand the concept of logical consequence or deductive inference can be introduced with the help of $\Rightarrow$ and the concept of logical truth. The assumption A is said to *entail* the logical consequence or *theorem* t just in case the conditional $A \Rightarrow t$ is logically true (tautologous). Whether or not this is the case can be checked by giving A and t all possible truth values. In other words, given A we can conclude t provided $A \Rightarrow t$ is not contingent upon the eventual reference of either A or t but holds in every circumstance, i. e. necessarily. We write '$A \vdash t$' for "A entails t" or "t follows from A". The entailment condition is then

$$A \vdash t \text{ iff } A \Rightarrow t \in L \qquad (1.12)$$

where 'L' stands for the (infinite) set of logical truths.

The premise A need not be asserted as true in order to entail the consequence t: it may be introduced just for the sake of argument, as a working hypothesis. What is required is that the conditional $A \Rightarrow t$ be logically true. Again: if we are to deduce t from A, A and t need be neither logically nor factually true, but $A \Rightarrow t$ must be necessary. Thus given p we can conclude $p \vee q$ with q arbitrary, because "$p \Rightarrow p \vee q$" is logically true; i.e., $p \vdash p \vee q$. The converse is false: $p \vee q$ does not entail p. But if p and q are assumed jointly it is because they were assumed separately to begin with, whence $p \wedge q$ entails both p and q separately. In brief, $p \wedge q \vdash p$, $p \wedge q \vdash q$. The most important inference pattern is the *modus ponens*: $p, p \Rightarrow q \vdash q$. (The widely used inference schema "$p, p \Rightarrow q \vdash p$" is fallacious.) Other important inference rules are: (a) *any-all:* $P(x) \vdash (\forall x) Px$; (b) *conditional proof:* $p, q \vdash r$ iff $p \vdash q \Rightarrow r$.

If only a consequence is given, the argument cannot be reconstructed mechanically; i.e. the premise(s) A cannot be uniquely determined from their logical consequences t. Thus in the case of the modus ponens, given q we cannot ascend to either p or $p \Rightarrow q$, where 'p' now stands for a definite proposition: in the above inference schema p is arbitrary — in particular p may be identical with $\neg q$. The consequence for theory construction is clear: low level theorems do not reveal high level hypotheses (in particular postulates); they can only refute them. The simplest refutation schema is this: if $A \Rightarrow t$ and $\neg t$, then $\neg A$ *(modus tollens)*. In fact, the only way in which $A \Rightarrow t$ can hold with t false is when A itself is false. Therefore, to say that NEWTON's laws of motion or

SCHRÖDINGER's equation are deduced from experimental data, is absurd: they can only be invented, then tested — and finally perfected. In brief empiricism, the belief that all knowledge is "abstracted" from experience, is at variance with logic (see POPPER, 1935, 1963a).

1.2.7. Alternative Logics. The logical theories we have peeped into are the propositional calculus and the *predicate calculus* (PC), which subsumes the former. (PC enriched with the theory of identity is abbreviated PC =.) These are the standard tools for analyzing statements, systems of statements, and deductive arguments; they constitute *ordinary* or classical logic. There are a number of alternative logics, as many as the logicians have cared to invent; the best known are intuitionistic logic, modal logics, and many-valued logics. Although interesting in themselves as mathematical frameworks, none of these exotic logics is used in contemporary science and mathematics either to analyze conceptual systems or to justify inferences.

The belief that in factual science one uses logical theories other than ordinary logic — e.g., "probability logic" (REICHENBACH, 1944, 1949) — is mistaken. (In particular, there is no such thing as a probability implication $p \underset{w}{\Rightarrow} q$, read '$p$ implies q with probability w'': '$p \underset{w}{\Rightarrow} q$' is not a new kind of implication but is short for the second degree statement "It is possible, with probability w, that the conditional '$p \Rightarrow q$' holds". But one rarely knows what the probability of a statement is.) And the claim (BIRKHOFF and v. NEUMANN, 1936) that quantum mechanics employs a logic of its own is wrong as well: physical theories have ordinary logic (PC =) built into them, as shown by the mathematics they use (see Ch. 5, 8). Ultimately there is no escape from ordinary logic save irrationalism. This does not mean that deductive logic is sufficient for scientific work: even in mathematics one does use a number of nondeductive (plausible) inference patterns, among them analogies and inductions of various kinds (see POLYA, 1954). But these inference patterns are lawless, they have no logic of their own; they must be analyzed but they cannot be justified because they depend on the nature of every case whereas formal logic does not. Let us now face the problem of content.

1.3. Semantics

Scientific semantics is neither the juggling with words nor logotherapy but a discipline dealing with concepts such as those of designation, reference, and truth. It was not systematically explored until the 1930's (see TARSKI, 1944 and KNEALE and KNEALE, 1962). Even now the semantics of physical theories remains to be built or rather rebuilt because most of it — and there is not much — is vitiated by the belief that every theoretical statement refers directly or indirectly to some experience — a belief fathered by the operationalist confusion between the meaning of a statement and the way it is put to the test (see 4.1.3.).

1.3.1. Designation. A sign is meaningless unless it is intended to point to something else — another sign, an idea, or a concrete object (thing, event, process). The simplest nonformal relation a sign can have to something other than itself is the relation of designation or *naming*. Thus the letter 'm' often designates or names mass — either a mass concept or the physical property the mass concepts are supposed to represent. In a fully formalized presentation of physics one would have to include, among others, the *designation rule* "'m' designates mass", or "$\mathscr{D}(m, \text{mass})$" for short. In general, the name relation shall be written: $\mathscr{D}xy$, where $\mathscr{D}$ is the binary and asymmetrical relation of designation, x is in the set S of signs and the individual y — the *designatum* of the sign x — is in some set O of objects. In particular, y may be of the same kind as x, i.e. another linguistic object. Ex.: 'H' is short for 'hydrogen'. In this case the sign abbreviates a longer expression — as is usually the case with definitions. In other cases the designatum is either an idea or a physical object.

When the two relata of $\mathscr{D}$ are signs, paradoxes can occur unless certain precautions are taken. For one thing $\mathscr{D}$ is asymmetrical and therefore irreflexive: $\neg \mathscr{D}xx$, i.e. x does not designate itself. Rather, 'x' names x, "x" names 'x', and so on. Next, if a sign belongs to a given language and we make a statement about that sign we are not making an intralinguistic but a metalinguistic statement or *metastatement* for short. Thus if we say that the sentence s expresses the proposition p that all bodies have inertia, we do not make a physical statement couched in what is called the *object language* of physics but a statement about a sentence occurring in this language. Hence the formula '$\mathscr{D}sp$' belongs to the *metalanguage* of physics: it refers to physics not to nature. Since any sufficiently explicit piece of physical discourse contains rules of designation all of which belong to the metalanguage, we see that the language of physics is a multiple level structure: it contains expressions about physical objects (physical statements), others about these expressions (metastatements), and so on. The most important metastatements are about law statements — e.g., the statement that MAXWELL's equations are Lorentz invariant: they will be called *metanomological statements* (BUNGE, 1961 b).

Designation rules are conventional, hence neither true nor false. But conventions need not be foolish — save with regard to units; in particular, rules of designation need not consecrate misnomers, i.e. names evoking the wrong partners in a name relation. Thus it is incorrect to speak of the spin-orbit *coupling* in reference to a single electron, since 'coupling' is a name for interaction, and there can be interactions among different physical systems not among properties of one and the same entity; it is better to speak of the spin-orbit *term*. Likewise it is incorrect to call *causality* the mere temporal precedence, since tradition has loaded that word with a far richer content, one involving production not just succession; it is preferable to speak of the *antecedence* condition (BUNGE, 1959a). Likewise DIRAC's bra-ket notation, though handy, severs a linguistic tie between physics and mathematics. Moral: when choosing a name take into account the existing associations between signs and their designata.

1.3.2. Reference. A deeper semantical concept is that of reference, in which conceptual objects rather than signs are involved. Unlike meaningful signs. *conceptual objects* — concepts, propositions, theories — do not just name objects but point to them or denote them. Thus a law of associativity refers to or concerns any three members of a set; the concept of meson occurring in a given meson theory refers to an arbitrary member of a certain class of mesons; and the concept of measured value of the electron charge refers to both electrons and measurement procedures (whereas the concept of electron charge refers only to electrons). In the first case the reference relation obtains within the conceptual level; in the second it holds among members of two different realms: concepts and things; in the third it relates a concept to both a thing which is supposed to be out there and a procedure applied by some subject.

The *reference relation* shall be written: $\mathscr{R}co$, read 'c refers to o' or 'c represents o'. In particular, if o is a physical system, say an electric network, and c is hoped to *model* or *mirror* o if only sketchily (as is the case of a network diagram), we shall write: $c \triangleq o$, read 'c models o' or 'c mirrors o'. Clearly, $\triangleq$ is a subrelation of $\mathscr{R}$. Both $\mathscr{R}$ and $\triangleq$ are relations from the set O of objects to the set C of conceptual objects, where $C < O$. (The philosophical not the popular sense of 'object' is herein involved). Within a given theory, $\mathscr{R}$ and $\triangleq$ are *functions* from O to C since, for each o in O, there is a unique conceptual partner $c \in C$. Thus any real electron is (partially) mirrored by the Lorentz electron theory. But since there are alternative electron theories involving as many electron concepts, in an open context such as that of the whole physics, $\mathscr{R}$ and $\triangleq$ are not functions.

Whereas a designation rule is conventional, a *reference assumption* does not stipulate anything but renders explicit what we think we are talking about — say, electrons. Reference hypotheses (often called correspondence rules) run the risk of criticism on the strength of either experience or alternative theories. Thus special relativity (SR) debunked the reference hypothesis that E and B stood for (referred to, represented) the elongations of oscillating aether particles. On the other hand SR did not modify the relations between E and B, i.e. MAXWELL's equations. In short, special relativity kept the syntax of MAXWELL's theory (it only made it more explicit) but it introduced some important changes in its semantics.

Reference relations serve the purpose of bridging the chasm between ideas and facts. Very often the context makes the reference relations apparent, but in a rigorous presentation of a factual subject matter an explicit statement of the reference hypotheses is mandatory: otherwise conceptual muddles can occur. Therefore when laying down the foundations of the basic physical theories (Chs. 3 to 5) we shall not neglect stating what our symbols are supposed to refer to. For example, in mechanics we shall not only say what the structure of a certain set is but shall add that the set is intended to represent a body — and we shall assign a velocity and other physical properties to the body not to the set.

1.3.3. External Reference. Concepts can be either *formal* (without an external or objective reference) or *nonformal* (with an external reference). A geometrical triangle is a formal concept while "a thin foil of triangular shape" is a nonformal concept for it referes to a concrete object. Between the two a reference relation can be set up: the mathematical triangle can be said to mirror the physical triangle as far as shape is concerned. Similarly the position of an event in spacetime may be represented by a quadruple of numbers, which does not entail that we identify events with quadruples. Physical concepts constitute a subclass of nonformal concepts.

Notice that we are not identifying "formal" with "abstract" except in the epistemological sense of remoteness from concrete experience. From an epistemological point of view the concept of magnetic permeability is abstract but semantically speaking it is not for it refers to a property of a kind of material. In mathematics an object is called *abstract* if it is devoid of properties other than those attributed to it by the postulated relations it enters into. Thus if we postulate that R is a reflexive, symmetrical and transitive relation in a set U of otherwise nondescript elements, we are specifying the formal or structural properties of R but are not determining whether R happens to be the relation of identity, or of equivalence, or of congruence, or of simultaneity. As soon as we assign R some such specific meaning (which will force us to say something equally specific about the hitherto unspecified basic set U), R ceases to be a semantically abstract concept to become an interpreted one. Even so it may still be a formal concept, i.e. one with no external reference. For example, if we lay down the designation rules "$\mathscr{D}(U$, the set of all statements)" and "$\mathscr{D}(R, \Leftrightarrow)$", R becomes more specific yet it remains with both feet in the realm of ideas.

Physics handles only concepts which, in addition to a definite form, have a physical meaning — although during the computation stage in a piece of research one may well forget what one's symbols refer to. Now the reference of physical concepts can be immediate (without intermediaries) or mediate (via other concepts). In experimental physics reference is often fairly immediate, which is not the case of theoretical physics, where all concepts have mediate referents — in other words they do not describe perceptible things but more or less faithful idealizations of real things, mostly imperceptible.

1.3.4. Immediate Reference. The referent of "ammeter" is any member of the class of instruments capable of measuring the intensity of an electric current. In this case the referent of the concept is not only concrete but also perceptible. Likewise the empirical (as contrasted to the theoretical) temperature, viscosity, and weight of a fluid are mirrored in a fairly direct way. Any such physical predicate refers to a property of an arbitrary member of a set of real and moreover perceptible physical systems. On the other hand certain physical constants refer to unique individuals, or to (practically) identical copies of an individual. Thus initially the concept of kilogram referred to a unique body deposited at Sèvres and was so specified. The formula "1 kg is the weight of the standard body at Sèvres" stipulates a relation between a concept, "1 kg", and a thing, the standard weight. The statement can be so symbolized: "$1 \text{ kg} \stackrel{rf}{=} W(s)$", read '1 kg equals, by *referition*, the weight of s', where $\mathscr{D}$ (s, standard body at Sèvres). Similarly, the value of the atomic unit of electric charge may be introduced thus:' $e \stackrel{rf}{=}$ Electric charge of an electron". In the former case the referent is a unique individual, in the latter an arbitrary member of the class of electrons. But in either case the statement establishes a concept-physical object relation and moreover such a relation is fairly direct and the statement involves a definite (unambiguous) description. We shall call such a semantical formula a *referition* and shall not mistake it for a *definition*, which is a sign-sign or a concept-concept correspondence rather than a concept-thing one. In particular, what are usually called "operational definitions" are refertions — often inadequate because ambiguous.

1.3.5. Mediate Reference. In theoretical physics reference is more or less roundabout. Thus "light ray" refers to no real thing since a light pencil, however narrow, has a nonvanishing width. Yet the concept does intend to refer, even though roughly, to a real narrow light beam: otherwise there would be no point in introducing it. Likewise the linear propagation law applies approximately to real narrow high frequency light beams: it holds exactly, and by stipulation, for light rays and it is nearly true for real beams under certain conditions. Moreover a light ray is, by definition, that which satisfies ("obeys" in anthropomorphic language) the laws of geometrical optics. (No operational definition can be given of a light ray.)

Accordingly we distinguish the *immediate referent* of a theoretical construct from its *mediate referent*. The former is a *theoretical model* or sketch, hence a conceptual model; yet not a formal idea nor a fiction, for it is supposed to represent, however coarsely, the corresponding intended or mediate referent, which in turn is supposed to inhabit reality, i.e. to exist by itself. Calling m the theoretical model (rigid body, light ray, Dirac electron, etc.) to which the conceptual object (concept, hypothesis, theory) c applies, and p the presumably real physical object (thing, event, process or property thereof) c is intended to refer to, we split the reference relation $\mathcal{R}$ into two: $\mathcal{R} = \mathcal{R}_i \cup \mathcal{R}_m$, where

$\mathcal{R}_i \, c \, m$: immediate reference of c to m (e.g., "w represents the vorticity in a nonviscous fluid")

$\mathcal{R}_m \, c \, p$: mediate reference of c to p (e.g., "w mirrors the vorticity in a low viscosity real fluid").

If we focus on the theoretical model (the one described exactly by the theory) we may forget that the goal of every physical theory is to find out the patterns (laws) of reality, whereas if we focus on mediate reference we may forget that physical theories cannot be expected to be completely accurate because they idealize their intended (mediate) referents. In a given theory the two reference relations are functions; but one and the same physical object p and one and the same ideal model m of it may be disputed by alternative conceptualizations — whence if all of the latter are taken into account neither of the two reference subrelations remains a function. Thus a fluid may be sketched as a continuous substance satisfying the laws of either classical or relativistic (special or general) mechanics.

1.3.6. Extension and Intension. The concept of reference allows us to introduce another semantical concept: the one of extension or denotation of a conceptual object. We define the *extension* $\mathscr{E}(c)$ of $c \in C$ as the collection of referents of c, i.e. as the set of objects $o \in O$ to which c refers or applies — whether accurately or not. In short, the extension of c equals the range of the reference relation satisfied by c: $\mathscr{E} = \operatorname{ran} \mathcal{R}$. Thus the extension of "carbon" is the set of carbon samples or, also, the class of things $\{y \mid C y\}$ satisfying the function(s) C that characterize c. And the extension of GALILEI's law of falling bodies is the set $P = B \times G$ of all body-gravitational field pairs $p = \langle b, g \rangle$ satisfying the law with reasonable accuracy. (Usually only concepts are assigned an extension but the notions of reference and satisfaction allow us to extend the extension concept.)

Suppose in a given context the following semantical information is supplied:

$\mathscr{D}('V', "volume")$ (a sign-concept relation)

$\mathscr{R}("volume", bulk)$ (a concept-physical object relation).

We shall then know how to interpret the symbol 'V' in that context. In fact the semantical statements, by establishing sign-concept and concept-referent relations, enable us to interpret signs. Yet knowing what is the concept a given sign stands for and what referent a concept points to, is insufficient to specify the meaning of that sign. To this purpose we need, in addition, some information concerning the properties of the concept itself; or, if preferred, we need its connotation or *intension* $\mathscr{I}(c)$. Thus the concept of equivalence relation is characterized by reflexivity, symmetry and transitivity in some set: these properties constitute the intension of "equivalence relation". Since the concept applies to (embraces) identity ($=$), propositional equivalence ($\Leftrightarrow$), congruence ($\sim$) and other relations, we say that these constitute part of its extension. In brief,

$\mathscr{I}$ ("equivalence relation") = {reflexivity, symmetry, transitivity}

$\mathscr{E}$ ("equivalence relation") = {$=$, $\Leftrightarrow$, $\sim$, ...}

where the leaders show that the extension is an open set.

If told what the intension of a concept is we can try to find out what it applies to: intensions are pragmatically determined by the corresponding extensions. The converse does not hold: the extension of a concept barely hints at its intension. Thus knowing that stars and men are bodies is knowing part of the extension of the body concept but does not advance us much in the way of determing the intension of "body": to this end we need physical theories concerning bodies; once we have the theories we can hope to determine which objects do in fact have the properties the theory assigns to or subsumes under the concept of body. In present-day physics something is called a body iff it is a physical system localizable in space and having a mass. In brief, $(\forall x)_P [B x \stackrel{\mathrm{df}}{=} L x \wedge M x]$. Consequently, $\mathscr{I}$ ("body") = {localizability, mass}. Since macromolecules are fairly well localizable and have a mass they qualify for the body status; electrons quite a bit less and photons not at all — at least according to current theories. In short, the intension and the extension of a physical concept are determined with the help of theory and experiment. Since neither is ever finished, we hypothesize that intensions and extensions vary alongside the growth of science.

1.3.7. Meaning. The above suggests that the meaning of a sign — say a physical symbol — is determined jointly by (*a*) the rule of designation that relates the sign to the corresponding concept and (*b*) the intension and the extension of the concept as determined by the available theoretical and experimental knowledge. In a compact way,

$$\mathcal{D}sc \Rightarrow [\mathcal{M}ean\,s \overset{\mathrm{df}}{=} \langle \mathcal{I}(c), \mathcal{E}(c) \rangle], \tag{1.13}$$

i.e. the *meaning* of a sign is the ordered pair intension-extension of the concept symbolized. (This is an analysis of the semantical not the psychological concept of meaning.) Whatever the formal structure and the domain of application of c, if c subsumes the predicates $P_1, P_2, \ldots, P_n$ (no matter what number of places they are), its intension and extension are given by

$$\mathcal{I}(c) = \{P_1, P_2, \ldots, P_n | P_1 c \wedge P_2 c \wedge \ldots \wedge P_n c\}, \quad \mathcal{E}(c) = \mathrm{ran}\,\mathcal{R}. \tag{1.14}$$

In the simplest case of a single unary predicate such as "body", $\mathcal{E}(c) = \{y | Py\}$. In the next simple case, of a binary predicate such as "is refracted by", the extension is constituted by the set of all light beam-transparent body pairs, i.e. $\mathcal{E}("\text{refraction}") = \{\langle l_1, b_1 \rangle, \langle l_1, b_2 \rangle, \ldots, \langle l_n, b_1 \rangle, \ldots \}$. In brief, $\mathcal{E}("\text{refraction}") = L \times B$. Similarly for higher degree predicates.

A sign will be condemned (or praised) as *meaningless* just in case it is both intensionally and extensionally void, i.e. if it evokes neither a cluster of predicates nor, as a consequence, a set of objects. Since something that fails to have both an intension and an extension is not a concept, we may also say that a sign is meaningless iff it designates no concept. In symbols: $\neg \mathcal{M}ean\,s \Leftrightarrow \neg (\exists c)_C (\mathcal{D}sc)$, which is just a complicated way of saying that $\mathcal{M}ean\,s \Leftrightarrow (\exists c)_C (\mathcal{D}sc)$. No idea behind a sign, no meaning.

We may now define the concept of equality of meaning. Since two ordered pairs are identical just in case the corresponding coordinates are the same, (1.13) leads to the following definition of *synonymity*:

$$\mathcal{D}sc \wedge \mathcal{D}s'\,c' \Rightarrow [\mathcal{S}yn\,s\,s' \overset{\mathrm{df}}{=} \mathcal{I}(c) = \mathcal{I}(c') \wedge \mathcal{E}(c) = \mathcal{E}(c')]. \tag{1.15}$$

All the formulas containing the concept of meaning must be qualifed (relativized). In fact intensions and extensions are *contextual*, i.e. dependent on the body of knowledge in which the given conceptual object occurs. Accordingly meanings are contextual as well. Thus the word 'photon' is meaningless in the context of mechanics, just as the word 'mass' is meaningless in the context of electromagnetism; consequently the hybrid 'photon mass' is meaningless as well. Similarly, the word 'temperature' does not mean the same in phenomenological thermodynamics and in statistical mechanics. In short, the preceding semantical formulas should be relativized to the language in which the symbol occurs. The corresponding notational changes are left to the reader.

Meanings are not only contextual but also *incomplete* except in trivial cases. Take the concept of electric charge: we do not know yet whether charges produce fields or conversely; we do not know whether the electron charge could be split; and we shall never care or be able to examine every individual making up the extension of "electric charge". Something similar applies to every other scientific conceptual object: if interesting it will be further examined, if further scrutinized it is bound to be modified in intension and/or extension. Another way of saying the same is this: every scientific idea seems to be intensionally and/or extensionally *vague* to some extent. To be sure the progress of knowledge decreases this blurring but it is unlikely to eliminate it completely; more likely, the construct itself will be transformed or even eliminated. Moral: There will always be room for further elucidations of scientific ideas — i.e. foundational workers need not fear unemployment.

1.3.8. Physical Meaning. A sign will be said to possess a physical meaning, or to be *physically meaningful*, iff it designates an idea which in turn refers to a class of physical objects — eventually a singleton. A sign system (= language) will be called a *physical language* iff all its nonformal terms are attached a physical meaning. And a theory will be physically meaningful iff all the nonformal terms of its formalism are physically meaningful — quite apart from the testability and a fortiori the degree of truth of the hypotheses of the theory. No external reference, no physical meaning. Thus the formula '$y = f(x)$' acquires a meaning if the domain (set of x's) as well as the mapping f are specified with reference to physical objects: prior to these semantical specifications the formula is devoid of physical content — it is an empty structure ready to receive a variety of contents. But the reference may be direct or indirect (via another sign). We stipulate that a sign s (simple or complex) has a *direct* physical meaning iff there exist at least one conceptual object c and one physical object p such that $\mathscr{D}sc \wedge \mathscr{R}cp$. And s will be said to have an *indirect* physical meaning iff s depends on other sign(s) s' such that s' has a direct physical meaning. Thus potentials have an indirect physical interpretation, via field strengths.

Now some cautions against certain crippling construals of the expression 'physical meaning'. (*a*) In order for a symbol to have a physical meaning it need not designate an idea referring to a thing: it may also symbolize a relation such as the duration of a process. Take, e.g., the concept of center of mass (*c.m.*) applied to a material ring: since that point does not fall inside the body it might be thought that '*c.m.*' is a nonphysical symbol. But a *c.m.* is always the *c.m. of* some body or other even if it is not part of the body. The referent is usually not indicated and therefore apt to be missed — but it is there. (*b*) The actual existence of physical objects of a given kind is neither necessary nor sufficient to ensure the physical meaning of the corresponding symbol: 'zero-width matter layer' has an intension in continuum mechanics and is therefore meaningful according to (1.13) although it has no physical extension; and "caloric" did have an intended real referent in the past. We require of physical symbols that they designate conceptual objects with a purely physical intension (physical predicates) and a hypothetical (not necessarily actual) physical extension. (c) The term *observer* meets the latter requirement since an observer is basically (though not just) a physical system; but it fails to meet the first requirement since 'observer' designates a concept connoting both physical and nonphysical (psychological and social) properties. Consequently the term 'observer' should be banned from the language of physical theory and kept only in the metalanguage — e.g., in order to be able to require the invariance of physical law statements with respect to changes of observer, and in order to discuss certain traits of experimental procedures. The same exile decree applies to 'observable' for being dependent on 'observer'. Observers and observables are worthy objects of psychology.

1.3.9. Operationalism. The thesis that only terms referring to ("defined by", in the incorrect jargon) concrete operations are scientifically meaningful, is the core of *operationalism* (MACH, 1883; EDDINGTON, 1924; HEISENBERG, 1927; BRIDGMAN, 1927 and 1955). Unmeasurables, such as the energy of a free photon traveling in empty space, or the location of every atom in a radiogalaxy, should according to operationalism be excluded from physics. It would certainly be very handy if meanings could be determined by reference to laboratory operations, even better by automata. But this is impossible if only because (a) laboratory operations bear only on observable objects — a small part of reality if our theories are not entirely mistaken; (b) even perceptible objects, such as bulky solids, are modelled in a schematic fashion; (c) everything measurable can in principle be measured in a variety of ways: there is no 1:1 correspondence between theoretical constructs and operations; (d) measurements can only sample the full domain and range of a function; (e) measurements presuppose theories.

Operationalism is not a possible interpretation of theoretical physics but its antithesis, for no theory is possible without *theoretical concepts*, i.e. constructs overreaching experience (and thereby making the explanation of experience possible). Our theories are supposed to span the whole expanse of spacetime quite apart from considerations of practical access. Limitations to the empirically accessible and the technically feasible arise only in connection with the tests of our theories and with their applications in technology: theory ignores such practical limitations, many of which can be overcome precisely because theory is not stuck to experience. Moreover, far from being experiment which determines the meaning of theoretical symbols, it is theories, and only theories, that enable us to interpret empirical operations. So much so that, when an empirical manipulation is not backed up by a set of theories, we regard it as nonsignificant or even magical. What must be required from theories is physical meaningfulness and testability.

1.3.10. Truth. The reference relation $\mathscr{R}$ may or may not be satisfied by a given couple $\langle c, o \rangle$ with $c \in C$ and $o \in O$. In case a given reference relation does hold between c and o, we say that this couple *satisfies* $\mathscr{R}$. In particular, if c is a proposition and there is an $o \in O$ such that $\langle c, o \rangle$ satisfies $\mathscr{R}$, we say that c is *true* or that what c predicates about o is true. Truth, then, is a case of satisfaction of the reference relation: it is the property some statements have of adequately pointing to their referents. Equivalently: a statement is true just in case it satisfies the $\mathscr{R}$ relation to its referent. The concept of truth is then seen to be semantical (TARSKI, 1944). The particular question whether a given $\langle c, o \rangle$ pair does satisfy the reference relation is not philosophical but scientific.

If o happens to be another conceptual object, the $c-o$ reference relation remains on the conceptual level and we get the notion of *formal truth* or adequacy of one idea to another (see 1.2.6). Ex.: $c = 3$ is prime, $o = 3$. Since the referent of c is in fact prime, the reference relation is satisfied by this couple and the statement "3 is prime" is true. Moreover it is formally true since it has no external reference. All logical and mathematical truths are formal. On the other hand if o is a physical object things are quite different: now there can be adequacy or adaptation of c to o only in a metaphorical sense since these relata of the reference relation are heterogeneous. Whereas the truth or falsity of a formal statement can be ascertained if at all by purely internal (formal) means, the truth value of a factual statement such as a physical law statement must be evaluated both internally (coherence with other statements regarded as true) and by resorting to empirical means such as measurement. In any case, in physics we must face the problem of *factual truth*, i.e. the case when a statement c is paired by $\mathscr{R}$ to a physical object $p \in P \subset O$. But the external reference of a statement may not be completely accurate or utterly inexact: factual truth can be partial.

1.3.11. Partial Truth. In ordinary logic and in mathematics a proposition is either true or false: no intermediate truth values are recognized as warned by the excluded middle law "$p \vee \neg p$". But in applied mathematics (e.g., numerical integration) and in physics, and consequently in every other factual science as well, we constantly use the concept of *partial truth*. Thus we say that 1.99 satisfies approximately the equation

"$x^2 = 4$", whence the statement "'1.99 is a root of $x^2 = 4$" is *approximately true* rather than outright false. Similarly with every nonformal statement of physics: we ought to know that, however abundantly confirmed, it is at best partially true. Moreover in many cases we can estimate the error of a statement, i.e. how much it deviates from what is taken as the truth. That is, we use the concept of partial truth and indeed a semi-quantitative one. Only, there is no accepted theory of partial truth enabling us to compute the truth value of compound statements out of the truth value of their components.

Logicians could be blamed for their reluctance to accept the very notion of partial truth just because it does not occur in pure mathematics. And philosophers of science are responsible for mistaking partial truth, a semantical predicate, with probability (a mathematical functor), degree of confirmation (a methodological concept), degree of acceptance (a pragmatic predicate), and even credibility or certainty (psychological concepts), not to mention relative truth (context-dependent truth) and value (subjective utility). Part of the difficulty logicians experience in accepting the concept of partial truth may stem from the mistaken belief that it would force them to give up two-valued logic, which is the skeleton of mathematics and science. If every statement were assigned a single truth value it would indeed be contradictory to handle it as plainly true (or false) at some steps in an argument and as approximately true (or false) at others. But the contradiction does not occur if every factual statement is assigned two different truth values, according to the function it discharges: (*a*) a logical truth value depending on its role in an inference (i.e., plainly true if functioning as a premise in a deduction or as a logical consequence of previously asserted premises) and (*b*) a factual truth value depending on its relation to its referent. If preferred, one and the same proposition concerning matters of fact may be regarded, for purposes of logical and mathematical processing, as if it were either true or false, and at the same time as being true or false to some degree depending on how closely it satisfies or fails to satisfy the reference relation. In the following we shall adopt this idea without committing ourselves to a definite full-fledged dualistic theory of truth (for which see BUNGE, 1963 a). This corresponds to the practice of research: thus we may assert the postulates of a physical theory for purposes of computation (deduction) but at the same time we mistrust them and are ready to declare that they are only partially true as regards fact. This enables us to use ordinary logic for unpacking their logical consequences, some of which are put to the test.

Let us now apply our universal tools to a logical and semantical analysis of physical concepts (Sec. 2), laws (Sec. 3) and theories (Sec. 4).

2. Predicates

Let us discuss physical predicates, the specific building blocks of physical statements. A *physical predicate* is a predicate referring to physical objects — i.e. a physical property or relation. Clearly, what is regarded as physical and what is not depends on the state of our science. Thus until recently "probability" was regarded as nonphysical and applicable only to states of mind (uncertainty, ignorance), whereas "absolute position" was taken to be physical — it still is but its extension has vanished. Even though the set of physical predicates varies with time, we must try not to admit in that set openly nonphysical predicates such as "uncertainty" and "observer" if we wish to prevent a relapse into anthropocentrism. On the other hand we should encourage the exportation of physical predicates to the higher level sciences; thus psychology can use some physics but not the other way around.

As regards their structure, some predicates are class concepts (e.g., "soluble"), others relation concepts (e.g., "soluble in water"), and finally others quantitative or metrical concepts (e.g., "solubility"). Although theoretical physics can dispense with none of theses kinds, it is characterized by quantitative predicates or quantities for short — also called magnitudes and variables. We shall concentrate on the latter.

2.1. Magnitudes

2.1.1. Variable and Fluent. The mathematical representative of a physical property is often called a *variable*. This name is misleading insofar as it suggests change in time as much as the Newtonian *fluent* does. If applied to the intensity of a field it is all right, but what about the fixed points in space, or time itself, which cannot flow? The pictorial conception of variables as describing some aspect of the flux of events died in mathematics with the arithmetization of analysis (WEIERSTRASS and DEDEKIND) but it still lingers among physicists, because it is often a powerful heuristic aid. If we employ this kinematical interpretation of variables we must recall (*a*) that not every physical variable is a fluent (time-dependent variable) and (*b*) that, logically speaking, a variable is not a changing thing but something that can take on at least one value: it is a blank (recall 1.1). Thus to say that x is a real variable means that x is an arbitrary member of the set R of reals. A variable is not a changing property but an unspecified member of a set. A variable does not change by itself: it is we who now choose a particular member of the set, now another. In brief, variables do not vary. On the other hand fluents do represent change in time — by definition of 'fluent'. Thus when we say that x is a function of time and write '$x = f(t)$', we mean that differences in t are mirrored in x; or, more exactly, that given a member t of the domain T of f, the corresponding member x of the range X of f is not arbitrary but is uniquely determined by the map $f: T \to X$. To the extent to which time and the referent of x are objective, and the map f faithful to reality, the corresponding changes in x are objective as well.

2.1.2. Structure and Content. The problem of the form and meaning of physical predicates has only been faced (CARNAP, 1926); it has not been studied carefully. We shall approach it via an example: the concept of mass in classical mechanics. The *mass* of a physical system as reckoned in a given scale-cum-unit system δ, is a nonnegative number m: $M(\sigma, \delta) = m$, with $\sigma \in \Sigma$, $\delta \in \mathscr{S}$, and $m \in R^+$, where 'Σ' designates the class of physical systems, '$\mathscr{S}$' the set of scale-cum-unit systems, and 'R^+' the set of nonnegative reals. By hypothesis, whereas m and δ are mental objects, σ names a real object: otherwise the mass concept would not apply (refer) to it and consequently it would not be a physical concept (see 1.3.3). We shall require of every physical predicate that it involves at least one $\sigma \in \Sigma$ as an argument if only tacitly.

For a given scale-cum-unit system δ we have a special concept of mass: it is a function that sends every pair $\langle \sigma, \delta \rangle$, where σ is fixed, to a nonnegative m. But there are as many δ's as we please: we can always rescale mass values by introducing an arbitrary continuous function f such that $m' = f(m)$, although we shall prefer the linear function as it maps equal m-intervals on equal m'-intervals. In order to get a general mass concept we must take the whole set $\mathscr{S}$ of scale-cum-unit systems. Let us then form all the ordered pairs $\langle \sigma, \delta \rangle$, i.e. the Cartesian product of Σ and $\mathscr{S}: \Sigma \times \mathscr{S}$. We now match every pair in the latter set with a given number $m \in R^+$ in such a way that no pair is assigned more than one number m, although any number of couples can have a single partner m. In other words we build the new couples $\langle \langle \sigma, \delta \rangle, m \rangle$ and take the whole bunch: we thus get the mapping M of $\Sigma \times \mathscr{S}$ into R^+. In brief, the structure of the general theoretical mass concept in classical mechanics is the function

$$M : \Sigma \times \mathscr{S} \to R^+. \tag{1.16}$$

This is an *analysis* or elucidation of the mass concept not a *definition* of it: in classical mechanics M is an undefined (primitive) concept notwithstanding the attempt to define it as an acceleration ratio (see Ch. 3). The domain of M is the half-real and half-conceptual set $\Sigma \times \mathscr{S}$ while its range is the wholly conceptual set R^+. Both sets are infinite, hence empirically inexhaustible. We can think of them as totalities (actual infinities) but we cannot effectively construct them, much less explore them experimentally in an exhaustive fashion: we can only sample them. In general, theoretical concepts are *nonconstructible*, whence physics pays no heed to the intuitionist and operationalist commandment to admit only constructible sets, i.e. sets all of whose elements can be exhibited by a finite number of human operations (see BUNGE, 1962a).

This shows also that physics does not deal only with what is the case (actuality) but with what is *possible* as well. (Conceptual not ontological possibility is herein involved: real possibility can enter through probability: see Ch. 2, Sec. 2.) Indeed, 'Σ' stands for the set of all physically possible material systems whereas '$\mathscr{S}$' designates the set of all available and thinkable scale-cum-unit complexes, and none of these sets is already out there. Were we to restrict Σ to the set of empirically accessible things and $\mathscr{S}$ to the set of available scales we would prohibit the application of the mass concept to new things as well as the introduction of new scales and units. Physics cannot limit itself to actual human experience, for such a limitation would stick it to the past. In brief, while experimental physics deals with actual facts theoretical physics concerns conceptually possible facts.

The fact that M and the other physical magnitudes are concepts involving nonphysical ideas such as $\mathscr{S}$ and R does not mean that they are fictions or that bodies have no mass and other properties. M is a concept that connotes a property of bodies. And merely assigning a numerical value m to a couple $\langle \sigma, \delta \rangle$ presupposes assuming σ to name a real body: Σ is the physical extension of M. What about its intension or connotation? As with other physical predicates, the intension of M must be inferred from its lawful relations with other predicates. In classical mechanics the numerical value m of the mass is regarded as a measure of inertia; but atomic physics teaches that the (proper) mass of a system is proportional to the number of heavy "particles" that make it up, for which reason mass is regarded as a measure of the quantity of matter in the system. We conclude with NEWTON that the inertia and the quantity of matter in a body are proportional to each other and both are measured by the (proper) mass. In short, $\mathscr{S}$ ("proper mass") = {inertia, quantity of matter}.

The previous analysis is at variance with the widely accepted analysis proposed by MACH (MACH, 1883), who tried to *define* mass as an acceleration ratio without realizing (a) that a law of nature, like NEWTON'S *lex secunda* — which underlies the inverse proportionality of mass and acceleration — is not a convention, and (b) that acceleration ratios are frame-dependent while in classical mechanics mass is an invariant. The root of MACH's destructive analysis of mechanics was his empiricist philosophy — the belief that science should yield to sensory experience rather than trying to enrich and understand it (see PLANCK, 1909; BUNGE, 1966).

2.1.3. Theoretical and Experimental Predicates. Physical predicates are assigned real or supposedly real referents but they are not all measurable. Thus wave phases and lagrangians are not measurable and consequently they do not reach experimental physics. On the other hand the temperature concept occurs in both theoretical and experimental physics — or, rather, there are theoretical and experimental temperature concepts. Let us cast a glance at them.

Let σ name a physical system and let m be a theoretical model of σ; briefly: $m \triangleq \sigma$ (see 1.3.2). Thus σ could be a volume of oxygen and m the corresponding ideal gas. Choose a scale-cum-unit system s, say KELVIN'S, to reckon the numerical values of the temperature function. The structure of this particular temperature concept is: $m \triangleq \sigma \Rightarrow T(m, \mathit{s}) = \vartheta_t$. Vary the theoretical model m or s, and the numerical value ϑ_t will vary accordingly: as many temperature values as m's and s's. To unify the various special temperature concepts we proceed as with the mass concept: we take the whole set $\mathcal{M}$ of models and the whole set $\mathcal{S}$ of scales and form their Cartesian product $\mathcal{M} \times \mathcal{S}$. We thus get the general theoretical concept of temperature: $T : \mathcal{M} \times \mathcal{S} \to \Theta$, where $\Theta \subset R$ is an interval of the real line. Again, this is an analysis not a definition of T. In phenomenological thermodynamics T is usually defined in terms of the energy E and the entropy S of the system: $T \overset{\text{df}}{=} \partial E / \partial S$.

So far theoretical or calculated temperatures. The empirical temperatures are on the other hand results of measuring T on real systems σ with the help of some feasible measurement technique t involving a temperature objectifier or index. For a given system $\sigma \in \Sigma$, a given scale-cum-unit system $\mathit{s} \in \mathcal{S}$, and a given measurement technique $t \in \mathcal{T}$, the outcome of a measurement of T on σ with t will be a statement of the form

$$\mu_t \, T(\sigma, \mathit{s}) = \vartheta_e \pm \varepsilon_t, \tag{1.17}$$

ε_t being the relative standard error associated with the given technique and the given temperature interval. The general concept of empirical temperature subsumes all these particular concepts: $\mu(T) : \Sigma \times \mathcal{S} \times \mathcal{T} \to \Theta$.

The difference between theoretical and empirical predicates is not just numerical but also structural and semantical. Were it not for these differences there would be no point in comparing calculated values (associated with m) with measured intervals $\vartheta_e \pm \varepsilon_t$ (associated with σ). Another difference between the theoretical and the empirical concepts of temperature is that whereas the former behaves orderly like length the latter exhibits an aberrant behavior. Thus a $1°$ C difference near the freezing point does not correspond, for real substances, to the same difference in thermal states near the boiling point. In short, the empirical temperature scales are not strictly linear. The reason is that T is not measured directly but indirectly, e.g. via volume changes (indicators or objectifiers: see 5.1.1), and the T–V relation is not strictly linear. (For this reason OSTWALD, 1902 held that empirical temperature is not a magnitude. But similar difficulties arise in relation with the objectification of other indirectly measured properties.) This is one of the reasons for not employing empirical magnitudes in the construction of general theories. General theories must be built with strictly theoretical concepts independent of the peculiarities of particular substances and experimental techniques: otherwise they cannot be expected to be general.

Temperature, mass and charge densities, field energy-momentum and angular momentum tensors and all other tensor densities are *intensive* magnitudes as opposed to *extensive* magnitudes such as length and the volume integrals of tensor densities. The latter are additive or at worst slightly subadditive. That is, if $\sigma_1, \sigma_2 \in \Sigma$ and σ_1 is juxtaposed to σ_2 to form the compound system $\sigma_1 \dotplus \sigma_2$ (physical addition), then if

$$P(\sigma_1 \dotplus \sigma_2) = P(\sigma_1) + P(\sigma_2) \qquad (1.18)$$

at least approximately, P is extensive — otherwise intensive. Energy and mass, being nearly additive, are extensive magnitudes. This does not mean that intensive magnitudes cannot be subjected to arithmetical and other operations, as is often claimed. For a given system σ it is correct to say that, if in a state s, $P_s(\sigma) = p$, and in a different state s', $P_{s'}(\sigma) = 2p$, then the P corresponding to the second state is twice the P of the first. In classical physics extensive magnitudes represent properties of wholes (e.g., the total mass of an aggregate) whereas intensive magnitudes represent either nonhereditary properties (e.g., temperature) or properties of mass points (e.g., densities). No such correspondence is found in quantum physics.

2.2. Constants

2.2.1. Reference. A variable that takes exactly one value is called a constant. A *physical constant* is a constant entering a physical law statement. From a logical point of view physical constants can be classed into numbers and numerical values of magnitudes. For example, BOLTZMANN'S constant k and PLANCK'S constant h are numbers not special values of a unique physical variable. On the other hand c and e are special values of magnitudes — velocity and charge respectively. This structural difference matches a deep semantical difference: the numbers refer to no physical system, they are *nonreferential* proportionality constants, whereas the special values of magnitudes are properties of classes of physical systems. Thus BOLTZMANN'S k is not the k *of* a class of substances or even of an arbitrary substance, while e is the charge of an arbitrary member of the class of electrons. In other words, unlike e, k is not a particular value of a function defined on a set of physical systems. (This difference, though, is contextual. Thus in BOHR's atomic theory h was the unit action of a periodic system.)

All nonreferential constants are universal but not all universal constants are nonreferential. The specific/universal dichotomy applies to the referential constants: a specific constant such as the electric resistivity of copper depends on the composition of the system it refers to; a universal constant is a nonspecific one. In short,

$$\text{Physical constants} \begin{cases} \text{Nonreferential } (k, \hbar) \\ \text{Referential} \begin{cases} \text{Universal } (e, R) \\ \text{Specific (specific heats, sound velocities)} \end{cases} \end{cases}$$

2.2.2. Fundamental and Derivative. Referential constants can in turn be classed into fundamental and derivative. *Fundamental* constants are those which are not further analyzed, *derivative* those analyzed in terms of fundamental constants. Thus $\alpha = e^2/\hbar c$ is a derivative universal constant. It may well be, though, that certain fundamental constants such as e and h become derivative in finer theories. And it is hoped that all specific constants will eventually become derivative, i.e. analyzed in terms of a few fundamental constants, because specific constants characterize macrosystems which we expect to understand in terms of fundamental theories regarding microsystems.

Fundamental theories contain no specific constants. Thus Newtonian mechanics contains no constant at all; Newtonian gravitation theory contains only G, and EINSTEIN's gravitation theory only $\varkappa$ (analyzable but not definable in terms of G and c); nonrelativistic quantum mechanics contains only $\hbar$ and quantum electrodynamics c, $\hbar$ and e. The reason for the absence of specific constants in fundamental theories is that they deal with comprehensive classes of physical systems, not with special systems like the CdS crystal or a position measurement device. Some universal constants are typical of macrophysics (e.g., k and R), othert of microphysics (e.g., e and $\hbar$), and finally others are level-independens (e.g., c and the gravitation constants G and $\varkappa$). In other words, some universal constants are more universal than others and therefore the theories characterized by them are more comprehensive than others.

2.3. Semantical and Methodological Status

2.3.1. Objectivity. Not all the symbols occurring in physics refer to some physical object. Thus the coordinates of a point in free space are just labels and PLANCK's constant is nonreferential (see 2.2.1), whereas the position coordinates of a physical system are physical predicates proper. Both sets of quantities are necessary in physics, but of course what characterizes our science is that it is a body of ideas referring if only hypothetically to physical reality (semantical side) in a testable way (methodological side). These two aspects are often mixed up. We must distinguish them and realize that testability is dependent on meaning not the other way around. For, before we can test for *adequate* external reference by means of empirical operations we must have at least hypothetical objective *reference*.

The invention of physical magnitudes, of predicates independent of the cognitive subject, was a momentous step in the quest for objectivity (subject-independence). Indeed objectivity is not attained by sticking to sense impressions, which are private, but by hypothesizing physical objects (autonomous entities). A few such constructs have sensorial correlates, most do not: most are empirically occult even if they are (indirectly) scrutable. Thus certain frequency bands of e.m. radiation correspond to colors — they elicit color sensations in certain animals — but the very concept of e.m. wave has no empirical correlate: we do not perceive light as a wave field, we can only think it this way. By conveniently combining such constructs we build physical object statements, the truth or falsity of which does not depend on our sense impressions but on the physical objects themselves.

This does not entail an evasion from ordinary experience but its enrichment and control: with the help of hypotheses relating magnitudes in a stable way (laws) we can explain some sense impressions and in particular we can find out under what circumstances they are reliable indicators of real but imperceptible events. The higher we climb the ladder of epistemic abstraction the less we ourselves appear in our picture of the world and the better we are at explaining our own experiences. On the other hand, by remaining close to the senses we will not transcend superficial, anthropocentric world views. In short, although experience is a test of our theories it is not the stuff our theories are made of or even the referent of physical theories: human experience proper is the subject of nonphysical sciences like psychology. These platitudes had to be stated on account of the widespread belief that in physics only observational predicates matter — a belief inherited from philosophies at variance with science.

2.3.2. Measurability. As regards their methodological status, magnitudes can be classed as follows:

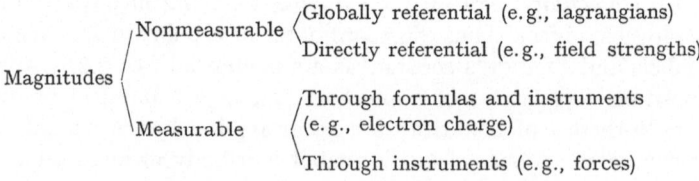

The unmeasurables of the first kind — e.g., hamiltonians and GIBBS' H — represent a system in a global fashion and are sources of magnitudes representing physical properties (e.g., $p = \partial L/\partial \dot{q}$, $\varrho = |\psi|^2$). The unmeasurables of the second kind, whether classical (wave phase) or not (spin), and whether they have a classical analog (q.m. momentum) or not (strangeness), are assumed to represent physical properties: they are directly referential. None of the measurables is directly measurable: every measurement requires both an experimental arrangement and a bunch of fragments of theories (mechanics, optics, etc.) whereby to plan and understand what one is doing (see Ch. 5). Some of the measurables are so only partially — e.g., the components of the angular momentum of a microsystem. In short most magnitudes are, though somehow scrutable, occult rather than manifest: they do not describe appearance even though they purport to refer to reality. In other words, theoretical physics contains no observables in the epistemological sense but it teems with hidden variables.

In every case measurability, rather than being an inherent property, depends on theory as well as on experimental technique. Thus in the past the aether wind was regarded as measurable and the electron spin as unmeasurable. And in any case measurables need not be and usually are not observable: thus energy differences, though often measurable, are never observable except in a Pickwickian sense of 'observable'. Also, measurability is not equivalent to physical existence, whence unobservability does not exclude physical reality. Thus there is little doubt that there are gravitational waves which interfere and diffract, but no one knows how to design a gravitational interferometer. Finally, the reducibility of the measurement of most magnitudes to length measurements (via theoretical formulas) does not prove that at bottom everything is spatial — not any more than the possibility of scaling electric energy meters in dollars proves that electric energy is nothing but money. These cautions are important in view of the lingering of an empiricist philosophy among theoreticians who rarely handle anything observable save pencil and paper.

2.4. Dimensions

2.4.1. Analysis. Physical symbols can be grouped into species and species into genera. Thus distances, lengths, wavelengths and mean free paths are as many magnitude *species* of the *genus* L. Similarly angular frequencies, frequencies and radioactive rate constants can be said to belong to the genus T^{-1}; equivalently, their reciprocals may be said to be in T. Every such genus is a set, whence we can write $\lambda \in L, \nu^{-1} \in T$, and so on. The reason for this is obvious: one and the same mathematical function, variously interpreted, is at stake in every genus: thus there is no essential difference in form between a wavelength and a mean free path. The usual expressions '$[\lambda] = L$', '$[\nu] = T^{-1}$' and so on for the dimensions of magnitudes can then be elucidated in the following way:

$$[\lambda] = L \overset{\text{df}}{=} \lambda \in L, \qquad [\nu] = T^{-1} \overset{\text{df}}{=} \nu^{-1} \in T, \quad \text{etc.} \tag{1.19}$$

Similarly for composite or derivative magnitudes:

$$[v] = L T^{-1} \overset{\text{df}}{=} (vt) \in L, \qquad [\varrho] = M L^{-3} \overset{\text{df}}{=} (\varrho V) \in M, \quad \text{etc.} \tag{1.20}$$

The analysis that exhibits the genera to which a set of magnitudes belong is called *dimensional analysis*. Once the dimension concept has been elucidated, dimensional analysis can be carried out by handling the sets L, T, M, etc. *as if* they were numbers and their juxtaposition *as if* it designated the arithmetical product. In every branch of physics dimensional analysis will lead to a family of irreducible or *basic genera*. The basic genera of mechanical magnitudes are of course L, T and M, since every mechanical magnitude can be analyzed in terms of them. In other branches of physics further basic symbols may be needed; thus in phenomenological thermodynamics a special symbol for the dimension of temperature occurs. But since every branch of physics can in principle be related to mechanics (though not reduced to it), it is in principle possible to reduce any dimensional formula to an expression of the form: $[Q] = L^{\alpha} T^{\beta} M^{\gamma}$ with α, β, γ fractions. It is only as long as such a connection with mechanics has not been efected that independent basic dimensions are needed — as was the case with thermological and photometric magnitudes. Once a basic family of dimension genera is adopted, dimensional analysis proceeds abiding by the rule of dimensional homogeneity: All the terms in a formula shall belong to the same magnitude genus. Any expression violating this principle is to be discarded as an ill-formed formula. Examples: $v + v^2$, $\sin t$, and e^t are sums of dimensionally heterogeneous magnitudes, hence ill-formed formulas.

Dimensional analysis is not only a good control but is apt to hint at physical meanings provided the physical object variable (usually a single σ) is not lost sight of. Thus the symbol 'E/c^2' has the dimension of a mass yet it is not a mass unless E refers to a body. And the gravitational potential has the dimensions of v^2 but it cannot be interpreted as twice the kinetic energy per unit mass, nor its square root as the velocity of something.

2.5. Scales and Units

2.5.1. Mapping Differences in Degrees into Numbers. In theoretical physics no scales and units need be specified: one works with general theoretical concepts. This does not mean that theoretical physics is scale-free and unit-free but that it subsumes all possible scale-cum-unit systems: it is scale-and-unit *invariant*. Take a magnitude of the form: $P(\sigma, \delta) = p$, where σ is the physical object variable, δ the scale-cum-unit variable, and $p \in R$. (If two different scales δ_1 and δ_2 are employed alternately then the relation between them is given by $P(\sigma, \delta_2) = p_2 = f(p_1)$ where f is a given continuous function.) For a given δ, any given interval of the p's will be divisible into smaller parts. If these parts are chosen to be equal, the basic intervals are called *units* and the scale will be said to be a completely specified and *uniform metric scale*. The scales of fundamental magnitudes are always uniform and metric. In this case one likes to write the barbarous formula ' $P(\sigma) = p u$ ', where 'u' stands for the unit interval of the scale or, equivalently, for the chosen unit of the range of P. This way of writing is symbolic — which usually means incorrect — as the multiplication of a number with a unit is not mathematically defined; but it is handy as long as it is not taken for a rigorous analysis of the magnitude concept. Different u's can be multiplied and divided symbolically but they cannot be added and subtracted: if $u_1 \neq u_2$, then ' $u_1 \pm u_2$ ' is not a well-formed formula (recall 1.1).

2.5.2. Standards. The materialization of a scale-unit complex is a task for experimental physics. When a uniform metric scale is adopted, the experimental problem boils down to the adoption of a standard serving as a referent or materialization of the given concept of unit. Sometimes nature supplies the standard, as in the case of the atomic units of mass and charge, but more often than not it is up to the physicist to devise and construct a standard. Clearly, it must be handy and technically feasible. Yet these conditions are not always fulfilled. For instance the international standard of electric current intensity adopted in 1946 was the intensity of a c.c. set up in two *infinitely* long wires interacting with a certain force. On top of this the committee mistook this referition (see 1.3.4) for a definition, as if the ampère were not definable in terms of the coulomb and the second. But there are worse confusions deriving from insufficient analysis — e.g., the frequent one between units and dimensions, and even the one between units and magnitudes. (For example, since a unit of force is definable in terms of unit mass and unit acceleration, it is often said that the concept of force is so definable.) Moral: Invest in FR what is overspent in committees on units.

2.5.3. Conventions at War. We cannot get involved in the War of Units in electromagnetism: life is too short and the war too confusing. Let us just point out the following cautions. (*a*) When setting up a system of units in some field one should not seize on an isolated law (e.g., Ampère's) but on the whole set of laws in it (e.g., MAXWELL's theory). For the choice of derived units is always theory-dependent. Otherwise one might prefer to adopt a fundamental unit of current rather than of density (or else charge), or of weight (rather than of mass) just because current and weight are more easily measurable. That is, by neglecting theory one might mistake the choice of units for the choice of standards. (b) One may well save a couple of 4π's in the basic equations but one will get the factor back in infinitely many solutions to those equations, so that the initial saving is illusory. (c) No matter how useful the MKSA system may be in engineering and even in experimental physics, it is senseless in theoretical physics as a whole and in particular in microphysics, because ε_0, the dielectric constant of the vacuum, is a particular value of ε occurring in MAXWELL's macroelectromagnetism; ε simply does not figure in microelectromagnetism (see Ch. 4, 1.1) and is therefore as meaningful as the Oedipus complex in it. (d) The ink-saving virtue of the so-called natural system of units in atomic physics (in which $c = 1$ and $\hbar = 1$) is as obvious as its disadvantage: the explicit occurrence of c and $\hbar$ are powerful heuristic aids in the interpretation of some formulas which are anyhow somewhat opaque. And after all, fundamental theories do not have so many fundamental constants that it becomes necessary to get rid of them.

Finally some comments on physical *conventions* in general. (a) Conventions, whether linguistic or concerning units, are freely chosen not dictated by nature. But they need not be silly: they should be practical (unlike the farad) and the introduction of every new convention should require a minimum readjustment in the body of accepted conventions — unlike the MKSA system. (b) The occurrence of conventions (notations, definitions, basic units) does not render science conventional but on the contrary it enables us to state matters of fact unambiguously, to argue about our statements and correct them whenever they fail to fit facts (see REICHENBACH, 1927).

3. Hypotheses

A *physical statement* is a proposition in which only formal and physical predicates occur (see Sec. 2). On this count, the statement "The eigenvalues of an observable Q_{op} are the only values an observer can read on a suitable instrument" is not a physical but a psychophysical statement. On the other hand "If Q_{op} represents the property Q of a physical system $\sigma \in \Sigma$, then the eigenvalues of Q_{op} are the only values of Q that σ takes on" is a physical statement.

From a methodological point of view physical statements can be so partitioned:

$$
\text{Physical statements}
\begin{cases}
\text{Incorrigible (conventions)} \\
\text{Corrigible}
\begin{cases}
\text{Hypotheses}
\begin{cases}
\text{General (e.g., laws)} \\
\text{Particular (supplementary conditions)}
\end{cases} \\
\text{Data}
\end{cases}
\end{cases}
$$

The only incorrigible statements accepted in science are conventions, i.e. stipulations: designation rules, definitions, and agreements concerning units and standards. Conventions are irrefutable because they say nothing about reality. Yet they need not be final: they can be replaced by more convenient stipulations (see 2.5.3) or they can become pointless — as when the associated idea is dropped. Any nonconventional statement referring to reality is corrigible by rational argument, whether or not this invokes experience: this is the thesis of rationalism *lato sensu*. We shall concentrate on a subclass of corrigible statements.

3.1. Assumptions

The corrigible statements expressing outcomes of empirical operations — observations, measurements, or experiments — are called *data*. All others are assumptions or *hypotheses* even after satisfactory corroboration (SCHLICK, 1935). When we affirm that the speed of a given e.m. pulse in empty space was found to be near $3 \cdot 10^{10}$ cm/sec, we state a datum — though not one of raw experience, for to produce it a number of theories were used. But when we jump to the conclusion that every e.m. perturbation propagates in vacuo with that same speed, then we state a hypothesis: we could not possibly examine every e.m. wave — and if we did we would find that the speeds are not quite the same. What characterizes hypotheses is that they overreach experience both ordinary and scientific, either because they are universal (all-statements) or because they refer to unobservable things or properties, or because they are both universal and nonobservational — which is the case of most physical hypotheses. And what is peculiar to scientific hypotheses in particular is that they are either testable or fruitful or both. Let us examine the main sorts of physical hypotheses.

3.1.1. Special, Subsidiary, and Comprehensive Hypotheses.

We shall call *special hypothesis* any statement that is assumed rather than inferred from experience and which accompanies one or more laws. Initial conditions, boundary values, special values of parameters, and special forms of hypotheses and operators (e.g., hamiltonians) are special hypotheses. Some special hypotheses are more conjectural and less accurately testable than others. Thus if we assume that a given field vanishes at infinity, we have no way of knowing for sure that this is the case. (Experiment may not tell the difference between the vanishing at infinity and the vanishing over a large sphere, for even though in the latter case certain sets of numbers will be denumerable not continuous, the separation between those numbers can be made as small as desired by taking a sufficiently large sphere.) Another example: the so-called Cauchy data are not data but either stipulated or conjectured, as they consist of the infinitely many values of certain field variables over a whole surface (usually a spacelike hypersurface, e.g. $t=$ const). In brief, many particular physical statements are hypothetical in the sense that they are not gathered from experience although they are empirically testable to some extent.

The special hypotheses are logically independent of the law statement(s) they accompany: they are prescribed (hypothesized) independently of the latter. For this reason it is often said that they are contingent or even beyond the reach of law. Epistemically contingent (not logically necessary), yes; physically contingent no for, after all, the physical conditions prevailing at a given moment are the outcome of some lawful process or other. What happens is that no actual process is characterized by laws alone: actuality can only be specified by a set of laws jointly with a set of data and subsidiary condition.

Next to special hypotheses come supplementary conditions, in particular constraints. A *constraint* is a particular conditions, e.g. a state, expressed by a restriction on some of the variables characterizing the problem concerned. Examples: "Temperature$=$const", "Interparticle distance$=$const", "$v_\mu v^\mu > 0$ (time-like character of v), "Hamiltonian$= 0$", "$\partial_\mu A^\mu = 0$" in CEM and "$(\partial_\mu A^\mu)\,\psi = 0$" in QED. No matter what the laws (equations of motion or field equations) "allow" the system to do, it is required that the system changes in a way compatible with the constraints. This is a hypothesis and by making it we draw some traits of a theoretical model supposed to be in turn a rough sketch of the intended real referent (see 1.3.5). In particular, *constitutive equations* like "$H = (1/\mu)\,B$" characterize ("define") ideal models of real materials of a kind (e.g., ferromagnetics). This does not mean that they are definitions proper: in fact in CEM the field strengths and the constitutive parameters are logically independent (not interdefinable). The example of constraints (in particular constitutive equations) shows that not every restriction on a set of variables counts as a law of nature. To be sure laws are, metaphorically speaking, prohibitions (POPPER, 1935), but not every prohibition is a law. Finally constraints, just as initial conditions, are logically independent of laws as proved by the fact that they can be modified while keeping the laws (and conversely). Thus the comprehensive laws of mechanics (the equations of motion) hold for rigid bodies as well as for elastic bodies, and MAXWELL's equations for material media hold no matter what the constitutive $E-D$ and $B-H$ relations may be.

Finally *comprehensive hypotheses* are assumptions so pervasive that we rarely notice them. Examples: the assumptions that spacetime is a continuum, that it has a metric, that the e.m. field in vacuum is representable by a tensor field, that particle trajectories are continuous. Some of the comprehensive hypotheses are so sweeping that they belong to protophysics (Ch. 2) rather than to any particular chapter of physics. In any case they are operationally meaningless: what operation could one possibly perform to measure the 5 *th*, let alone the 105*th*, order derivative of a position coordinate? The main reason we keep continuity hypotheses is that they enable us to use analysis. Another reason is that, if we proposed alternative hypotheses — e.g., that spacetime is full of holes — we would have to give some reason for them and should expect some observational consequences.

3.1.2. Laws. If a physical hypothesis is universal or nearly so, and in addition systemic (belonging to a system of hypotheses) and corroborated (yet not final), we call it a physical *law statement* — or law for short and for confusion. If it is logically very strong — if it is pregnant with many lower level laws — we promote it to the rank of *principle*. Thus HAMILTON's variational principle entails LAGRANGE's equations, which are high level laws yet not peaks, and which in turn entail low level laws such as GALILEI's, which finally subsumes infinitely many substitution instances — one per initial position-initial velocity-gravitational acceleration triple. Schematically: Principle $\vdash$ High level law $\vdash$ Low level law $\vdash$ Substitution instance.

Turning from form to content we get a different systematics of law statements. Some typical genera are:

Reaction schemata such as $K_2^0 \overset{.}{\to} \pi^+ + \pi^-$, where the letters designate individuals of natural species (kaons and pions) while '$\to$' and '$+$' designate the physical concepts "decays into" and "adds physically".

Jump equations: relations between values of magnitudes across surfaces of discontinuity (e.g., interfaces).

Equations of evolution without spatial motion, such as those referring to a black box.

Equations of evolution with spatial motion: laws of motion and nonstationary field equations.

Laws of force, e.g. Coulomb's.

A reaction schema does not describe a process such as it takes place in space and time but seizes on the input and output of such a process. Jump equations are sorts of exception laws. It is desirable though not always possible to deduce them from equations of evolution in conjunction with information concerning the frontier. Equations of evolution contain either time derivatives (no memory) or time integrals (materials with memory). Laws of force have been placed apart because in most theories they are logically independent of the laws of evolution. This situation changes with general relativity, whose field equations entail some laws of motion, and with elementary particle theories (different laws of evolution for different kinds of particle as characterized by charge, spin, isotopic spin, mass, etc.).

Let us now briefly examine the ontological and epistemological status of law statements.

3.2. Law Statements

3.2.1. Law and Pattern. Since physical laws are general physical propositions, they refer — with some degree of accuracy or other — to pervasive traits of nature: they depict, symbolically to be sure, the *patterns* of physical reality, i.e. the unchangeable structure of a world in flux. These patterns or stable modes of being and behaving are supposed to be objective (*realistic* thesis: see AMPÈRE, 1843). Otherwise we would not care to search for, test and correct the statements referring to them. We have then objective patterns (nomic structures) or $laws_1$ on the one side, and their various conceptual reconstructions — the law statements or $laws_2$. The relation between the two is the semantical relation of modelling: $law_2 \triangleq law_1$ (see 1.3.2). Every law_1 can be represented in an unlimited number of ways: the modelling relation is a one-many relation. In other words, the mesh of objective patterns is representable by a variety of conceptual nets of law statements (theories).

A law statement concerns an arbitrary member of a whole set of facts rather than a specific fact; equivalently: it refers to every *possible* fact of a kind. Which possibility will be actualized will in general not be indicated by the law statement: actuality is determined jointly by laws and circumstances (see 3.1.1). In brief, a law statement does not say what is the case but what is possible. Thus a field equation determines the possible modes of a field (standing waves, cylindrical travelling waves, etc.). If the law is stochastic (probabilistic), it will refer to possibilities even after the circumstances (e.g., initial distributions) have been specified: these special hypotheses and data will only select a subset of possibilities. Which possibility will be "chosen" by nature can only be determined with the help of empirical data. (The "choice" or "decision" is of course metaphorical.)

3.2.2. Law and Experience. According to radical empiricism law statements do not express objective patterns but are empirical relations: (*a*) they are equations between the measurable elements of phenomena (MACH, 1883) and (*b*) they are inferred (induced) from observed cases (e.g., REICHENBACH, 1951). This view may describe some of the generalizations of ordinary knowledge but is inadequate for science: (*a*) 'phenomenon' means "fact as perceived by a subject", and the point of physical theory is to go beyond phenomena, to the things themselves — as thought by man, to be sure, but not as sensed by him; (*b*) physical $laws_2$, even the lowest level ones, are about factual patterns, and these are not observable (we do not see laws$_1$) — consequently laws$_2$ cannot be "abstracted from experience"; (c) every high level law statement contains indirectly measurable and often nonmeasurable predicates (see 2.3.2); (d) some of the low level generalizations are suggested by scientific experience — but not unambiguously and then that experience is enlightened by theory. This does not prevent laws$_2$ from being related to experience: they are tested empirically and they guide new experience.

Law statements are introduced in a variety of manners: generalizing empirical relations (high grade induction), noting resemblances to known patterns (analogy), deriving consequences from general principles (deduction) and, above all, by invention. Neither induction nor deduction introduce radically new ideas: they only relate available concepts in new ways. Radically new ideas, and particularly strong ones, require acts of creation similar to artistic creation (see BUNGE, 1962a). Admittedly one does not start from scratch but from the available body of ideas and one uses analogy and heuristic recipes; nonetheless, radical novelty does not arise by shuffling a pack of old items. Thus in the case of the simplest of the ideal gas laws, "$pV = nRT$", the concepts "p", "V" and "T" were initially available in a coarse state. By immersing them into theories (mechanics, thermodynamics and kinetic theory) they were refined to the point of becoming full-fledged theoretical concepts which do not overlap with their raw ancestors ("push", "bulk", "warmth"). Moreover, p and T are not directly accessible but require indicators or objectifiers involved in further law statements. As to n, it was from the start a theoretical concept with no experiential counterpart: thus putting 1 mole of O_2 equal to 32 g involves the hypothesis that oxygen is diatomic.

The simple and orthodox idea that laws$_2$ are a posteriori summaries of data is, like nearly every other orthodox simplicity, false. To begin with, before we can join concepts to build statements, the concepts must be there. Next, one does not get new concepts by looking hard but by thinking hard. Then, scientific observations are planned and interpreted with the help of law statements. Also, the gist of a law statement is that it transcends particulars — circumstances and experimental arrangements — whereas measurement can yield only particulars. Indeed, every empirical operation is performed by some operator under definite circumstances and with a given technique; but none of these items occurs in a law statement. For these reasons laws are not discovered by cooking data — but data are produced with the aid of laws. Laws must be hypothesized — notwithstanding which the inscription in the Pisa campanile informs the visitor that GALILEI "*legibus motus detectis*".

One of the functions of experiment is to subject hypotheses, in particular laws$_2$, to test. Even the so-called empirical laws (e.g., the "empirical curves") must so be tested since they extrapolate sets of data. If a hypothesis is corroborated it is assigned a truth value in a given domain — the extension of the proposition. Rarely does this extension overlap with the whole range of facts of the kind; in other words, the extension of law statements is usually limited. Thus experiments shows that the simple gas law breaks down at extreme values of pressure and temperature. These deviations disconfirm that particular law statement and call for its correction; moreover, if read within the context of the kinetic theory the discrepancies themselves suggest definite modifications, such as blowing up the point-like molecules to small spheres and gluing them with a molecular force field. That is, one will try to find reasons for the deviations and will eventually come up with a new, usually more refined and complex theoretical statement. Empirical relations are the starting point not the culmination of physical research.

Let us finally discuss two tricky sets of laws$_2$.

3.3. Variational Principles

If a variational principle stands at the apex of a physical theory then the theory exhibits its unity and its invariants. But unless philosophical soberness is kept, the introduction of variational principles can result in confusion: remember MAUPERTIUS' and PLANCK's linking extremum principles with teleology and even theology.

We shall deal only with integral variational principles or *action* principles. An action principle is a statement of the form "$\delta S = 0$" where 'δS' designates the first variation of a functional S of a set of physical variables $q, q', q'', \ldots$, referring to a physical system and such that (a) S is a definite integral, (b) $\delta q = 0$ at the integration limits, and (c) "$\delta S = 0$" entails physical laws — say equations of motion or field equations. The variations δq are conceptually possible differences. (In general virtual displacements are physically impossible: BANACH, 1951.) An action principle says then that these possible differences combine in such a way that the actual S value of the system is not affected: S is always either a maximum or a minimum. It does not say that the system strives to conserve S but just that it keeps a constant S throughout its changes: no teleology is implied. Nor does a variational principle indicate which, among all the courses of nature limited by the condition that they pass through the given extremes q_1 and q_2, will actually be realized: the actual course of events must be charted with the help of the laws of motion and the supplementary conditions.

In most cases a variational principle is mathematically equivalent to the laws of motion (or field equations) it entails. But this does not mean that the two are identical: in fact given a set of laws of motion we can usually manufacture infinitely many action principles that will entail them. For example, if a lagrangian density $\mathscr{L}$ leads to the correct laws of motion of a continuous system, so does $\mathscr{L}' = \mathscr{L} + V \cdot \mathscr{A}$, where $\mathscr{A}$ is an arbitrary vector vanishing on the boundary of the system. (On the other hand certain magnitudes derived from $\mathscr{L}$, such as momenta, may not be invariant under this transformation.) Consequently an action principle is stronger than the set of laws it entails. It is correspondingly further removed from experience. Indeed, with the exception of the simplest of all, which is FERMAT's principle, the integrand of an action principle is nonmeasurable. Moreover, the precise form of the lagrangian is immaterial as long as it has the prescribed transformation properties and leads to the desired equations of motion or field equations. Consequently variational principles are not empirically testable in a straight-forward way: what can be tested are some of the solutions of the laws of motion or the field laws they entail.

Given a set of such laws there is no guarantee that a corresponding variational principle will be found. If a law is expressible as a self-adjoint differential equation then it is derivable from a variational principle. This sufficient condition is not necessary: less restrictive conditions can be found for casting a set of differential equations into the Lagrangian form (see HAVAS, 1957). But then not all laws are differential equations: some are integro-differential equations and some are algebraic relations such as the commutation and anticommutation relations in quantum field theory. All of these fall outside the scope of action principles, which is therefore large but not unlimited.

The reasons for wishing to set up action principles are the following (HELLINGER, 1914; CORSON, 1949; LANCZOS, 1949; BUNGE, 1957). An action principle (a) compresses into a single formula the central state-ments of a theory and consequently (b) the compatibility of the resulting Euler-Lagrange equations is automatically ensured; (c) when the equa-tions of motion or the field equations are unknown, it helps finding them for it requires only very general conditions — e.g., Lorentz invari-ance — on a single functional S; (d) it transmits its invariance properties to its immediate offspring, the Euler-Lagrange equations; (e) by NOETHER's theorems it entails conservation laws, and correspondingly the transformation groups that preserve the principle are easily recogniz-ed. Which brings us to our next subject.

3.4. Conservation Laws

3.4.1. Conservation-Like Equations. If the rate of change of a function is either zero or is balanced by some other function, we can call the whole a *conservation-like equation*. Any fluent may thus be balanced, whether or not it represents an actually conserved physical property. In fact given an arbitrary scalar ϱ and an arbitrary magnitude $\mathscr{A}$ (scalar, vector or tensor), it is always possible to find a flux density function or functional $i[\mathscr{A}]$ and a source density function or functional $s[\mathscr{A}]$ such that in any region V of space

$$\frac{\mathrm{d}}{\mathrm{d}t}\int_V \mathrm{d}^3 x \cdot \varrho \mathscr{A} = \oint_{\partial V} \mathrm{d}^2 x \cdot i[\mathscr{A}] + \int_V \mathrm{d}^3 x \cdot \varrho \, s[\mathscr{A}] \qquad (1.21)$$

(TRUESDELL and TOUPIN, 1960). If the functions occurring in this formula are smooth enough, the integral statement is equivalent to a differential equation, otherwise it is stronger than a local equation. In particular, for $\mathscr{A} = 1 \wedge i = 0 \wedge s = 0$ (corresponding to a closed system without sources and sinks), we have the familiar continuity equation: $\dot{\varrho} + \varrho \, V \cdot \dot{x} = 0$.

The mathematical skeleton (1.21) is physically meaningless as long as $\varrho, \mathscr{A}, i$ and s remain physically meaningless (uninterpreted). And upon attaching these functions a physical meaning we get a physical statement but not necessarily a true one. In other words, not every conservation-like equation models a genuine conservation law. The best way of finding conservation laws is to derive them from basic laws, be it integral varia- tional principles or their immediate logical consequences. We shall stipu- late that only the conservation-like equations accompanying laws of mo- tion or field equations deserve being called *conservation laws*. Several reasons underly our stipulation: (*a*) any general statement entailed by a law$_2$ is itself a law$_2$ not just a mathematical skeleton; (*b*) only laws con- cerning the flux of events can say what is preserved in the midst of change; (*c*) the bewildering (actually infinite) multitude of conservation- like equations occurring in the theories of continua and fields, particularly in CEM and in GR is thus automatically shrunk to a manageable set. (In GR genuine conservation laws are called *strong* and our conser- vation-like equations are called *weak* conservation laws.)

3.4.2. Conservative Systems. The most important conservation laws have the form "Div $T = 0$", where T is the energy-momentum-stress tensor of the system and 'Div' abbreviates the four-divergence. A system with conserved T is called a *conservative system* — not to be mistaken for a conservative force, i.e. one deriving from a scalar potential. Most of physics is concerned with conservative systems. And if a given system is not conservative one hopes that it can be embedded into a larger system which has this property exactly or to a good extent. This hope is unful- filled in two important cases. One is the class of short-lived elementary particles: since they decay spontaneously — meaning in the absence of known external disturbance — their Div T should depend explicitly on space and time coordinates (PROCA, 1943). Unfortunately this lead does not seem to have been followed up: such is our conservatism. The second exception is the gravitational field: only locally, i.e. in the flat space approximation, does one obtain the classical conservation laws. More on this in 3.4.3.

Quantum mechanics (QM) is sometimes said to *violate energy conservation* because energy need not be conserved in the transitions between intermediate or virtual states under the action of an external perturbation: the over-all process is conservative but at any given step the system can either borrow or lend some extra energy. Yet the general theory does involve energy conservation for time-independent interactions, hence the nonconservation in the case of virtual transitions introduces an inconsistency. This contradiction is usually dodged with a pinch of subjectivist (operationalist) philosophy: since virtual states are unobservable — being in fact intermediary between states that are wrongly supposed to be observable — no violation can be detected, and what cannot be measured is meaningless and what is meaningless has no existential import. This solution — which curiously enough acknowledges that QM handles unobservables — is like holding that it does not matter whether honesty rules are broken as long as nobody notices it.

To an objectivist the solution is straightforward: he refuses to attach a physical meaning to every step in a calculation. In particular, the intermediate states introduced by perturbation theory should not be attached objective referents: QM in its present state accounts for quantum jumps in a wholistic way, and in any case perturbation theory is not a physical theory but a computation method. The same holds for the virtual quanta of QED. And the two cases are parallel to the infinitely many possible resolutions of a force into components and to the infinitely many possible series and integral expansions of a function: they are formal devices.

3.4.3. Symmetries.

Euclidean and pseudoeuclidean spaces have certain symmetry properties not shared by other differentiable manifolds such as Riemannian spaces. The symmetries characterizing the spacetime E^{3+1} of special relativity are particularly important since deviations from flat space are significant only in the large. Certain transformations of E^{3+1} onto itself, such as infinitesimal translations along the coordinate axes and spatial rotations about a fixed point, are associated with invariants that are both physically meaningful and important. Thus energy conservation corresponds to the invariance of the lagrangian under a shift of the time origin. Something similar holds for all other classical conservation principles: every one of them is mathematically equivalent to a symmetry principle concerning the lagrangian of the system (NOETHER, 1918). Thus the conservation of linear momentum is equivalent to the invariance of the lagrangian under infinitesimal spatial displacements — i.e. place is irrelevant as long as there are no fields.

This shows that the properties of physical systems and those of space-time cannot be separated although they are distinguishable. I.e., in every case we should know whether a property is dependent on the structure of spacetime or not, for if it does not there may be no reason to adopt one metric rather than another. Moreover the relation between physical systems and the spacetime in which they are embedded — short for the net of relations of the given system to everything else — enables us to probe into the spacetime structure by watching the behavior of the high level representatives of matter and fields (lagrangian densities and the associated energy-momentum tensors). Indeed every conservation law that can be linked to a symmetry property (via an action principle) can be used to test the spacetime theory presupposed by that conservation law. The failure of a conservation hypothesis may indicate that the assumed spacetime lacks the corresponding symmetry and thus suggests modifying it. For example, in GR the classical conservation laws do not hold in the large because Riemannian spacetime has no inherent symmetries: those laws hold only in the small (flat space approximation) or for a fictitious single isolated system, for which spacetime is Euclidean at infinity (TRAUTMAN, 1962). Ever since the law of conservation of energy was shown to hold only regionally (HILBERT, 1917), other conservation laws$_2$ were shown to be only approximately true — so much so that the misprint *conversation laws* has already occurred.

The relation between dynamical properties and spacetime becomes less intimate at the microphysical level: a number of important invariants, such as electric charge, isotopic spin (kind of nucleon) and baryon number (No. of baryons — No. of antibaryons) are associated to no known symmetry operations in ordinary spacetime. Thus charge conservation corresponds to the invariance of lagrangian densities — which are bilinear or 4-linear forms in ψ and ψ^+ — under gauge transformations of the first kind: $\psi \rightarrow \psi\, e^{i\varepsilon}$, ε being the parameter of the one-parameter group. In other words, in today physics these are intrinsic properties. This suggests modifying the hypothesized structure of spacetime or adding on intra-system space (J.-P. VIGIER, in BUNGE, ed., 1967d). Anyway while some invariants are spatiotemporal symmetries others are so far independent of spacetime.

3.4.4. Symmetries and Observers. The association between some of the invariants and spacetime symmetries is often stated with reference to measurement operations, e.g. thus: "If a system is isolated then the results of a linear momentum measurement on the system do not depend on where the measurement is performed." Similarly for energy (irrelevance of "when") and angular momentum (irrelevance of the direction of observation). But neither the basic physical theory nor group theory refer to measuring devices and observers; hence these operationalist formulations of symmetry properties involve an *ad hoc* interpretation of the formulas, which refer to the bare system not to the system-apparatus-observer complex, as shown by disclosing the variables involved. Moreover, the attempt to read every conservation law in operationalist terms is self-defeating for, if the energy of an isolated system does not depend on the time at which energy measurements are made, then the property is clearly an objective one and the operator is supernumerary. It is like saying "No matter who looks at the Mt. Blanc he sees it covered with snow". The same holds for all the other symmetries: they can be restated as laws of indifference with respect to the observer, which shows that the latter is supernumerary in physical theory (see 4.1.6 on semantical closure). To be sure, the observer has got to design the experimental arrangement, operate it and interpret its readings (with the help of theories), but he must withdraw from the final picture if this picture is to be objective.

3.4.5. Conservation and Change. It is occasionally believed that the existence of conserved quantities and the associated symmetries is incompatible with a dynamic world view. This is mistaken: invariants are traits that remain unchanged through change — they are invariants (or symmetries) under certain transformations. They are not substances but patterns of a changing world. Physics is still closer to Heraclitus than to Parmenides.

4. Theories

A theory, e.g. MAXWELL-BOLTZMANN's statistical mechanics, is a *hypothetico-deductive system*, i.e. a set of hypotheses glued by the relation ⊢ of deducibility or entailment. In a theory no formula is isolated: every statement is either a basic assumption (= axiom = postulate) or a logical consequence of previously asserted formulas — unless it happens to be a definition. What is peculiar to *physical* theories is that they contain semantical assumptions (interpretation hypotheses) conferring a physical meaning upon its basic symbols: physical theories are physically interpreted formalisms, i.e. formalisms with an intended objective reference. For a physical theory to be scientific not speculative it must be testable, in addition to having a reasonably correct formalism or structure and a definite content or meaning: it must be capable of matching empirical data and other theories covering adjoining fields. Briefly, a *scientific physical theory* is characterized by these traits: mathematical formalism, physical meaning, and testability. In this section the first two traits will be handled; testability will concern us in Sec. 5.

4*

4.1. Form and Content

Every hypothesis belonging to a physical theory, whether it is postulated or deduced, refers to reality and is corrigible. But the relation among a set of premises and its logical consequences is rigid: unlike the highly ambiguous relation between a set of facts and the corresponding theory, the relation of entailment obtains or it does not within a given body of statements (and with a given underlying system of logic). In other words, while no set of data points unambiguously to a given theory and no theory, if factual, is more than partially true (see 1.3.11), entailment is an all-or-none affair. In fact for a theorem t to be deducible from a set A of assumptions, the conditional $A \Rightarrow t$ must be logically true: it must hold regardless of fact (see 1.2.6 and 1.3.10). Unless a set of formulas has the rigid structure given by the entailment relation it does not count as a theory.

4.1.1. Logical Structure. Every well-formed statement is a formula of the predicate calculus with identity (PC=). Disregarding the fine mathematical structure, a theory is, as regards its form, a set F of formulas of PC=. This is necessary but not sufficient: a structureless set of statements is not a theory. If we want a theory, the whole set F of its formulas must be *closed* under deduction — i.e. deduction within F must yield no formula outside F. In brief, a theory is a *relational structure* $T = \langle F, \vdash \rangle$, where the relation $\vdash$ is the one of deducibility (see 1.2.6), which has the gross properties of the relation $\leq$ of partial order.

The relation $\vdash$ that orders the set F of formulas of a scientific theory T is characterized by the rules of inference of PC=. It is not permissible to break the logical unity of science by proposing a theory employing some nonclassical system of logic. If a logic other than PC= were to underlie one scientific theory, all other theories would have to be reformulated on the basis of the same exotic logic, for otherwise it would be impossible to apply them jointly to the explanation of facts and the design and interpretation of experiments, as each of these procedures summons a number of different theories. In other words, the theory with an extraordinary logic would remain isolated: inapplicable and untestable. Therefore the claim that QM has its own logic is a joke (see Ch. 5, 8).

To return to serious matters: historically and psychologically almost anything can spark off a theory: the wish to explain a fact, the urge to bring order to a chaotic mass of data, the desire of unifying hitherto separate fields, etc. But logically the starters of a theory are certain strong *initial assumptions* which suffice to generate the whole set F of formulas by sheer logic. Clearly, for this to happen, i.e. for a set of initial assumptions to generate a whole theory, they must meet certain conditions: (*a*) they must talk about the same thing (the universe of discourse of the theory), (*b*) they must somehow dovetail, and (*c*) at least some of them must be universal. Otherwise there could be neither deduction nor subsuming of particulars under generals. A set of data (singular propositions) does not qualify as a theory starter, nor does a set of universal but altogether heterogeneous propositions concerning disjoint universes of discourse, such as atoms and experimenters: in neither case can deducibility relations be established — except trivial and irrelevant ones such as $p \vdash p \vee q$ and $p \vdash q \Rightarrow p$, where p is given and q is an arbitrary statement (e.g., $p =$ OHM's law, $q =$ God is almighty).

The subset $A \subset F$ of logical (not historical) generators of all the remaining formulas of a theory T is the *axiom base* of T. The set F of an axiomatizable or actually axiomatized theory is the collection $Cn(A)$ of all the logical consequences of A: $F = Cn(A)$. In turn, $Cn(F) = F$ (closure under deduction). In particular, $Cn(\emptyset) = L =$ set of logical truths. An axiomatizable theory looks like a network or a tree and is in fact so represented. Look at a finite part of the tree representing an imaginary theory with two axioms and six theorems:

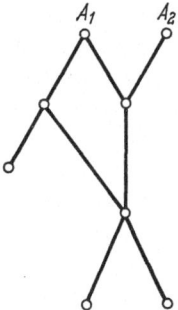

The calculus of deductive systems (TARSKI, 1956) studies both the inner structure of theories and the inter-theory relations. Among these two stand out: formal identity and formal inclusion. Let A_1 and A_2 be the axiom bases of the sets of formulas F_1 and F_2 of two theories T_1 and T_2 respectively. Then the following theorems can be proved:

DS 1 T_1 is (formally) *identical* with T_2 $(F_1 = F_2)$ iff $A_1 \Leftrightarrow A_2$.

DS 2 T_1 is (formally) a *subtheory* of T_2 $(F_1 \subseteq F_2)$ iff $A_2 \Rightarrow A_1$,

where A_1 and A_2 are the conjunctions of the respective postulates. In words. (*a*) If two axiom bases are equivalent they generate the same theory: one speaks in this case of two different *formulations* of a single theory. But notice that the equivalence at stake is formal not semantical: two formally identical theories may have different physical meanings. Thus two different physical theories may be brought under the same Lagrangian formalism even if they refer to altogether different facts. And FEYNMAN's spacetime formulation of QM, although mathematically equivalent to the usual theory, is semantically different from it since it postulates that the systems concerned are point particles with definite though random trajectories — so much so that the ψ at a spacetime point is a sum over all paths reaching that point from the past. Two theories will be said to be *completely equivalent* iff they are both formally and semantically equivalent — i.e. provided they are just different ways of saying exactly the same things. (*b*) If T_1 is a subtheory of T_2 then T_1 is also said to be (formally) *reducible* to T_2, which is in turn called an *extension* of T_1. Examples: statics, which used to be an independent science, has been reduced to dynamics (notwithstanding D'ALEMBERT's claim), just as geometrical optics is a subtheory of wave optics. Reducing one theory to another amounts to proving the weaker one upon asserting the stronger theory. Methodologically things are the other way around: the stronger theory gains support from its subtheories if the latter pass the test of experience.

4.1.2. Mathematical Structure. Every theory presupposing mathematics has not only a logical but also a mathematical structure that logic is much too universal to discern. Even logical theories have a mathematical structure; thus the mathematical structure of the propositional calculus is Boolean algebra. The logical together with the mathematical structure of a theory make up its *formal structure*. Any theory presupposing analysis has, in addition to its logical structure, what may be called an *analytic structure* given by its transformation properties. Thus one theory is invariant under space inversions and time reversals (severally or jointly), the other under canonical transformations, and so forth. These invariant properties of a theory are expressed by metasentences such as 'Every local field theory invariant under Lorentz transformations is invariant under the combined charge, time, and space inversions' (the PCT combined parity theorem). The set of statements exhibiting the transformation properties of a theory is called the associated *transformation theory*. A transformation theory is a *metatheory*, a theory about a theory, that characterizes the analytic structure of the object theory — just as the calculus of deductive systems analyzes the logical structure of theories.

The metatheoretical statements concerning transformation properties do not speak directly about nature: their concern is a theory. In fact, establishing whether or not a given theory has certain transformation properties is a purely conceptual operation. Moreover, this operation may fail to have a factual counterpart. Thus we cannot reverse the direction of time: all we do is to change t into $-t$ in certain equations, or to reverse the motions. Consequently it is hopeless to try to give "operational definitions" of all such transformations as well as of the corresponding symmetries or invariances (see 3.4.4).

A theory with a known formal structure is a grown-up mathematical theory or *formalism* that must be attached a physical interpretation if it is to count as a physical theory. Let us then help ourselves some filling.

4.1.3. Abstract Theory and Model. An *abstract* theory is a deductive system containing only uninterpreted symbols apart from the logical ones. Examples: Boolean algebra, lattice theory, group theory. One of the simplest abstract theories is the one of partial order (P), which contains two undefined and uninterpreted specific symbols: U (a set) and $\leq$ (a binary relation). The pair $\mathsf{P} = \langle U, \leq \rangle$ is subject to the sole conditions (postulates)

$P\,1$ For every x in U, $x \leq x$ (reflexivity).

$P\,2$ For every x and every y in U, if $x \leq y$ and $y \leq x$, then $x = y$ (antisymmetry).

$P\,3$ For every x and every y and every z in U, if $x \leq y$ and $y \leq z$, then $x \leq z$ (transitivity).

By interpreting the basic (primitive) symbols of an abstract theory it acquires a meaning. Every such interpreted theory is called a realization or *model* of the theory if in fact it satisfies the axioms of the theory. There is no limit to the number of models of an abstract theory. But they must be true interpretations, i.e. they must respect both the structure of the concepts and the axioms. Thus if in P 'U' is interpreted as the set of atoms and '$\leq$' as "less patriotic than", a set of meaningless statements is obtained. And if 'U' is interpreted as the set of girls and '$\leq$' as "uglier than", a false interpretation of P is produced, as the relation of ugliness is not transitive. On the other hand the following interpretations of the primitives of P produce models of P:

$$M_1(\mathsf{P}) \begin{cases} I(U) = \text{set of all statements of a kind} \\ I(\leq) = \vdash \end{cases}$$

$$M_2(\mathsf{P}) \begin{cases} I(U) = \text{set of all points on a straight line} \\ I(\leq) = \text{to the left of or coincident with} \end{cases}$$

$$M_3(\mathsf{P}) \begin{cases} I(U) = \text{set of bodies} \\ I(\leq) = \text{lighter than or as heavy as.} \end{cases}$$

These are, respectively, a logical, a geometrical and a physical model of the abstract theory P. (Physical, not operational, for if '$\leq$' is interpreted as "was found to weigh less than or equal to", $P\,3$ is falsified every time a scale is employed which can discriminate between a and c but not between a and b, and b and c.) In other words, the *semantic systems* $M_1(P)$, $M_2(P)$ and $M_3(P)$ are models of P within logic, geometry, and physics respectively. The first two are *conceptual models* since U and $\leqq$ are made to correspond to further concepts: in other words, the first two models are interpretations of P in certain theories (logic and geometry). On the other hand, since $M_3(P)$ has a factual referent, it will be called a *factual model* of P. Every physical theory is a factual model of its underlying abstract skeleton.

Whether conceptual or factual, a model $M_n(T)$ of a theory T must respect the structure of the basic concepts of T and satisfy its axioms. Conceptual models are as definite as the interpreting concepts, which can be specified by further formal (but not abstract) theories. Such models are studied by *model theory* (ROBINSON, 1963). On the other hand all factual models are somewhat hazy because the objects they point to are extraconceptual and therefore at best half-known. In particular, the physical meaning of any given formalism is somewhat vague: even if all the interpretation assumptions are stated explicitly, they do not determine exhaustively the meaning of the symbols concerned. In factual science meanings are never fully determined unless they are trivial (see 1.3.7, 4.2.5 and 4.2.6). In short, physical theories are *partially interpreted* hypothetico-deductive systems. The reason for this semantical vagueness is not that some of the theoretical concepts cannot be defined operationally (none can), but that their referents are extratheoretical and are known gradually. This vagueness, far from excusing sloppy interpretation jobs, calls for utmost care in discussing the specifically physical ingredients of a physical theory — a discussion that may well be as endless as the growth of our science.

In conclusion, a physical theory is a formalism endowed with an interpretation. The formalism is a set of scraps of mathematical theories and is therefore referentially noncommittal: it is the physical interpretation that coordinates some of the mathematical symbols with properties of a physical system. This interpretation must be distinguished from the means whereby the truth value of the theory is ascertained: thus, long before neutrino detectors became available theoreticians had to attach a definite content to neutrino theories — otherwise the very search for those elusive entities could not have started.

4.1.4. Meaning and Testability. A theory has a *physical meaning* iff it contains interpretation assumptions that assign physical correlates to its basic concepts. These correlates (referents) need not be and in general are not perceptible. But they must be *scrutable*, i.e. they must show up as observable effects even if remotely, i.e. at the end of a long chain accounted for by other theories. The phenomenalist claim that science ought to eliminate unobservables derives from a confusion between *unobservability* — unavoidable in most cases — and *inscrutability*. This claim is just an instance of the widespread confusion of *test* with *reference*. For a theory to be testable it need not refer to phenomena or to human operations. Moreover, by definition a hypothetico-deductive system referring to human experience is a psychological not a physical theory.

The reference to observable or measurable traits is unnecessary for meaning as well as for testability; what is needed for the latter is the existence of further theories capable of bridging unobservables to observables (more in 5.1.4). A physically meaningful theory can moreover be untestable by itself. Thus the electrostatics of a single isolated spherically symmetrical charge is perfectly meaningful. But the theory is obviously untestable, as every real test requires the existence of at least one more charged body — if only a test body. On the other hand every empirically testable theory must be interpreted before we can hope to test it: we must know what it is about — and this is all that is meant by 'meaning'. In short, meaning is necessary though insufficient for testability, and the latter is sufficient though unnecessary for meaning.

Yet the popular doctrine is that meaning and testability are equivalent. This doctrine, popularized by the Vienna Circle under the name of *verifiability theory of meaning*, has long since been abandoned by philosophers and it only lingers among scientists, many of whom take it as an article of faith. The doctrine involves a confusion between semantics and methodology, rooted in turn in a confusion between reference and test. Worse, it endangers every deep theory, as every such system is untestable without the assistance of further theories: just try to test a field theory without the help of mechanics. Relativistic thermodynamics, for instance, is sometimes condemned as meaningless on the strength of that tenet, because one does not know yet how to test, say, the transformation formula for the temperature. Yet the theory can be tested indirectly, namely through its consistency with other theories that are less indirectly testable. Moreover, given the rest of relativistic physics, relativistic thermodynamics follows logically from the former and is therefore necessary, whence if it is condemned all its premises ought to go as well. Moral 1: Do not mistake testability (a methodological requirement) for meaning (a semantical condition). Moral 2: Do not convict scientific theories on grounds of philosophical prejudice. Moral 3: Keep your philosophy up to date.

4.1.5. Meaning and Reality. An interpretive hypothesis, such as "e is the electron charge", involves the assumption that there are certain physical objects — e.g., electrons — that is, certain things out there, independent of the mind. The referents of a theory constitute the latter's *reference class*. It is assumed that every such reference class is nonempty, i.e. that at least one member of it exists. But this is an assumption that may turn to be false. Therefore one speaks of the *hypothetical* or *intended* referent of a theory — in the philosophical not the psychological sense of the word. Nonetheless a physical theory does talk, even though hypothetically, of real entities: total fictions are left to literature (e.g., HERTZ, 1894; BOLTZMANN, 1905; PLANCK, 1909; EINSTEIN, 1949).

If the theorist assumes reals, the experimentalist takes many of them for granted either because he manipulates them (e.g., his apparatus) or because he employs certain reality hypotheses without questioning them all simultaneously. Thus when scanning a set of tracks in a nuclear emulsion he assumes that the tracks were produced by some entities, often imperceptible, in a way that the theorist is supposed to explain in detail. Moreover the experimenter will assume the existence of entities that leave no trace at all on his plates, such as those originating forks that seem to emerge out of the blue: not being a continual creation cosmologist, he will attribute those forks to the disintegration of neutral particles. He will thereby use the principles of conservation of mass and charge, which are not put to the test in this case but on the contrary render the whole operation meaningful — just as optics renders the use of microscopes intelligible and therefore justifies it. In no case does the experimenter remain content with recording, interrelating and reporting on observations: in every case he hypothesizes real things as sources of appearances.

Whether or not the reality hypotheses contained in a theory are true must be inferred from the way the theory matches the body of accepted knowledge and from the performance of the theory in describing and predicting events amenable to empirical control (always with the assistance of further theories). This is a fallible not a fool-proof criterion of physical reality: some or even all the entities hypothesized by the theory may eventually be shown to be unreal. But no other criterion is known. And in any case, when a set of low level theorems of a theory are confirmed, one "concludes" (assumes) that the dressed formulas not the mathematical skeletons of the theory are (partially) true. Experience cannot prove theories but can render some of them pointless, others false, and finally others likely — and whenever the latter happens conventionalism and instrumentalism are refuted. The most popular criterion of reality is the one of measurability: "To be is to be measurable." Like most popular beliefs about science, this one is inadequate: measurability is very often indirect and is always dependent on theory — remember that our predecessors made accurate measurements of properties of nonthings, such as the caloric, and that hundreds of our contemporaries are probably measuring properties of fictive particles.

In any case, every physical theory presupposes the *philosophical hypotheses* that there are physical objects (mind-independent things), that most of them are imperceptible (HERTZ, 1894), and that some of them are knowable if only in part (THOMSON, 1963). Should these hypotheses be dropped we would turn to introspection and mysticism. In addition to those general existence hypotheses, every physical theory makes *special existence hypotheses* — e.g., that there are e.m. fields. These assumptions are part of the semantics of physical theories; they occur right at the start of our axiomatizations in Chs. 3—5. As any other hypotheses, these are fallible; but without them there would no point in building physical theories. And no possibility either, for a physical theory is, by definition, a theory referring (even if wrongly) to physical (observer-independent) objects.

4.1.6. Formal and Semantical Desiderata. A good physical theory satisfies a number of conditions that are prior to its experimental check; some of them are formal, others semantical. The formal desiderata are:

Internal consistency (mandatory): every theory should be free from contradictions, for otherwise it will entail anything — though nothing validly. (In fact: a false A entails anything, as $A \Rightarrow t$ is then logically true.) Consistency can be unwittingly violated upon trying to extend a given theory by adding some hypothesis. Thus by trying to extend dynamics to cover dissipative systems, certain laws concerning friction are occasionally adjoined which clash with the mechanical axioms. Conversely, a contradiction can be removed by dropping or modifying one or more assumptions: inconsistency, though sinful, can be atoned for. Relative consistency *test*: exhibit a contradiction-free model — e.g., an arithmetical interpretation.

External consistency (mandatory): compatibility of the given theory with the accepted noncompeting theories in the same field and in contiguous fields. A theory is automatically consistent with the theories it presupposes (is based upon). Therefore progress in factual theories cannot lead to refuting any of the formal theories they presuppose, say ordinary logic and probability theory. Truth tests include tests for external consistency: a thoroughly off-beat theory that contradicts every other theory is not even considered. Such a theory could not even be subjected to empirical tests, as these are always backed up by a number of "square" theories. The various correspondence principles of physics — e.g. ,"Locally GR→SR" — fulfil the function of testing for external consistency. Caution: this requirement should not be used to discredit partial revolutions, i. e. genuine novelties that do not clash with everything (see 5.2.3).

Primitive independence (highly desirable): the basic (= undefined = primitive) concepts of a theory should be logically independent (not interdefinable). Otherwise they cannot all be basic and one will not know of some formulas whether they are definitions or hypotheses. *Test:* reinterpret the primitives of a theory one at a time and check whether the axioms are still satisfied — i. e. whether the reinterpretation of one primitive after another makes any difference. Caution: do not mistake *logical* independence for either mathematical or semantical independence. Two concepts are *mathematically independent* in a given theory iff the theory does not relate them; they are *semantically independent* in a given theory iff they have different referents in it. Thus mass and charge are logically but neither mathematically nor semantically independent in the electrodynamics of massive bodies; and the metric and the matter tensor in GR are logically and semantically independent but mathematically interdependent.

Postulate independence (desirable): no basic assumption should be deducible from any other hypothesis in the theory. Independence *test:* drop or negate the postulates one at a time and check whether a consistent system remains.

These formal desiderata must be supplemented by the following semantical desiderata, the fulfillment of which make up the *semantical unity* of a theory (BUNGE, 1967b):

Unity of reference (mandatory): the theory should have a definite reference class, i. e. it should concern the members of a definite universe of discourse — e.g., bodies, or particle-field couples. Otherwise no logical relations among the formulas can be established. (Just try to conclude validly from the Schrödinger equation to states of mind or conversely.) Every physical theory has among its basic or primitive concepts one or more sets whose elements are assumed to mirror physical objects, and every specific statement of the theory is about members of that set — usually designated 'Σ' in this book. Only the generic statements, e.g. those concerning space and time taken as unanalyzed concepts, and those which have no physical content at all, will not refer to members of Σ.

Conceptual connectedness (mandatory): the basic concepts of the theory should be interrelated: not every postulate should concern a single primitive. Otherwise the axioms won't dovetail and deduction won't start. Thus from "Every A is a B" and "Every C is a D" nothing can be concluded unless a bridge is introduced — e.g., "Some B are C".

Semantical closure (highly desirable): no predicates other than those admitted at the start or introduced by definition should be allowed in the theory — i.e., no newcomers. Thus, if thy formulas refer to things physical, thou shan't interpret them in terms of things human such as uncertainty, predictability and observability. Otherwise anything can be concluded from any given statement by virtue of the inference rule "$t \vdash t \vee u$" with u arbitrary. Also, irrelevant conditionals can be correctly derived unless the restriction of semantical closure is observed: thus from "Hydrogen is diatomic" we validly but irrelevantly conclude "If an observer is looking then hydrogen is diatomic". (In fact, $p \vdash q \Rightarrow p$.) In a semantically open context arbitrary shifts of meaning can occur whereby a formula is sometimes assigned one interpretation and at other times another. Thus in QM standard deviations are sometimes interpreted as intrinsic random fluctuations, at other times as the disturbance caused by an apparatus on the system, or even as our subjective uncertainty concerning the latter. No such shifts of meaning can occur in a semantically closed context (see Ch. 5).

Semantical homogeneity (desirable): the predicates of the theory should belong to the same semantical family — e.g., entropy, a macrophysical property, should not be attributed to individual nuclei. Otherwise queer hybrids are born, such as the idea that an atom will jump from one state to another just because we read a gauge after the latter has done its job — an assumption actually occurring in v. NEUMANN'S theory of measurement (v. NEUMANN, 1931).

The above logical and semantical requirements are easier stated than tested for and satisfied, but anyway they characterize a well-built factual theory. For internal consistency, primitive independence and postulate independence there are definite tests elaborated in metamathematics (HILBERT-BERNAYS, 1934, and STOLL, 1963). Yet these tests are often hard to apply even if the theory is well organized (axiomatized). Thus a consistency proof — seldom supplied in physics — can be quite a feat. (The real number system and Euclidean geometry are so far only believed to be consistent.) External consistency can be even harder to establish and so is semantical homogeneity. On the other hand unity of reference, conceptual connectedness and semantical closure are automatically enforced by axiomatization.

4.2. Physical Axiomatics

Scientific theories are reared by logic but are not born from it: they are conceived in odd ways, with the help of analogies, heuristic clues, and metaphysical principles. Their structure and content emerge only gradually, as they mature. At some point they become ripe for axiomatization: this happens when the background and the essential components of the theory have been recognized. The *background* of a theory is the set of ideas the theory borrows and takes for granted or presupposes. And the *essential* (indispensable) components of a theory are the hypotheses that characterize it and that could not be changed without getting an entirely different theory — e.g., MAXWELL's equations in CEM. Every statement of a well organized theory T is either an essential initial assumption A of T, or a logical consequence t of essential assumptions, previously derived consequences, and pieces B of the background of T: $A, B \vdash t$.

4.2.1. Background. Save in elementary logic, which so far is background-free, there is no theorizing in a vacuum. Except logic, which follows from nothing, every theory is built on the basis of some old ideas — logical ones to begin with. Such fundamental ideas occur not only as heuristic clues — as was the case of hydrodynamics in building electrodynamics — but also as proper ingredients of the new theory. Only they shall count as the proper background of a theory: after all, heuristic guides need not be kept after the theory has been built.

The background of a physical theory consists of two sets of ideas: formal and nonformal. The *formal background* of a physical theory consists of all the logical and mathematical ideas it employs. Thus geometrical optics (GO) presupposes ordinary logic (PC=), elementary set theory, topology, analytic geometry, and analysis: this is the formal background of GO. The *material background* of a physical theory consists of all the generic and specific theories it presupposes. Thus the systems theory, Euclidean physical geometry, and a theory of time occur in the material background of GO. This theory presupposes only generic physical theories — or protophysical theories as we shall call them. Other theories have a richer and more specific background; thus SR presupposes CEM. When axiomatizing a theory we need not display its background but may just list its chief items — otherwise any paper on physical axiomatics would acquire book size. Once the background of a theory has been dug up, its basic units shall be displayed: they constitute the primitive base.

4.2.2. Primitive Base. The fundamental concepts of a theory are the building blocks of its postulates and definitions: no bricks, no building. They are primitive or undefined in the theory under consideration although they can in principle be defined in alternative theories. They are not therefore unanalyzed and indeterminate: the postulates characterize them both formally and semantically (the latter only in outline).

The set of primitive or undefined technical concepts of a theory is called its *primitive base* P. All the so-called independent variables of a theory are undefined in it; the converse is false: dependent variables can be primitives. The set(s) representing the object of study of the theory will be in the family P; the chief aim of the theory is to characterize this object of study (reference class). For example, the primitive base of GO is:

E^3 (Euclidean physical space, with points x)

Σ (set of light rays)

Σ' (set of transmitting media or optical systems)

n (function mirroring the refractive capacity of the medium).

GO does not worry about the first primitive: it borrows it from protophysics (Ch. 2). The proper business of GO is with the *specific primitives* Σ, Σ' and n. The Cartesian product $\Sigma \times \Sigma'$ constitutes the reference class of GO: in fact, every specific statement of GO is about a pair light ray-medium. Whenever the reference class of a theory is a set of individuals, we call the theory *unitary*. If, as in the case of GO, the reference class is made up of two mutually independent classes, then the theory will be called *dualistic*, as it postulates two mutually irreducible "substances"; CEM is dualistic, and GR is *pluralistic* as it concerns any number of mutually irreducible "substances": gravitational fields, e.m. fields, bodies, and anything else we may care to pour into the "matter" tensor.

Once the primitive base is available all other concepts of the theory can be defined — and they must be so introduced in an axiomatic theory. (For example, we can introduce the lagrangian function $L \overset{\mathrm{df}}{=} n\,\mathrm{d}s/\mathrm{d}t$.) In other words, the total stock C of technical concepts of a theory T consists of its primitive and defined concepts: $C = P \cup D$. A technical concept which is neither in P nor in D does not belong by right to T. Thus if the concept of observer does not occur either in P or in D, then it is an intruder. Once the basic concepts of a theory are on hand its basic statements can be stated.

4.2.3. Axiom Base. The basic statements or initial assumptions of a theory determine the nature (structure and meaning) of the primitives. Before advancing such basic statements we have, at least officially, no accurate information concerning the primitives — this being why the brief description of the primitives of GO in the last subsection was enclosed in parentheses as extrasystematic remarks. Those basic (logically fundamental) statements are the *axioms* or *postulates* of the theory. The postulate set A of a theory is called its *axiom base* or *axiomatic foundation*. (There is a catch in these expressions: they suggest that, from a logical point of view, the basis or foundation of a theory is a set of assumptions not a bunch of data, much less a sequence of operations.) Ideally, all other statements of T are entailed by A in conjunction with whatever logical, mathematical and protophysical premises may have to be roped in. Therefore the total set F of formulas of T is the union of A and the set $\{t\}$ of all conceivable (derived and derivable) theorems: $F = A \cup \{t\}$.

In modern metascience 'axiom' means *initial assumption* not self-evident pronouncement. There need be nothing intuitive and there is nothing final in an axiom; so much so that axioms are often tried for the sake of argument, i.e. to see what they entail and whether what they entail is approximately true (BUNGE, 1962a). Just as primitives are definers and builders, so postulates are provers: they should be pregnant with all the derived statements of the theory — even though the actual proof of a theorem may be a hard piece of midwifemanship. This holds for axiomatics in any science. What distinguishes *physical axiomatics* from axiomatics in logic and mathematics is: (*a*) whereas formal axioms are up to a point conventional, being chosen by their fertility, unifying power and perhaps beauty alone, physical axioms are supposed to be maximally true to fact; (*b*) whereas in mathematics one can admit almost anything as long as it does not conflict with logic (and provided it is not boring), and then prove everything else, in physics nothing save formal science can be taken for granted and no conclusive proof can be supplied either — except the proof that a given statement or a given theory (which can be false) does in fact entail certain other statements.

For the primitive base to be adequately characterized by the axioms, there must be one batch of axioms for every primitive concept. Any set of primitive statements sufficient to characterize the primitive base of a theory will be called *p-complete*. P-completeness is necessary but not sufficient to get all the desired statements of a kind — i.e. in a given field of inquiry. A *p*-complete set of axioms will be called *d-complete* if it is necessary and sufficient to derive all the desired formulas (rather than any given statement) in a given field. (But given a stray formula written in the symbolism of the axiom base it may be impossible to recognize whether the formula follows from that base: there may be no possible decision procedure for settling this question, i.e. for ascertaining whether the formula belongs to the given theory. Worse: all physical theories are *undecidable* in this sense, as they presuppose undecidable mathematical theories such as PEANO's arithmetic: see TARSKI *et al.*, 1953.) Obviously, *d*-complete axiom systems are desirable. Clearly, also, one will keep changing any such axiom base as some of its statements are shown to be false or inconsistent with statements regarded as true, and as some other statements are shown not to be part of the theory —

i.e. as the latter is shown not to be d-complete. For example, no axiom system for CM will be accepted if it fails to yield the usual conservation theorems: these are essential though not fundamental components of the theory. (Our requirement of d-completeness is both weaker and more practical than any of the completeness conditions given in metamathematics.)

4.2.4. Example: Axiom Base for GO.

A possible axiom system for ray optics is this:

GO 1 (*a*) E^3 is the Euclidean three-space. (*b*) E^3 maps physical space.

GO 2 (*a*) Σ is a nonempty set. (*b*) Every member σ of Σ represents a narrow light pencil.

GO 3 (*a*) Σ' is a nonempty set. (*b*) Every member σ' of Σ' mirrors a specimen of a kind of optical media.

GO 4 (*a*) n is a function from $\Sigma \times \Sigma \times E^3$ to $\{n \mid Re\, n \geqq 1\}$. (*b*) For every fixed pair $\langle \sigma, \sigma' \rangle$, n is first order differentiable and integrable over any region of E^3. (*c*) The value of n at (σ, σ', x) represents the bending propensity [= refrangibility] of σ' for σ at the place represented by x. (*d*) For every pair $\langle \sigma, \sigma' \rangle$ and any couple $\langle x_1, x_2 \rangle$,

$$\int_{x_1}^{x_2} n(\sigma, \sigma', x) \cdot ds = \text{extremum}.$$

This set of 10 axioms is p-complete in the sense that it characterizes every one of the primitives of GO. And it is d-complete as well in the sense that it covers the field of optics in the zero wavelength limit and for time-independent refractive index. Usually the first three axioms are taken for granted and only *GO* 4d is stated explicitly. This is the central axiom and the only law statement among them: all the other basic statements prepare so to say the stage for the entry of FERMAT's principle. The latter entails the eiconal law and all the lower level hypotheses of GO, such as HERO's minimum principle, the law of rectilinear propagation, and the reflection law. In the intuitive theory all these theorems are derived by using *GO* 1 — *GO* 4 tacitly.

The preceding axiomatization of GO has been made in terms of set-theoretical predicates: indeed the four specific primitives of GO are either sets or mappings among sets. This illustrates the strategy of axiomatizing a theory within set theory (SUPPES, 1957). This strategy will be adopted throughout in this book. Needless to say, using set theory to build a physical theory does not render the latter a part of set theory: a physical theory is a semantic system not a syntactical or formal one (see 4.1.3). An alternative type of axiomatics, one in terms of mappings, is conceivable since every set-theoretical predicate is definable in terms of mappings (LAWVERE, 1964). Such a strategy (categorial axiomatization) would deepen the foundations of physics as regards its structure and would reinforce the focus on function at the expense of substance but it would presumably keep the physical content fairly unaltered. In any case, the important point to note is that there is not a single possible approach to axiomatics.

The preceding it not the sole axiomatization of GO. Thus we could also have added time as a primitive and taken a certain function L, instead of n, as the central primitive, subjecting it to HAMILTON's principle. We would then introduce refrangibility as a derivative concept: $n \overset{\text{df}}{=} L \, dt/ds$. This procedure has the advantage of placing GO within the generous matrix of Lagrangian "dynamics" (a protophysical theory). But nearly every advantage has its price. In this case the disadvantage is that L does not represent a physical property but is a source of properties. Needless to say, this alternative axiomatization would constitute a different theory. Indeed, if upon tampering with the primitive base and the axiom base a different total set F of formulas results then a different theory is obtained since $T = \langle F, \vdash \rangle$. If changes in the primitive base and/or the axiom base result in the same total set F of formulas, then we have to do with different *formulations* of the same theory (recall 4.1.1).

4.2.5. Interpretation. In contrast to mathematics, in physics we must characterize not only the structure and the interrelations of the primitives but also their meaning. This is the function *semantical axioms*, i.e. reference relations, discharge. For example, all the three postulates called (*b*) in the first three groups of assumptions in the preceding axiom system are semantical hypotheses concerning the reference of the various primitives to physical entities of properties. This is not a mere question of nomenclature as would be 'n is short for "refractive index"', which is a conventional sign-concept relation, i.e. a designation rule (see 1.3.1). Semantical hypotheses lay down concept-physical object relations, i.e. reference relations (see 1.3.2). And these relations are somewhat vague (see 1.3.7) and may not be satisfied by real objects — i.e. the semantical axioms may prove false or at least inaccurate.

A semantical axiom such as GO 4c does not fully determine the meaning of the primitive 'n' but just delineates it. In general the semantical or interpretive postulates of a theory trace only the semantical profile of it. Psychological consequence: if one is given the preceding axiom system for GO with no previous knowledge of elementary optics then one is not likely to make much sense of it. The meaning of a symbol, say 'n', is given both by its extension (in this case the set of all pairs light ray-medium, i.e. $\Sigma \times \Sigma'$) and its intension, and the latter is determined by the set of statements in which that concept occurs — both the law statements and the experimental statements regarding the bending of narrow light beams. (Recall the definition of 'meaning' in 1.3.7.)

Physical meanings are then only sketched not fully determined within every single theory: they are determined by the whole of physics. Accordingly meanings do not jump but emerge gradually with the growth of science. This has always been so, but it is more markedly so since the birth of GR and QM, which have shown how much it pays to play with semi-interpreted mathematical frameworks. (So much so that in many cases the resemblance of the theoretical model to its referent is unintentional: DIRAC's theory of holes intended to portray protons not positrons, and YUKAWA's mesons were finally shown to be pions not muons as first thought.) The physical meaning of the formalism of QM was cleared up (and simultaneously muddled) only gradually and there is no reason to believe that its interpretation poses no longer tough problems (see Ch. 5). Indeed, the popular idea that QM was just an inductive synthesis of spectroscopy and some scattering experiments, and that it is now cut and dried, is a fable. This holds a fortioti for relativistic QM: thus of its 16 linearly independent bilinear forms ($\psi\gamma^\mu\psi$, etc.), in the beginning only $\psi^+\psi$ and $\psi^+\gamma \times \gamma\psi$ were attached physical meanings; the interpretation of the remaining densities took two decades and is still controversial. Only utterly false and rejected theories cease to change and are fit to be cut and dried.

4.2.6. Interpretation Procedures. There are four main ways in which the physical interpretation of a theory is worked out: by deriving general theorems, by applying the latter to special cases, by reading the formulas in experimental terms, and by finding analogies with other theories. While the first two are legitimate the last two can be misleading because *ad hoc*. The safest way of finding the physical meaning of a theory is to work it out and apply it to paradigmatic cases — even if purely "academic" like the linear oscillator, found nowhere in nature, much less at the atomic level. Thus in a hamiltonian theory we may not know exactly what the generalized coordinates and momenta stand for until their time derivatives have been computed; in a field theory we will refrain from interpreting the square of a field amplitude (or some more complex bilinear or 4-linear form) as a density unless we have proved that it satisfies a conservation law; in "elementary" particle theories we shall interpret certain operators as spin operators provided they do not depend on position coordinates and that, added to angular momenta, they yield conserved quantities. In addition to finding general theorems we must apply them to simple examples, not because simplicity is the seal of truth but because it is understandable.

5*

In any case physical interpretations are not given from the start —
except in an axiomatic reconstruction — but are "discovered" as
certain logical consequences are drawn and additional specific assumptions are introduced. Interpretation, then, far from being alien to logic
is dependent on it. It is only in axiomatic theories that interpretation
hypotheses are introduced at the start rather than as afterthoughts,
but even so they only sketch the semantical profile of a theory: the
semantics of a physical theory becomes more precise or even alters
radically as the theory grows. Unlike some mathematical systems,
physical theories are not born axiomatic.

The discussion of experimental situations in the light of basic theories
is unlikely to be enlightening because experimental situations are so
specific and complex that a realistic account of them requires several
scraps of different theories in addition to purely descriptive statements.
Most empirical interpretations of theoretical concepts are therefore
phony. Thus the interpretation of 'n' in GO as the result of measuring
incidence and refraction angles, computing their sines and finally dividing
the latter, is *ad hoc* because the foundations of GO are not about gonio-
meters and mathematical operations. We surely expect to check SNELL's
law, but before we start performing the necessary empirical operations
we must know what the law statement states: otherwise we might
start feeding rats or counting unhappy people.

As to analogy as a gate to meaning, the temptation to use it is so
compelling that we tend to forget its pitfalls. If an otherwise opaque
expression is formally similar to an expression having a definite meaning
in another context, we feel driven to transport meaning from one
container to the other. After all, this is how electromagnetic waves
were read in MAXWELL's equations and how wave mechanics was
born. But of course the trick fails more times than it works. It props
the imagination to speak of the phase *fluid* in phase space — after
all, LIOUVILLE's theorem "says" that the "fluid" "moves" *as if* it
were an incompressible fluid; but since the fluid is fictitious, this is
not what LIOUVILLE's theorem can "say". Similarly we may well
speak of field oscillators if the field hamiltonian is formally similar to
the hamiltonian of an aggregate of independent mechanical oscillators — but again those oscillators are fictitious, so that the analogy
fails to supply a correct interpretation of the theory; worse: it encourages
nonsense, like speaking of the mass of field "oscillators". Analogies are
not meaning suppliers but double-edged psychological crutches that
must be handled with care (BUNGE, 1957, 1962a).

4.2.7. What is Gained by Axiomatizing? The axiomatization of a
theory does nothing but organize and complete what had been a more

or less disorderly and incomplete body of knowledge: it exhibits the structure of the theory and makes its meaning more precise. Yet axiomatization is still resisted among physicists. Some fear that axiomatization may eliminate physical content — which will indeed be the case if the semantical axioms are left out as usual. Others fear that upon axiomatization theories may become rigid — an irrational fear for any precisely stated idea is rigid anyway and cannot evolve without changing into another idea. It can be argued that axiomatization, though insufficient, is necessary for attaining full maturity (BUNGE, 1967c), as only by axiomatizing a theory (a) its structure becomes perspicuous — in particular, the status of its formulas (postulates, theorems, definitions) becomes clear; (b) we do not go beyond the assumptions or, if we do wish to go beyond them, then we change them accordingly; (c) we realize that the intuitive or natural theory contained tacit assumptions that were either false or redundant or just unexplored, and anyhow were not kept under control because unstated; (d) the essential concepts of the theory, those which cannot be eliminated or defined, are spotted; (e) the essential or central hypotheses are identified as well as those which provide their proper setting; accordingly (f) one avoids trying to define and prove everything; (g) one can explore what differences would it make to drop or change certain basic assumptions; (h) one avoids regarding as well proved theorems what are stray hypotheses; (i) one does not inadvertently inject at the level of theorems variables missing in the postulates; (j) one can handle if not the whole theory at least its generators and by so doing (k) the weaknesses of the theory can be traced back to its assumptions; (l) one can spot the growth tips of the tree: the assumptions that must be worked out or can stand generalization in the "spirit" of the theory; (m) by attempting to ground certain assumptions on still stronger hypotheses one "deepens the foundations" (HILBERT, 1918); (n) axiomatics discourages woolly philosophical vagaries.

Yet if improperly used axiomatics, just like everything else, can be misleading. For example, it can obscure the problems that gave rise to the theory — which is particularly sad if the theory fails to solve them. And by presenting the theory in its gala garb it can make us overlook its ugly spots. Finally, by exhibiting only the bare essentials it may lead us to forgetting that the actual solution to any particular problem within a theory requires additional premises — special hypotheses and data — that are not among the postulates. Thus we may well forget about coordinate systems in laying down the foundations of a theory but, unless the theory is unconcerned with changes of place, we will have to marry a definite coordinate system when solving the basic equations. Axiomatics, as every other human creation, has pitfalls but they can be avoided with some intuition or common sense and with small doses of history and philosophy of science which, if critical, are the best antifreezers.

4.3. Theory Construction

In order to better understand, work out, criticize and evaluate a theory it is convenient to axiomatize it. But axiomatization is in physics a redecoration job: a theory must be there in a natural or intuitive state before it can be axiomatized. And how do we get a good natural theory? Answer: by hitting on a good problem and getting hold of a good background and a better head. There are no rules for building theories from scratch: there are only some more or less loose heuristic guides, constraints, and desiderata: theories are invented not ground by pure logic or distilled from data. In view of persistent opinions to the contrary we shall give a quick review of how theories are not built and what the invention process does not supply.

4.3.1. Theories do not Come out of the Blue. Theories can be born and killed by inspiration but they are built with pre-existing materials. To begin with, any physical theory has a fairly rich background, or set of presuppositions (see 4.2.1), which enables one to state and work out any specific physical idea. Then, theories are not born spontaneously but in answer to some problem posed in a body of available specific knowledge — e.g., the problem of organizing a heap of facts (phenomenological theory), or of explaining their mechanism (representational theory), or of joining two or more given systems (unifying theory), or of deducing a given theory from fewer and stronger assumptions (deepening theory). In short (a) every new physical theory contains fragments of available knowledge, both formal (logicomathematical) and physical or protophysical, and (b) the motivation for proposing a new theory is always some unsatisfactory trait of the available knowledge — whence an analysis of our stock is apt to suggest the need for new ideas and even some fresh ideas. In any case the subjectivist recipe for evolving theories out of general epistemological principles alone (EDDINGTON, 1939) does not work: there must be a vacuum leak for some new theory to emerge.

4.3.2. Theories are not Derived from Experience. Since so many theoreticians proclaim the slogan "Stick to experience", it should prove instructive to pause and recall how the great physical theories were born. Classical dynamics was not born until observable motions were taken as the thing to be explained *(explanandum)* rather than as the explainers *(explanans)*; elasticity theory and the elastic theory of light dealt from the start with imaginary models and moreover with propensities (dispositions) such as inner stresses rather than with actual properties; statistical mechanics postulated the existence of unobservable entities responsible for the overt behavior of macrophysical systems; classical electromagnetic theory postulated invisible fields and so did general relativity. With quantum theory the invention of transobservational concepts, hypotheses and models became even more conspicuous although people tried to disguise theoretical concepts as empirical ones by calling them 'observables' — only to produce a carnival effect. The origin of every great physical theory has been speculative in the sense that it did not emerge by processing data or even analyzing special cases — although data can motivate theory construction and good examples can guide it — but from trying new bold hypotheses involving unobservables. Observation generates problems and tests theories but does not exude them. Not even phenomenological theories are just data-fitting devices: they all contain theoretical concepts that do not occur in the output of experimental arrangements (BUNGE, 1963 b, 1964).

This is not just a matter of historical record, which after all might change in the future: it is logically impossible to deduce theories from facts. Indeed, the concepts needed to describe and explain facts do not come attached to them but must be created. Also, a fact is expressed by a particular (singular or existential) proposition, and no amount of particulars amounts to a theory, which is a system containing universal propositions that moreover hardly ever refer to observable facts. The actual process is the other way around: new singular statements concerning facts can be derived from general statements (e.g., laws$_2$) in conjunction with old singular ones (e.g., initial conditions).

Take a typical problem in theoretical physics, such as setting up and solving a second order differential equation with initial and/or boundary conditions. The latter are either hypothesized, as is the case of most boundary values (see 3.1.1), or they are obtained from measurement. Even if all such items were fully determinable by measurement, they would be insufficient to determine the differential equation itself if only because the equation contains terms of higher differential order than those figuring in the supplementary conditions. Furthermore the basic physical laws contain no empirical data at all, whence they cannot be uniquely determined by data. This holds for all sorts of theories but particularly for the deeper ones: those which, far from establishing input-output relations, contain a model of the mechanism underlying the net effects. Finally, even if a computer could be designed to produce theories out of data (which is logically impossible hence technically unfeasible), it would have to apply other theories to gather those data. In conclusion, the belief that experience is the source and basis of theory rests on a faulty understanding of both scientific theory and scientific experience.

4.3.3. Theories Have no Observational Content. Our account of physical theory differs from the widely accepted analyses proposed by some important philosophers (CARNAP, 1939, 1966; BRAITHWAITE, 1953, 1959; HEMPEL, 1952, 1965). According to them a scientific theory serves essentially the purpose of systematizing data and it consists of two parts: an abstract calculus and a set of correspondence rules (e.g. operational definitions) conferring an empirical meaning upon all or most of the specific symbols of the calculus. The terms that are not so linked to sensory experience (like lagrangians and spin operators) are declared to fulfil a merely syntactical function: they are said to be physically meaningless or nearly so, and any theory containing such terms reluctant to empirical anchorage is said to be only partially interpreted — for a different reason then than the ones given in 4.1.3 and 4.2.5. Some go as far as affirming that deduction can eliminate them all, leaving only the observational components: for example, by solving a set of Lagrange equations for a body one comes up with a set of trajectory equations, from which the lagrangian is absent.

Let us start with the latter contention, namely that deduction can eliminate all theoretical concepts. This is logically impossible for the simple reason that the whole theory is made up of theoretical concepts alone: even seemingly empirical concepts such as those of position and momentum are theoretical; and even seemingly observational statements like trajectory equations concern theoretical models of the system in question — and moreover they can be expressed as functionals of higher level concepts such as the one of lagrangian. Deduction reshuffles the available set of primitives (all of them theoretical), it can eliminate none, but on the other hand it can introduce (by the principle of addition: $p \vdash p \lor q$) new concepts, which have to be controlled by the principles of semantical homogeneity and closure (see 4.1.6). Only the converse operation is logically possible: in principle observational concepts could be defined in terms of theoretical ones since all primitives (definers) are theoretical. But the connection between theory and experiment should not be sought at the expense of either: they are irreducible but also intimately related: experiments are planned and interpreted by theories, which are in turn controlled by experiments.

And now to the first contention of the official view. While it is true that from a formal point of view a physical theory is a calculus or formalism (see 4.1.1 and 4.1.2), it is false that this skeleton acquires a meaning only in contact with experiment and remains otherwise uninterpreted (meaningless). Indeed (*a*) although physical theories have hardly an empirical content (interpretation), they do have a physical meaning as shown by analyzing their primitives, which — except for a few universal constants (see 2.2.1) — have a definite objective reference — even if only an intended one; (*b*) this physical meaning is sketched by the basic assumptions of the theory, particularly the interpretive postulates, which link constructs to physical objects and properties — e.g., "$\psi(\sigma)$ represents a state of $\sigma \in \Sigma$"; (*c*) it is theories that interpret experiments not experiments that confer a meaning upon theories (see 5.1.5). Rather than trying to live without theories or to convert them into nontheories we should be proud of the privilege of being able to raise above spontaneous experience and ordinary knowledge thereby eliminating the subject-dependent features from the picture (PLANCK, 1909). Scientific experience is not needed to build theories but to supply problems that will motivate the theorist and to find out how close or far theories are from being true. To this task we now turn.

5. Theory Checking

A physical theory should be tested for logical and semantical unity (see 4.1.6) and for truth. Truth tests are conceptual and empirical. The former consist in ascertaining the compatibility of the given theory with the bulk of available knowledge (if only correspondence-wise), while empirical tests consist in finding out how the theory fits empirical data, both available and accessible. If the theory fails to pass the test of external consistency — consistency with accepted theories (see 4.1.6) — then it is declared speculative or even crackpot — as is the case with the cosmological speculations that do not tally with physics (BUNGE, 1962c). But if the theory agrees reasonably well — perhaps in some correspondence limit — with the antecedent knowledge and in addition predicts new facts, then it should be examined as to empirical worth.

If a theory is empirically confirmed it is regarded as verisimilar, as containing a grain of truth — yet as fallible. Its value with respect to competing theories, if any, will be estimated with the help of further criteria, among them its heuristic and unifying powers. The truth value assigned a theory will therefore vary in time. The theory will always be watched for failings but as long as none show up it will be used both to judge other theories and to understand particulars. When a grand old theory begins to show failings it becomes worthwhile trying to correct them even at the price of introducing *ad hoc* hypotheses — provided the latter are in principle independently testable (case of PLANCK's quantization conjecture). Eventually the theory may die on the surgery bed or it retires to a more modest domain. In any case the search for a new

theory will start immediately and it will involve some amount of philosophizing concerning the nature of good theories. In the present state
of our science this philosophizing is done on an amateur level: the evaluation criteria are hardly stated in a clear way, let alone analyzed and
examined for consistency and fertility. Thus people speak of simplicity
as if it were a simple concept and they demand predictive power but
propose no measure for it. As usual, philosophy is easier abused than
studied — but it is always there, in the midst of research. It is up to
us to use it or to be misled by it. Let us now cast a glance at some of
the metascientific problems that come up in the checking of a theory.

5.1. Testability

Before subjecting an idea to actual tests we must see whether it is
testable at all: it might well prove immune to criticism on the strength
of experience, either because it is not a statement but something else
(a concept, a proposal, etc.), or because it is purely formal, or because
it is excused by some other idea (a shielding hypothesis), or because
presently feasible experiments are too coarse. In any case testability
requires company: an isolated idea is untestable — moreover it is
hardly meaningful. Before an assumption can be tested a context must
be supplied, i.e. the hypothesis must fit into some body of ideas some
of which make contact with experiment. This holds for every hypothesis
but is more obvious in the case of hypotheses concerning unobservables:
these remain inscrutable unless they are logically associated with some
observable by means of a link that often lies beyond the theory to which
the hypothesis belongs. In short, indices or objectifiers must be introduced.

5.1.1. Objectifiers. In objectifying temperatures we may employ a
length-temperature relation (law statement) belonging to the theory
of heat propagation; in objectifying electric current intensities we may
use an angle-intensity relation included in electrodynamics; and in
comparing masses we may use the momentum conservation law deduced
in mechanics. All these are unobservable-objectifier relations and every
one of them belongs to some theory or other. Unless they did, there
would be as little ground for using them as for regarding hand lines as
objectifiers of one's destiny. In other words, the bridges between unobservables and observables, by means of which physical hypotheses
and theories are put to empirical tests, are theoretical themselves.
Occasionally the bridge belongs to the theory under test, at other times
it belongs to alternative theories acting as conceptual instruments. The
progress of experimental physics consists largely in multiplying such
bridges.

In any case these are not "operational definitions", correspondence rules, or interpretive hypotheses: they are physical law statements that should be independently checked. Their function is methodological not semantical: they do not supply meanings but ways of indirectly manipulating the unobservable and force it to manifest itself. And they are systemic not isolated. Very often they are the offspring of the mating of several theories — just think of the formulas that go into the design and operation of a thermoelectric couple or of an interferometer, as well as of the number of ways the Avogadro number and the electron charge can be measured. In brief, systemicity is favorable to testability: the more numerous the relations a construct holds the better scrutable it is.

5.1.2. Testability and Systemicity. The testability or exposure of a statement is the greater the more friends it has. Thus if the linear momentum is conserved in a collision, then the kinetic energy is conserved as well; therefore an independent checking of momentum conservation is evidence for the kinetic energy conservation. The trouble with many newly introduced magnitudes — such as the baryon number and the strangeness — is that they figure in few statements and have therefore few chances of being exposed.

Another good example is the old hypothesis that there are single magnetic "masses", i.e. magnetic monopoles. Although it still lingers in textbooks nobody believes it, not just because it does not hold for large scale magnets, but because it is inconsistent with MAXWELL's CEM. In other words, CEM constitutes indirect evidence against the hypothesis "$\nabla \cdot B = 4\pi\varrho_m \neq 0$". If ϱ_m did not vanish exactly it would be impossible to set $B = \nabla \times A$ and, by replacing this expression into the field equations, to obtain the basis of radio engineering, i.e. $\Box A = (4\pi/c)j$. Now if we wish to test the unorthodox hypothesis that there are magnetic monopoles, one must first expand it into a theory constituting an extension of MAXWELL's CEM and predicting some manifestation of monopoles which does not conflict severely with the well corroborated consequences of CEM. The merit of DIRAC's theory of magnetic monopoles (DIRAC, 1948) is precisely that it constitutes such an expansion and makes some new predictions, e.g. that the binding energy of a monopole is ca. 500 MeV. This is a precise indication for experimenters to look for monopoles in high energy events. The search has so far been unsuccessful but this result, which considerably decreases the likelihood of the theory, does not discard it completely. (Notice the writer's effort to remain impartial.) In any case the testability of the monopole hypothesis was remarkably increased upon being systematized. In general, systemicity enhances testability, wherefore experimental work is apt to stimulate theorizing.

5.1.3. Experiment Stimulates Theorizing. The popular philosophy is that, by keeping close to experience, one does not risk proposing untestable ("metaphysical") hypotheses. One could rejoin that, in the absence of theory, experiment becomes pointless and even impossible. In any case the popular doctrine is that theory is a *pis aller* and that, if one is forced to theorize at all, then one should either play with observational concepts alone or, if nonobservational ones are unavoidable, they should be regarded as mere useful intermediaries between observables and as lacking any real import, i.e. as representing no physical objects. This is the philosophical ground for the widespread preference for phenomenological (black box) theories over "mechanistic" (representational) ones. It is an argument the energetists of the turn of the century (MACH, OSTWALD, DUHEM) invoked against the atomic and molecular theories. Quite apart from the questions of depth and richness, the argument ignores that, by suitably increasing the number of adjustable parameters in a phenomenological theory, it can be rendered unassailable — witness the phenomenological theories of nuclear forces (BUNGE, 1964). A system of input-output relations with no indication of the process whereby inputs are converted into outputs is not only mysterious but can be so accomodating as to become a fixed framework to pour data into rather than a network of assumptions to be checked by data (BUNGE, 1963 b).

Another popular tenet is that comparisons between theory and experiment require the elimination of theoretical concepts proper. Some philosophers have gone so far as to give recipes for detheorizing theories, on the curious assumptions that one can get rid of theoretical concepts by stretching the deductive chains far enough (see 4.3.3), or by giving "operational definitions" of constructs (see 1.3.9). The truth is, empirical tests require first of all working out the theory to cover particular situations, and secondly joining it to further theories. Thus any partial test of general relativity (GR) requires obtaining a special set of solutions of its field equations, which necessitates making additional assumptions — e.g., a definite distribution of matter and nongravitational fields, simplifying assumptions making the computation feasible, and the choice of a convenient coordinate system. The empirical test itself will require the design of an objectifier of the effect referred to by the given solution (see 5.1.1); and this design will call for any number of theoretical relations not included in GR. In short, empirical testing increases theoretical activity rather than putting an end to it: it involves an expansion of the body of theory as well as the introduction of further, auxiliary theories.

5.1.4. Auxiliary Theories. Deep theories, like CEM and QM, are untestable by themselves: their test requires hypotheses that are extraneous to them. They are untestable because they do not talk about observable facts but about facts that must be inferred deviously — e.g., e.m. wave propagation in empty space. Moreover no theory lays down the conditions of its own test even though it may talk symbolically of observable facts. Thus CM does not say how the motion of bodies can be recorded and QM is not about spectrometers. Deep theories are untestable in isolation because they do not refer to phenomena — very complex events as they appear to some observer. And phenomenological theories are inherently untestable as well because they are partial: they concentrate on certain traits of facts while phenomena are many-component, manysided macroevents that must be accounted for by a large number of theories. For one or the other reason no theory is testable by itself: the test of any theory requires additional theories. (Even if a single theory did suffice to design and interpret its own tests, the latter would constitute no control at all: i.e. they would be as worthless as the identification of accuser and defendant.) These additional theories needed to test a given theory will be called *auxiliary theories*.

Auxiliary theories are the conceptual partners of the material instruments used in empirical tests. If a set of scraps of auxiliary theories explains how an apparatus or a whole experimental set-up works, it is called a *measurement theory*. Examples: theories of the scale, the galvanometer, the Geiger counter, and the cyclotron. All these are strictly physical theories: they do not cover real observers with their peculiar perception and interpretation mechanisms: they are not concerned with what goes on in the observer's mind when he takes a reading (see Ch. 5, 7). Measurement theories, whether classical or quantal, are applications of fundamental physical theories and they constitute the basis of experimental physics. (The official religion is of course the opposite doctrine.)

5.1.5. Experimental Physics. The experimental physicist takes so many auxiliary theories for granted that he may occasionally forget that he uses them all the time (DUHEM, 1914). While he may think of himself as an untheoretical operator, the good experimentalist is actually a man full of ideas: he is an applied theoretical physicist with a flair for finding objectifiers of unobservables and skill in designing feasible material counterparts of theoretical models, computing in terms of orders of magnitude, putting aside irrelevant details, and making plausible inferences. Without theories his manipulations would be gadgeteering or even white magic and his data would not count as evidence for or against some idea. No hypothesis, no evidence relevant to it: "evidence" is a relational not an absolute item. This suffices to render experimental physics as dependent on theory as conversely.

The theory-experiment interdependence ruins the empiricist thesis that theories are just summaries of observations. In advanced science there is no theory-free observation. Think of an experimenter who shells atomic nuclei with electrons: does he gather data without the help of theories and to no purpose of interest to the theoretician? Clearly not: far from limiting himself to recording observations, he hypothesizes that his cathode rays consist of electrons and that his target contains nuclei; he assumes that they can interact and that the interaction may depend on the incident energy; he uses a lot of theoretical formulas to operate the beam and the counting devices as well as to perfect them; and he sets himself the goal of finding some indirect evidence concerning, say, the transparency of nuclei to electrons. He handles unobservables as much as the theorist does; but he has the added burden of linking them to observables *via* auxiliary theories which the theorist can often afford to ignore. In short, the experimenter has no use for pure, theory-free experience: he knows his data are insignificant unless they can somehow be accounted for (explained). And explanation, in advanced science, is made in a body of theory.

5.2. Explanation and Prediction

H. A. LORENTZ (1909) explained the emission of light and the "normal" Zeeman effect by means of MAXWELL's CEM and three additional hypotheses: (a) that matter consists of atoms, (b) that atoms contain charged corpuscles, and (c) that some of these particles oscillate harmonically thereby emitting radiation of the same frequency. Characteristically enough, LORENTZ proposed his radiator model not in response to new experimental data but because he wished to understand a fact known of old: he wanted to disclose the mechanism of light emission. And he did it one year before J. J. THOMSON conjectured that cathode rays were electron beams and measured the electron specific charge. The quantum theories of radiation are sophisticated versions of that classical model: they supply more refined explanations of the same as well as of additional facts.

In every case, within advanced science and as far as logic is concerned, *explanation* is deduction within a theory. In a *physical* explanation the premises are pieces of physical theories, supplementary hypotheses, data, and mathematical laws. If the conclusion is general and refers to an objective pattern, one has explained a law; if the conclusion is singular and refers to a circumstance, one has explained a fact. In the latter case the explanation is called a *prediction*. But in every case explanation has the structure of a logical tree. Look at the explanation of TORRI-CELLI's theorem:

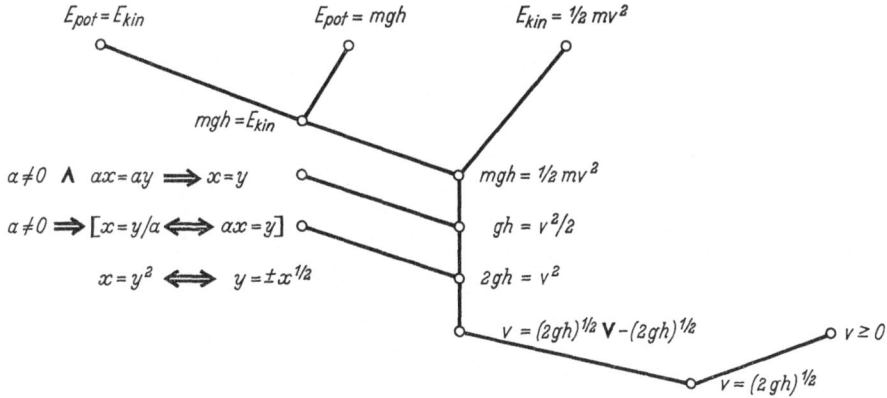

5.2.1. Subsumptive and Interpretive Explanation.

Variational principles explain laws of motion in the sense that they entail them, not in the psychological sense of making us understand how the laws of motion come about. Indeed, that inference does not intimate what mechanism, if any, results in this particular pattern. On the other hand a statistical explanation of a thermodynamic law, and a solid state theory explanation of a macroproperty, do say something about such a mechanism and are thefore scientifically deeper and psychologically more satisfying than a mere subsumption of one statement under a set of premises. Phenomenological or black box theories supply explanations since they are hypothetico-deductive systems; the kind of explanation they afford will be called *subsumptive explanation*. On the other hand a "mechanistic" theory — e.g., a field theory that explains the formation of wave packets by interference of wavelets — supplies what is often called an interpretation (rather than a description) of facts; we shall call *interpretive explanation* the kind of explanation afforded by nonphenomenological (translucid box) theories. Interpretive explanations are the deeper and the more challenging. They also satisfy our wish to understand, a wish which classical positivists — still surviving in quantum mechanics — considered sinful.

Explanation alone, without accuracy, is of course worthless: we want our theories to supply *accurate* explanations. A theory accounting accurately for a large mass of facts may be said to have a large *coverage* — whether its explanations are epidermic or deep. And a theory which on top of having a large coverage explains in a deep way can be said to have a high *explanatory power*. Thus a quantum theory of superconductivity has a higher explanatory power than the corresponding electron theory. While phenomenological theories can achieve large coverages — the larger the more specific traits they disregard — only deep (translucid box) systems have a high explanatory power.

But not everything can be explained in a given context: some premises must be accepted however critically and tentatively if conclusions are to be inferred from them. Thus a theory of electrons will perhaps (not necessarily) explain the trajectory of an electron in a given field but it will assume that there are electrons and fields. Yet some physicists complain that we do not understand why there should be electrons, i.e. systems with a given mass, charge, and spin. They seem to believe that everything ought to be explained — an illogical demand because to explain is to deduce from given premises. The existence of electrons could be explained only in the wider context of a theory of unstable "particles" accounting for electrons as decay products. In general, what is taken for granted in a given context may become a subject for explanation in a wider body of theory. This process has a limit: it is pointless to try to explain within physics the existence of the universe: this must be taken as a datum unless we wish to turn physics into theodicy — or worse, science fiction cosmology.

5.2.2. Prediction. A forecast is a singular statement concerning an unexperienced future or past fact. If made in a scientific context, i.e. derived from scientific premises (laws, data, etc.), it is called a scientific prediction; other kinds of forecast are prophecies and expert prognoses. To have a scientific prediction it is sufficient that it follows from general theoretical premises in conjunction with singular statements (data) that can be fed into some theory. That is, if {Scientific theory, Data} $\vdash$ Singular statement — then this conclusion is a scientific prediction. When a full-fledged theory is not on hand, a set of candidates to law statements will do even if they are taxonomic rather than kinetic or dynamic — but we cannot settle for less.

Data alone are insufficient to generate predictions, even with the help of computers, because they refer to a cross section of actual events — e.g., to the present state of the system concerner. And laws alone are insufficient as well because they indicate possible situations and possible changes rather than actual ones. We need both laws and data (real or conjectured) if we wish to anticipate or retrodict the most likely outcome of a given trend originating or terminating in a given state of affairs. This ontological argument has a logical counterpart: since a prediction is a particular statement (in general a conjunction of singulars), it can be deduced only from a set of general statements (e.g., laws) supplemented by singular propositions (e.g., data). Example: {For all t,

$dy/dt + ky = 0$, $y(0) = a$} $\vdash$ [For all t, $y(t) = a \exp(-kt)$]. And, by specification, $y(1) = ae^{-k}$ (a substitution instance).

The occurrence of the time variable is necessary (but insufficient) to predict the time or the time interval at which an event will or did happen, but is otherwise unnecessary. Thus the strangeness conservation law entails that two particles with strangeness values $S_1 = 1$ and $S_2 = -1$ can be produced out of a nonstrange system $(S = 0)$. No matter what the kind of law and the kind of theory, they will allow making predictions of some kind, even if "only" statistical — provided the necessary data are found (or at least hypothesized) and fed into the appropriate hypotheses. This condition is not easily satisfied. Thus strictly speaking a field theory, even if nonstochastic, does not lead to precise predictions concerning the field evolution, because this requires knowing the field values over a spacelike surface. But it is practically impossible to gather all the infinitely many information bits required for this; and even if it were possible, it would be physically impossible to transmit them simultaneously to the computer performing the calculation. Similarly, the effective prediction of the behavior of real materials with memory would require a detailed knowledge of their whole past history, which is again inaccessible.

Therefore predictions performed with field theories and mathematically similar theories are often bogus forecasts in the sense that they employ hypothetical data rather than genuine empirical information. This does not entail that they are unphysical theories: the predictions computed in that way can be checked if only partially, i.e. at selected points in space and time. Nor does it mean that those theories are indeterministic, just as PM is not indeterministic even though a small uncertainty in the knowledge of the initial conditions can, after a while, blur the whole picture of the evolution of the system. A theory may be deterministic yet its application to specific circumstances may fail to supply precise predictions due to lack of empirical knowledge. If a stochastic theory like QM has a high truth value we can infer that it adequately covers the random aspects of its referents. But we are not logically justified in concluding that its referents have only random traits or that these could never be deduced from nonstochastic hypotheses. The claim that nature is irreducibly random, on the strength that QM is so, presupposes that QM is a final, infallible theory. It is as dogmatic a claim as its negate, classical determinism. Both are ontological hypotheses which may never be conclusively established or disproved. In any case, since predictability is a human ability not an objective trait of nature, it is necessary to distinguish *ontological determinacy* (= determinacy *in re*) from *epistemic determinacy* (= complete knowability). We would never attain full epistemic determinacy even if nature were objectively determinate and if chance, far from being a mode of being, were just a mode of ignoring.

5.2.3. Measures of Coverage and Predictive Performance. The concepts of coverage and predictive performance of a theory can be quantified in terms of the concept of partial truth. Let T be a physical theory and $\{t_i\}$ the (finite) set of tested (not just testable) consequences of it. The truth value of every single t_i will be estimated on the strength of T itself (t_i might be an approximate solution), of the antecedent knowledge A, and of the body E of empirical evidence relevant to it. If we believed the popular fable that T can entail by itself observational consequences we would say that T covers E iff $E \subset F$, where F is the set of formulas of T. Unfortunately, things are more involved: F does not include E but, from F and A (which includes all necessary auxiliary theories), a number of testable statements t_i can be derived which, upon certain modifications, can be compared with experimental statements $e_i \in E$. If t_i is equivalent to e_i to within the accepted experimental error, we declare t_i to be confirmed. (For details see BUNGE, 1967b.)

But we can do better than that. Indeed, call $V(t_i | K)$ the degree of truth of t_i relative to the body of knowledge K, and agree to confine the range of V between -1 (complete falsity) and $+1$ (complete truth). We define the actual *coverage* of T up till the Nth test has been performed as

$$C(\mathsf{T}) = (1/2N) \sum_{i=1}^{N} [V(t_i | \mathsf{T}) + V(t_i | A E)]. \qquad (1.22)$$

Extreme values: (*a*) all-round confirmation with experimental error ε and theoretical error $\delta : C(\mathsf{T}) = 1 - (\varepsilon + \delta)$; (*b*) all-round refutation with experimental error ε and theoretical error $\delta : C(\mathsf{T}) = \varepsilon - \delta$. Recall that (1.22) is a measure of coverage not of explanatory power, which involves the somewhat imponderable dimension of theory depth (see 5.2.1). Let us now turn to predictability.

Although the prediction of a fact is logically identical with its explanation, the predictive power of a theory does not coincide with its coverage because in the former the element of novelty (serendipity) is essential. While coverage is matching with whatever knowledge is available, predictive performance is agreeing with fresh experience and disagreeing to some extent with tradition: it is made up of the differences $V(t_i | A E) - V(t_i | A)$ between the *posterior* truth values of the t_i and their *prior* truth values, or verisimilitudes. If the two cancel pairwise the theory is all right but it says nothing that was not contained in A: it is a traditionalist not a revolutionary theory. We define then

$$\Pi(\mathsf{T}) = (1/N) \sum_{i=1}^{N} [V(t_i | A E) - V(t_i | A)] \qquad (1.23)$$

as the *projective performance* of T up to the Nth test. If we now decide to assign new experience as much value as tradition, i.e. if we set

$$V(t_i | A E) = \tfrac{1}{2}[V(t_i | A) + V(t_i | E)], \qquad 1 \leqq i \leqq N \qquad (1.24)$$

then we get

$$\Pi(\mathsf{T}) = (1/2N) \sum_{i=1}^{N} [V(t_i | E) - V(t_i | A)]. \qquad (1.25)$$

Extreme values: (*a*) all round empirical confirmation of predictions contradicting the relevant antecedent belief (revolutionary theory): $\Pi(\mathsf{T}) = +1$; (*b*) all-round empirical refutation of predictions agreeing with antecedent knowledge: $\Pi(\mathsf{T}) = -1$; (*c*) all-round empirical confirmation of predictions tallying with antecedent knowledge (conformist theory): $\Pi(\mathsf{T}) = 0$.

Since we care for originality we shall elucidate this concept. We define *originality* as a sort of complement to compliance with tradition:

$$O(\mathsf{T}) = (1/2N) \sum_{i=1}^{N} [V(t_i | \mathsf{T}) - V(t_i | A)]. \qquad (1.26)$$

Adding $C(\mathsf{T})$ to $O(\mathsf{T})$ we obtain

$$C(\mathsf{T}) + O(\mathsf{T}) = \tfrac{1}{2} \Pi(\mathsf{T}) + (1/N) \sum_{i=1}^{N} V(t_i | \mathsf{T}). \qquad (1.27)$$

If all the t_i belong to T (no approximations made in deducing the t_i) we infer

$$\Pi(\mathsf{T}) = 2[C(\mathsf{T}) + O(\mathsf{T}) - 1]. \qquad (1.28)$$

That is, projective performance is made of coverage and originality. But clearly the two cannot be maximized at the same time. If maximum coverage is cherished, a cautious phenomenological theory will be preferred; if maximum originality is wanted, a representational theory will be sought without caring for coverage. A reasonable policy seems to strike a balance between these two extremes by maximizing predictability.

5.3. Rival Theories and Programmes

The evaluation of scientific theories is done by means of a score of metascientific criteria such as internal consistency, semantical unity, coverage, and predictability (BUNGE, 1961a, 1963a). In no case are theories gauged by their empirical support alone. Experience, though important, is not more important than plausibility on the strength of certain nonempirical criteria such as consistency (internal and external), depth (recourse to mechanisms and fundamentals), and objectivity (observer-independence). It could not be otherwise, for there are no such things as theory-free evidence and philosophy-free theory (on the other hand there are evidence-free theories and theory-free philosophies). Data, theories, and philosophy are distinct but intertwining threads of physical research.

Nonempirical criteria such as continuity, Lorentz invariance, and depth, occur in the evaluation of programmes (research lines) as much as they do in assaying available theories. Take the recent revival of phenomenological theories in "elementary" "particle" physics — e.g., S-matrix theory. The following reasons have been given in support of the exclusive exploration of theories of this kind and against field theoretical attempts. (a) The deeper theories, involving the field concept, have failed to solve the problems of strong interactions, e.g. the forces that keep the nuclei together *(true)*. (b) Since the particular field theories so far proposed in this domain are defective, their very approach must be wrong, i.e. the very programme of explaining the behavior of matter in terms of fields is doomed to failure *(untestable)* and it must be shelved once and for all *(authoritarian)*. (c) Predictability is a test of every physical theory *(true)*. Hence the ability to predict accurately is the sole requirement that should be imposed on theories: these should be regarded as handy predictive boxes not as more or less true models of reality *(crippling)*. (d) So far we have but two ways of gathering information concerning "elementary" "particles": scattering experiments and disintegration measurements *(true)*. All that will ever be known is what these operations can yield *(conceited)*. (e) Theories should be testable *(right)*. Hence no theory should state anything that cannot be directly checked by currently feasible experiments *(dangerous)*. (f) The black-box approaches have given good results (nuclear bindings, pion-nucleon interaction, etc.) there where more detailed theories have failed *(true)*. Hence one must continue along this path and discontinue the exploration of any others, particularly those involving a detailed spacetime description of events and interaction mechanisms *(wrong and dangerous)*.

The above fashionable *non-sequitur* chain endangers the future of physics by disavowing the more difficult and ambitious lines of research. This is not to imply that field theories will ultimately prevail: possibly a *tertium quid* is needed. In any case it is no wonder that the more superficial approach should be the more rewarding in the short run, as it tackles more modest problems and therefore attracts more people. Any attempt to disclose hidden mechanisms, e.g. to go beyond the kinetics of reactions, is bound to be more difficult and therefore less successful, in the short run, than the phenomenological approach. Yet certain failures are more instructive than some successes: at least they show which conceivable mechanisms nature does not care for. This is not to suggest that phenomenological theories are worthless: though insufficient they are indispensable if only to know what is it that deeper theories are expected to entail (BUNGE, 1964). But a successful superficial theory may bar progress and in any case it hardly supplies an understanding of how things work, which is after all what we are after (HEISENBERG, 1966).

In any case research programmes and their end products are evaluated with the help of philosophical insights and prejudices. Therefore it is convenient to tame one's philosophy and train it to furthering the search for truth rather than hindering it. But no one should be denied the pleasure of being devoured by the wild philosophy of his choice.

This completes our quick review of some of the metascientific tools that will be used in the following in taking apart and reassembling certain basic physical theories.

Chapter 2

Protophysics

No physical theory is free from presuppositions: every one has a background constituted by a set of ideas it assumes without questioning — although they may become and should become subject to criticism in other contexts (see Ch. 1, 4.2.1). The body of ideas shared by every set of physical theories but investigated by none in particular falls into two classes: the set of logical and mathematical theories constituting the formalism of physical theories, and a quaint collection of nonformal yet generic and important principles and theories, which may be christened *protophysics*.

There are no treatises on protophysics and this for the following reasons. Firstly, the very existence of such a nonformal generic background goes usually unnoticed or, if noticed, it is either taken to be trivial or to belong to some specific field — often mechanics. Secondly, due to its generic character protophysics leads nowhere in particular though without it we get nowhere: thus the theory of additive magnitudes (which the psychologists call 'measurement theory'), applied to the energy concept, is necessary but insufficient to build energetics, just as in order to build mathematics the mathematician must go beyond its foundations. Thirdly, protophysics is fragmentary rather than unitary: it is not a single theory or even a set of contiguous theories but an heterogeneous assortment of principles and theories; consequently it has by itself no great deductive power.

Yet it would be wrong to dismiss protophysics: precisely what is taken for granted should be scrutinized once in a while. Remember the stories of absolute time and of parity. Trivialities are harmless provided they are kept under control. Let us then cast a glance at some chapters of the unwritten book of protophysics. Many of them, such as general field theory and general stochastics (the statistical "mechanics" of aggregates of unspecified entities), will have to be left out of this account.

1. Zerological Principles

An assumption occurring in the body of a theory will be called *constitutive* while one serving as a guide or constraint in theory construction will be named *heuristic*. (For the constitutive-regulative distinction see KANT, 1781—1787.) This is not a dichotomy: some principles are both constitutive and heuristic. For example the principle "All interactions vanish asymptotically" is a generic hypothesis occurring wherever forces and fields figure and is both a partner of law statements and a heuristic tool for discarding unphysical solutions of basic equations. On the other hand the principle that transition probabilities must be independent of the choice of representation and observer is solely heuristic: it does not concern the transitions themselves, which are the physical objects, but our way of handling certain ideas about them: it is metatheoretical not theoretical.

Whether constitutive or heuristic or only the latter, such pervasive principles are not the property of a single theory. Being generic and fundamental in a genetic though not in a logical sense, they deserve being collected in a set of their own: the set of *zerological* principles. Let us sample it. No order save the one of shrinking extension shall be followed because anyhow these principles constitute no system.

Uniform convergence (McKINSEY, SUGAR and SUPPES, 1953): "All series whose terms represent physical objects (e.g., forces) shall be uniformly convergent on the given interval". This requirement does not refer to nature but to certain ideas of ours, whence it cannot belong to a theory about nature. Yet its motivation is clearly physical: since the way the terms of a series is arranged is not dictated by nature, the order should make no difference to the value of the series — a condition met by uniform convergence. We may violate the principle only when the individual terms of a series do not represent physical objects, e.g. when we fourier analyze a function without claiming that every fourier component can be detected, say, by a spectrometer. Extension of the principle: physics.

Classical relativity (GALILEI): "The basic laws of physics shall be invariant under transformations of coordinates representing changes of inertial (Galilean) frames." Notice that (a) the requirement concerns laws not facts, i.e. it is metanomological, and (b) only the fundamental equations (the highest level laws) are subjected to this covariance principle, not their deductive consequences (integrals of motion). Intended extension: physics; actual extension: classical mechanics. The Galilei group was eventually embedded in the Lorentz group, in turn included in the group(oid) of arbitrary coordinate transformations. The motivation of all three covariance principles is one and the same, i. e. the goals of objectivity and universality, attainable at the level of principles (basic laws), not always at lower levels. More on relativity principles in Ch. 4.

Antecedence ("causality"): "If A and B are time-dependent properties and A determines B (whether causally or stochastically), then B at time t is a functional of A at all times t' prior to and up to t". Particular formulation:

$$B(t) = \int\limits_{-\infty}^{t} dt'\, K(t', t)\, A(t').\qquad(2.1)$$

This principle is usually mistaken for the principle of causality but is actually independent of it: 'A' need not be interpreted as the cause of 'B' and cannot be so interpreted if the memory kernel K is a random function; moreover, the principle holds also in the limiting case $B = A$, whereas the causal relation is irreflexive. But antecedence is compatible with causality and moreover an ingredient of it (BUNGE, 1959a, 1963a, 1963b). Extension of the principle: the whole of physics except some controversial versions of electrodynamics (see Ch. 4, 1.7).

Contiguity (nearby or local action): "If A and B are place-dependent properties and A determines B (whether causally or stochastically), then B at the place x is a functional of A at all places x' surrounding x". Particular formulation:

$$B(x) = \int\limits_{-\infty}^{\infty} dx'\, K(x', x)\, A(x')\qquad(2.2)$$

where the value of the kernel K represents the weight of the local contribution of A. Extension: continuum theories, including field physics.

Spacetime contiguity ("strict local causality"): "Only field regions connectible by a field disturbance (hence separated by a timelike interval) can interact." The precise mathematical formulation of this principle depends on the theory: in classical field theories it can be expressed in integral form, as a mere spacetime generalization and merger of the two preceding principles (BUNGE, 1963a); and in quantum field theories it is expressed as a set of commutation relations among field variables at different points in spacetime. In either case the designation 'causality condition' is a misnomer (BUNGE, 1959a, 1962b, 1963a). Extension: relativistic physics.

Material indifference or objectivity (NOLL, 1959; TRUESDELL and TOUPIN, 1960; NOLL, 1963; TRUESDELL and NOLL, 1965): "The constitutive equations must be invariant under coordinate transformations." Rationale: the behavior (response) of a material is something objective, i. e. independent of the observer and his choice of representation. Implementation: by writing the constitutive equations either in a coordinate-free fashion (e.g., the dispersion formulas in classical electron theory) or in tensor form (e.g., the relation $\mathfrak{G} = \frac{1}{2}\chi F$ in phenomenological CEM). In either way the frame-dependent features are discarded. From a logical point of view this principle is a case of the general covariance principle. But material indifference must be required even if the equations of motion are not required to be generally covariant — as is the case of CM. Intended extension: theories of matter; exploited in continuum mechanics. Close relative: the Neumann principle of crystal physics, demanding the rotation and reflection invariance of physical properties — a restriction that can be imposed on every tensor quantity characterizing a material medium, such as, e. g., temperature and magnetic permeability (POST, 1962).

Equipresence (TRUESDELL, 1966): "An independent variable present in one constitutive equation is so present in all". Rationale: macro-events are manysided (mechanical, electromagnetic, thermal, etc.), whence realistic theories of materials will take all possible variables into account. This principle of open-mindedness is a metanomological statement of the prescriptive (normative) species (BUNGE, 1961 b). Intended extension: phenomenological theories of materials; exploited in thermomechanics (COLEMAN, 1964).

The principles of uniform convergence, relativity, and material indifference are of an epistemological nature: they are conditions of objectivity and their heuristic value consists in that they help demarcate the objective or physical from the subjective or operator-dependent features of a theory. They are constraints rather than propellers. On the other hand the principles of continuity (contiguity, antecedence, and spacetime locality) are of an ontological nature: they exclude jumps in space or time, their heuristic value lying in that they stress the weight of the immediately preceding and the close by, thereby suggesting the use of differential equations. Finally the equipresence principle is purely heuristic; it often leaves hardly any traces. The construction of any physical theory employs some of the above zerological principles as well as certain specific desiderata — e.g., that the basic equations be of no higher than the second differential order. The set of guiding principles can be completed in such a way that a single theory results — as was the case of GR. This does not mean that all such theory construction principles are kept in the theory: some of them do not entail anything but are just prescriptions — e.g., that every property be represented by a hermitian operator. In comparatively young fields like GR and QM the heuristic components are not always distinguished from the constitutive ones, which results in gay confusion; it will be instructive to disentangle them (see Chs. 4 and 5).

2. Physical Probabilities

Many nonphysical pitfalls, particularly subjectivism, can be avoided in statistical mechanics and QM upon realizing that the calculus of probability (CP) is nowadays a formal theory devoid of any specific factual content, and precisely for this a framework apt to be used almost anywhere. The formal nature of CP is best seen by axiomatizing it, not because axiomatization leaves only clean carcasses but because it discards heuristic scaffoldings. Although there are more refined axiomatizations (e.g., POPPER, 1963 b), to our purposes it will suffice to discuss the bare essentials of the elementary CP (KOLMOGOROFF, 1933).

The presuppositions of elementary CP are ordinary logic, set theory, and real functions; the advanced theory presupposes also measure theory — it is actually an application of the latter. The specific primitives of CP are a set U and a function P. The members of the reference class or universe U are nondescript: U is an abstract set that can eventually be interpreted as a collection of events, data, statements, or what not. Call $\mathscr{P}(U)$ the set of all the subsets of U. The probability function P maps the power set $\mathscr{P}(U)$ into the unit interval. That is, if A is in $\mathscr{P}(U)$ and p is in $[0, 1]$, then $P(A) = p$. The latter is the form of an elementary probability statement such as "The probability of casting a six is $1/6$". (Physicists tend to focus on the numerical variable p, i.e. to regard probability as a number rather than as a function. This amounts to take the range of P for the whole thing and is a remnant of Pythagorean mathematics with its emphasis on number rather than on form.) After these informal preliminaries we lay down the following *axiomatic definition of probability:*

P is a probability (measure) iff it satisfies the following axioms:

$P\,1$ U is a nonempty set.

$P\,2$ P is a nonnegative function on $\mathscr{P}(U)$. [I.e., For every A, if A is in $\mathscr{P}(U)$, then $P(A) \geqq 0$.]

$P\,3$ P is normed. [I.e., $P(U) = 1$.]

$P\,4$ P is additive. [I.e., for every A and every B in $\mathscr{P}(U)$, if A and B are disjoint then $P(A \cup B) = P(A) + P(B)$.]

These assumptions are p-complete as they characterize the two primitives of the theory (see Ch. 1, 4.2.4). And they are also d-complete for they are necessary and sufficient to derive all the theorems of elementary CP. Mark the formal character of CP: it is not said that U is a sample space, or that its elements are events or that they are statements, nor that the values of P are frequencies. (KOLMOGOROFF himself did employ a misleading terminology but it is as supernumerary as the observer and the ideal experiment in QM: the formulas remain and they become more perspicuous and general if the heuristic appendages are extirpated.) Coins, urns, observations and random samplings will occur in informal inferences and in applications of CP not in its foundations.

In order to apply CP in physics we must turn it into a factual theory — a theory of physical probability. This amounts to interpreting the primitives (U and P). In the old times it was supposed that there was a single correct interpretation of CP. The logicists held that U could only be a set of propositions and $P(A)$ the likelihood that A be true. The psychologicists, that U is a set of beliefs or some other psychological states and $P(A)$ the credibility or degree of certainty or rational degree of belief we feel compelled to attach A. And the empiricists, that U can be nothing but a collection of experiences or of data and $P(A)$ the observed frequency of A in U.

With the development of semantics, and particularly model theory, it became clear that these are just so many interpretations of U and P, yielding each a possible model of the abstract theory CP (see 4.1.3). It was also realized (*a*) that there are many other possible interpretations — in principle infinitely many, and (*b*) that some interpretations of CP may not be true, i.e. they may not produce genuine models of the theory. For instance, U might be interpreted as the set of one's friends and $P(A)$ as the likelihood that the subset A remains faithful to us; or U could be interpreted as a set of techniques, or of operation rules, and $P(A)$ as the a priori efficiency of the class A of procedures. But we are here interested in physical interpretations of CP, i.e. in introducing semantical postulates that will turn CP into a physical theory — no doubt a generic or protophysical one.

We shall say that CP is assigned a *physical interpretation* iff its primitives U and P are assigned physical referents (see Ch. 1, 4.1.3). These assignments of meaning will be entrusted to additional assumptions. Once they are added CP ceases to be a purely mathematical theory to become a protophysical system, one that will stand or fall with the specific physical theory it backs. The chief physical models of CP are given by the following interpretation postulates:

M_1 *Propensity interpretation (quantified objective possibility)*

$I(U)$ = set of possible physical events of a kind

$A \in \mathscr{P}(U) \Rightarrow I[P(A)]$ = the natural disposition of event(s) A to happen

M_2 *Randomness interpretation (objective chance)*

$I(U)$ = set of random (mutually independent) events of a kind

$A \in \mathscr{P}(A) \Rightarrow I[P(A)]$ = the objective odds of event(s) A in the space of events U

M_3 *Statistical interpretation (relative observed frequency)*

$I(U)$ = Population of empirical data of a kind

$A \in \mathscr{P}(U) \Rightarrow I[P(A)]$ = the relative frequency of the sample A of observations in the context U.

The extension of these semantical hypotheses to conditional probabilities is left to the reader. Caution: in M_3, (a) the arbitrariness of A is implemented by making reasonably certain A is a random sample of U, and (b) the values of P are rational numbers included in the unit interval. Empirically found frequencies are therefore bound to differ from calculated probabilities.

Similar considerations apply to the richer CP employed in most of physics, namely the system based on three primitive concepts: a space U, a continuous distribution function ϱ, and a family of random variables X. The differences in interpretation are transmitted from the basic to the derived concepts *via* the respective definitions. Thus in M_1 and M_2, '$\langle X \rangle$' will mean "average X" whereas in M_3 it will mean "measured average of X" or rather "average of measured values of X" — never "expectation value of X", which belongs to a psychological model of CP. And '$\varDelta X$' will be interpreted in M_1 as the average scatter of X around $\langle X \rangle$; in M_2, as the mean random fluctuation of X about $\langle X \rangle$; and in M_3 as the mean dispersion of data around the mean measured value. M_1 and M_2 involve unobservables and occur in theoretical physics whereas the abode of M_3 is experimental physics. The three models are needed in their proper places: M_1 and M_2 to explain and predict, M_3 to test such predictions by means of data. M_1 and M_2 are semantically independent of M_3 in the sense that their meaning does not depend on the content of M_3; but the test of any theory involving M_1 or M_2 requires M_3, whence M_1 and M_2 are methodologically dependent on M_3. Thus the sentence 'The probability of the reaction A is p' means that the reagents in question have, under specified circumstances, a tendency or strength p to react according to the pattern A: the sentence does not mean "The relative frequency of A in an observed sequence U is p", but the statement becomes testable if so interpreted. The union of the three models M_1, M_2 and M_3 constitutes what shall be called *physical probability theory*.

Notice that adopting an interpretation of CP in terms of observed frequencies does not entail *defining* p as the long run relative frequency of a kind of events. Such a reduction not only restricts CP unnecessarily but is also mathematically impossible, if only because probability values are real numbers whereas frequencies are rational, and because whereas the former are stable the latter depend on the population size. In short, while the frequency theories and definitions of probability (e.g., MISES' and REICHENBACH'S) are wrong and were given up by mathematicians 30 years ago (BUNGE, 1956a) the frequency interpretation of the formal CP is a genuine and indispensable model. Indeed, the empirical test of a theoretical statement of the form " $P(A) = p$ " requires reinterpreting A as a possible outcome of an empirical test and p as close to the long run relative frequency with which A actually happens.

The logical and psychological models of CP are irrelevant to physics although some of them are legitimate models of CP. In physics we are not interested in propositions for their own sake nor in states of mind — e.g., expectations and uncertainties — but in what a physical system may do and in the frequency with which it manifests its potentialities under experimental control, i.e. when we examine a random sample of the set of all its physical possibilities. In other words, in physics probability statements must be interpreted in a strictly objective way, whether as stating objective (but potential) properties or as expressing traits of runs of observations rendering those dispositions manifest. Otherwise they cannot be counted as physical statements — by definition (see Ch. 1, Sec. 3). On the other hand we are allowed to use psychological probability when speaking about physical statements — as when we say that it is probable that the basic difficulties in statistical mechanics will be solved — but for this we do not need the CP.

Let us now scotch a few popular myths.

First myth: "The probabilities occurring in statistical mechanics and QM are subjective or psychological in the sense that they represent the strength or weakness of our beliefs concerning the behavior of a system, or else the degree of our ignorance about it." That this view is false can be seen (*a*) by unearthing the arguments of the corresponding probability functions — no cognitive subject occurs among them — and (*b*) by recalling that physical probability statements must be objectively testable — e.g., by counting frequencies not by asking a physicist how uncertain he feels. (Further arguments in SMOLUCHOWSKI, 1918; BUNGE, 1955b, 1956b and POPPER, 1957a, 1957b, 1959.) Granted, the assignment of a probability to an event is an act of knowledge — but so is the assignment of definite single values to any other function. If physical probability were subjective then (*a*) we would have to compute it in the context of psychological theories and (*b*) we would be hardly able to correct probability statements on the strength of objective experience. Yet the experimental physicist will tell the theoretician whether his calculated (in this sense subjective) values match the experimental ones. There is no more subjectivity in a Maxwell velocity distribution or in a position distribution than in a distribution of radioactive decay times. Since the myth that all probability is subjective underlies the Bohr-Heisenberg interpretation of QM, this becomes suspicious (see Ch. 5).

Second myth; "Any theory containing probability statement employs some system of many-valued logic, perhaps CP itself, rather than ordinary logic: probability reasonings are themselves probable not conclusive." Rejoinder: (a) a logical analysis shows that CP presupposes ordinary logic, whence (b) every proof in CP uses two-valued logic alone, and (c) any given probability statement can have the same truth value as a suitably chosen nonprobability statement. The real problem is elsewhere: when the mathematical CP is interpreted in physical (or biological, or psychological, etc.) terms, its statements acquire a factual truth value on top of the logical truth they had to begin with — and the two may not coincide, which shows the need for a dualistic theory of truth (see Ch. 1, 1.3.11).

Third myth: "'Probable' amounts to 'verisimilar' *(wahrscheinlich)*, and CP is the theory of partial truth." Criticism: (a) the probability that a statement be true cannot be equated with the degree of truth of the statement, for the former assumes that the statement is either completely true or completely false, not half-true; (b) the attempt to define truth in terms of probability is circular: the concept of truth, which probability is alleged to define, reappears in the definiens, namely thus: "Truth value of $s =$ Probability that the truth value of s equals unity." The concept of truth is logically independent of the probability concept and much more pervasive than the latter. A theory of partial truth may use CP though without the wrong identification of probability and verisimilitude (POPPER, 1963 a) or it may dispense with probability altogether (BUNGE, 1963 a).

3. Chronology

Every physical theory employs some time concept, none makes it its particular subject. Even statics, thermostatics and ray optics, though atemporal, take it for granted (a) that the facts they are concerned with happen "in space and time" (i.e. in the midst of a network of things and events), and (b) that the statements they make hold for every place and at all times — even if vacuously. Consequently it is desirable to catch and tame that most elusive of protophysical beasts. And since the structure and meaning of a concept are best elucidated by building a theory about it (BUNGE, 1963 a), it is convenient to have theories of time. Each such theory will characterize one concept of time; since there are in fact several theories of time and several others are conceivable, there exist various time concepts. The set of all theories of time may be called *chronology*. We shall restrict ourselves to a part of physical chronology, the set of theories of physical times.

About the simplest statement that is made about physical time is that it is the independent variable of physics. But this is an ambiguous characterization since most functions occurring in physics are smooth enough to be invertible and thus make the dependent/independent variable distinction a thin one. Moreover the statement is incorrect, as the concepts of physical system and of place occur no less conspicuously and with similar mathematical roles. Usually the next step is to state that time is a parameter and more precisely a real variable. But this is clearly insufficient: if this were all then it would be a task of mathematicians to study everything concerning time. But then recurrent phrases, such as 'x occurs at the instant t', would make no sense, because mathematics does not deal with events. We must then say more and more precise things about time: we must build theories of physical time, which none of the preceding statements are.

Historically the first attempt to characterize the physical time concept seems to have been NEWTON's doctrine of *absolute time*. It boils down to the statements that (*a*) time is absolute, i.e. independent of everything else, (*b*) the measure of duration is a numerical variable, and (*c*) time flows at a constant rate from past to future. The first statement leaves time hanging in the midst of nothingness, the second is incomplete, the third metaphorical; moreover, the last is circular, for the concept of rate involves in turn the time concept. In short the microtheory of absolute time is unsatisfactory and was so recognized by critics of Newtonian mechanics at various times (MACH, 1883).

Among the more satisfactory theories of time, the simplest formalize in a qualitative way the prescientific time concept. They are instructive exercises in symbolic logic but quite useless in science, where quantitative concepts of time are needed. A reasonable theory of physical time should satisfy the following desiderata: (*a*) it should refine and work out our intuitions about physical time; (*b*) it should purge those intuitions of subjective elements; (*c*) it should tally with important physical theories; (*d*) it should be quantitative; (*e*) it should be well organized, and (*f*) it should include the proper semantics, in the sense that it should deal with a certain relation among objective happenings rather than with a ghostly absolute time. Few theories of time satisfy all these requirements. Those which do are all *relational theories* of time in the sense that they work out the insight of PLATO, ARISTOTLE, LUCRETIUS, AUGUSTINE, SPINOZA, LEIBNIZ, MACH, BOLTZMANN, and a few others, that time is a relation among facts not a self-existent (absolute) object —

whence: no events, no time or, if preferred, time is neither *ante rem* nor *post rem* but *in re*.

The first exact relational theory of time was proposed only recently (NOLL, 1967) — which goes to show the sad state of neglect of foundations research. (For previous unsuccessful attempts see CARATHÉODORY, 1924 and REICHENBACH, 1924.) It is a theory of *universal time* in the sense that it assumes a single time for every place and every reference frame; it will be abbreviated UT. This theory specifies a relational not an absolute time concept because it analyzes time as a relation among pairs of events on the one hand and numbers on the other. Now since time lapses not only ride on events but are relative to reference frames, UT may hold for every single frame but it fails to state the timing relations among events referred to different frames. We shall therefore relativize it further, namely to the frame concept, building along its lines a theory of *local time* (LT).

LT has three primitives: the set E of events $x, y, z, \ldots$, the set K of frames $k, k', \ldots$, and the time map τ. The purpose of LT is to characterize τ in terms of E and K with the help of some formal concepts. This characterization, which may be regarded as an axiomatic definition of the local time concept, is effected by means of the following axiom system.

LT 1. (*a*) E and K are nonempty sets. (*b*) E represents the set of actual and possible point (elementary) events. (*c*) K represents the set of physical reference frames.

LT 2. τ is a function that maps the set $E \times E \times K$ of pairs of events relative to a frame, into a segment T of the real line:

$$\tau : E \times E \times K \to T, \qquad T \subseteq R. \tag{2.3}$$

[I.e., for every $x, y \in E$ and every $k \in K$, there is a unique $t \in T$ such that $\tau(x, y, k) = t$.]

Since a particular frame $k \in K$ is usually referred to, it is convenient to agree on

Df. 1. $\tau_h(x, y) \overset{\text{df}}{=} \tau(x, y, h)$

which allows us to compress the remaining postulates as follows:

LT 3. For every $x \in E$, every $k \in K$ and every $t \in T$, there is a $y \in E$ such that

$$\tau_k(x, y) = t.$$

LT 4. For every $x, y \in E$ and every $k \in K$,

$$\tau_k(x, y) = - \tau_k(y, x).$$

LT 5. For every $x, y, z \in E$ and every $k \in K$,

$$\tau_k(x, y) + \tau_k(y, z) = \tau_k(x, z).$$

Since a particular event $a \in E$ is usually taken as a reference point, it is convenient to adopt

Df. 2. If $a, x \in E$ and $k \in K$, then

$$k\text{-time coordinate of } x \equiv t_k(x) \stackrel{\mathrm{df}}{=} \tau_k(x, a).$$

Clause (*a*) of *LT* 1 warrants that the axiom system be nonvacuous: it is the conjunction of two statements of physical existence. Clauses (*b*) and (*c*) contribute to delineating the meaning of LT. If LT were regarded as a closed system then these two clauses would be designation rules not semantical assumptions: it is only in the wider context of physics that the expressions 'point event' and 'reference frame' acquire a physical meaning. But this is the case with every physical theory: in isolation no theory is fully interpreted (see Ch. 1, 4.1.3, 4.2.5).

LT 2 makes LT a *relational* not an absolute time theory, as it states that time is not a self-existence object — e.g., something flowing somewhere — but a relation among facts. This relation is not one to one: *LT* 2 does not preclude the existence of two or more pairs of events which, in a given frame, are matched to a single real number. Nor is the time relation onto: *LT* 2 does not state that, for any given real number, there is a triple $\langle x, y, k \rangle$ such that $\tau_k(x, y) = t$. Consequently as far as *LT* 2 is concerned there may well exist numbers to which no pairs of events are assigned; but such numbers will not be interpretable as time lapses for they do not fall under the relation τ. Notice also that the measure t of a time interval or duration is a real number, which excludes the narrow operational interpretation of the values of τ as only results of time measurements with clocks of some sort or other, for their readings are rationals. In other words, LT intends to be an objective (operator-free) theory. True, the value of the time function depends not only on the pair of events but also on the reference frame; but a reference frame is a material system not a sentient observer (TRUESDELL and NOLL, 1965; TRUESDELL, 1966). Not even the origin of time need be observer-dependent: any distinguished event may serve to characterize the zero of a time scale; usually the beginning of a process is chosen — for example but not necessarily the beginning of an observation. Finally, the inclusion of K makes LT not only a relational theory but also a theory of *relative* time in the sense that duration is here frame-dependent.

LT 3 states that, no matter what the event, the frame and the number, there is at least one other event such that the time lapse between the two equals the preassigned number. In other words, if something ever happens then something is always happening, i.e. the set of events is continuous — a bold assumption and one that cannot be put to the experimental test, which cannot discriminate between events arbitrarily close in time. Moreover, LT 3 states that if anything has ever happened at some time past then something will occur at any ulterior time, even infinitely later — and conversely. Since something is now happening (datum), LT 3 entails that there has always been and there will always be something going on, and moreover uninterruptedly. That is, the universe has always existed and will always exist in some form or other — both before the last big bang and after the next collapse if any. Consequently LT is inconsistent with creationist cosmologies, whence those who insist on the nonphysical ideas of creation and annihilation (instead of transformation) ought to propose their own theory of time — one involving nonphysical (supernatural) hypotheses. This trait of LT justifies the usual practice of writing integrals in which the integration variable represents time and runs from $-\infty$ to $+\infty$, as well of imposing asymptotic conditions on fields at $t = \pm\infty$.

LT 4 states that duration is an oriented interval: it expresses a trait of the past-future anisotropy. But it has nothing to do with irreversibility, let alone with entropy increase. In LT the flow of events is postulated independently of irreversibility; moreover, we cannot even state a theory of irreversible processes unless the time concept is on hand. Finally the triangle equality LT 5 expresses that time is a one-dimensional manifold as long as it is restricted to a given reference frame.

The concepts of temporal order in (relative to) a frame can now be defined exactly: If $x, y \in E$ and $k \in K$, then x is said to be *earlier* than, *later* than, or *simultaneous* with y relative to k, according as $\tau_k(x, y) > 0$, $\tau_k(x, y) < 0$, or $\tau_k(x, y) = 0$. If two events x and y are simultaneous in $k \in K$, we shall call them *k-simultaneous*; in symbols, $x \underset{k}{\sim} y$. This is an equivalence relation just as those of congruence and of belonging to the same species. Consequently $\underset{k}{\sim}$ induces a partition of the set of events into disjoint subsets t_k of k-simultaneous events. Every such class of k-simultaneous events may be called a k-instant. And we may say that the event x happens at the k-instant t_k, or at the instant t relative to the frame k, iff $x \in t_k$. We may finally pick any convenient event $a \in E$ and call it the *initial event*, without thereby implying that there was no previous event. And we may take the corresponding k-instant $t_k(a)$ at which a occurs as the *initial instant*. By LT 4 and Df. 2, $t_k(a) = 0$. The details of the construction of the simultaneity concept are given in the next paragraph. What matters from a philosophical point of view is, firstly, that in this reconstruction or elucidation of the time concept the instant is not an unanalyzed individual but an equivalence class of events (RUSSELL, 1914). Secondly, the qualitative (and relative) concept of simultaneity is defined in terms of the quantitative concept of duration. Thirdly, once we have introduced the concept of instant we may, for all practical purposes, forget how it was defined and handle it as an individual, thinking of time as a continuous line whose points are instants — which explains the success of the naive view that time is a self-existent or absolute one-dimensional continuum.

(We now justify the identification of an instant with a class of simultaneous events. The set of all pairs of k-simultaneous point events is an equivalence relation on E, i.e. one that induces a partition of E into equivalence classes $t_k = \{x, y \in E \mid x \underset{k}{\sim} y\}$ of simultaneous events. Every t_k is called a k-instant. $LT\,3$ applied to simultaneous events $(t = 0)$ ensures that t_k is nonempty, i.e. that every event has at least one partner in t_k. Moreover, the subsets t_k are mutually disjoint: no event belongs to two different sets of k-simultaneous events. Therefore the family $T_k = \{t_k\}$ is a partition of E, i.e. each event belongs to a single subset t_k of T_k. Correspondingly if $\{k\}$ stands for the set whose sole member is k, then $E \times E \times \{k\}$ is exhaustively covered by $T_k \times T_k$. The latter is the quotient space of $E \times E \times \{k\}$ over the simultaneity relation $\underset{k}{\sim}$. In symbols: $T_k \times T_k = E \times E \times \{k\}/\underset{k}{\sim}$. In other words, any pair of events is assigned a single pair of k-instants. The function p_k: $E \times E \times \{k\} \to E \times E \times \{k\}/\underset{k}{\sim}$ that effects this correspondence is the natural projection of $E \times E \times \{k\}$ into its quotient space. Define now $\bar{\tau}_k : T_k \times T_k \to T$, with $T \subseteq R$, as follows: If $x \in t_k$ and $x' \in t_k$, then $\bar{\tau}_k(t_k, t'_k) = \tau_k(x, x')$. This is a "well defined" function since it is independent of the chosen representatives of the classes of simultaneous events. In fact take two different representatives y and y' of t_k and t'_k respectively. Then $\tau_k(y, y') = \tau_k(y, x) + \tau_k(x, x') + \tau_k(x', y')$ by $LT\,5$. But $\tau_k(y, x) = \tau_k(x', y') = 0$ by $LT\,4$ since $x \underset{k}{\sim} y$ and $x' \underset{k}{\sim} y'$. Hence $\tau_k(y, y') = \tau_k(x, x')$. Therefore $\bar{\tau}$ is an extension of the map $\tau_k : E \times E \times \{k\} \to T$ over the quotient space $E \times E \times \{k\}/\underset{k}{\sim}$; that is, the restriction of $\bar{\tau}_k$ to the original domain equals the time function: $\bar{\tau}_k \mid E \times E \times \{k\} = \tau_k$. In short, $\bar{\tau}_k$ assigns a real number t to every pair $\langle t_k, t'_k \rangle$ of instants: $\bar{\tau}_k(t_k, t'_k) = t$, and by definition this lapse equals the lapse between the representative events $x \in t_k$ and $x' \in t'_k$.)

The preceding theory of local time reduces to the theory UT of universal time if every mention to reference frames is dropped: in this case UT $\subset$ LT. It might be thought that this is in fact what happens with our axiom system for LT, as in no case was the reference frame changed, and after all a variable that is held constant is not a variable. Yet although no relation between $\tau_k(x, y)$ and $\tau_{k'}(x, y)$ was laid down, none was explicitly excluded either, so that LT is a useful framework for thinking of time as a relation between events and frames. LT is made more specific in relativistic theories, whereas nonrelativistic physics presupposes UT. So far, the quantum revolution has not affected LT: in QM t is a "c-number" not an eigenvalue of an operator representing a dynamical variable; and constants of the motion, i.e. magnitudes M such that $dM/dt = 0$ for all t, are as important in QM as in classical physics. But this situation may change — given time (RANKIN, 1965).

It will be noticed that τ was not made to correspond to a special physical object: this would have been as impossible as trying to find an independent physical correlate for the relations of being to the left of, or heavier than, beyond the bodies for which they hold. What has physical reality are neither the bare events nor the bare frames nor the time relation but the whole set of events and frames ordered by the time function. Nor have measurement procedures and devices been mentioned in the preceding, for this would have amounted to taking finite subsets of the set of events and denumerable subsets of the set of reals: and this would have tied not only chronology but also chronometry. Moreover, measurements can say nothing about the time function itself: if they could determine it, time theories would be unnecessary or they would have been proposed long ago. On the other hand theories of time do say something as to how to take time measurements and how to process the results of such operations. In short LT is an objective and general theory, one which is supposed to hold even if no one is making time measurements and which can apply to any of the many time-keeping devices.

Note also that LT retains the useful fiction of the *point event*. We seem to know this is a fiction: real events last some nonvanishing time; and if the 4*th* Heisenberg relation turned out to be true (see Ch. 5, 6), a point event would require an infinite energy. The simplification could be dropped by making extended events the stuff of time. That is, instead of starting with the set of point events we could try and start with the set of processes. A process could in turn be regarded as an ordered set, either continuous or piece-wise continuous, whose elements could be interpreted as events. Something like a nesting of intervals in the style of WHITEHEAD (WHITEHEAD, 1919) might be tried, although this is not clear. But it must be done if the relational theory of time is to be given an even more realistic basis — which desideratum is controvertible, as anyhow no theory can afford to take reality exactly as it is, without making brutal simplifications. Moreover, no such refinement is needed for measurement purposes: in measuring we always make the pretence that the initial and terminal states of a process are point-like. In other words, time reckoning is so sophisticated that it need not refine the oversimplified assumption that there are point events. The foundation of time theories on the more complex concept of process is necessary, if at all, for philosophical reasons.

Let us finally hit at three widespread misconceptions.

First myth: "The time order is reducible to the causal order, in the sense that the two are logically equivalent: event c is prior in time to event e iff c causes e." False: the assumption that c precedes e does not imply that c causes e; to assert that it does is to commit the *post hoc ergo propter hoc* fallacy. What is true is that, if c does cause e, then c is earlier than or simultaneous with e in any frame. The converse not being generally true, the equivalence does not hold. Moreover, as made clear by LT, it is perfectly possible to have time in a random world and even in a lawless world. In short the causal theory of time is a gross oversimplification of the relational theory of time and it rests on an incomplete analysis of causality.

Second myth: "The 2nd law of thermodynamics, supposed to be satisfied by the irreversible processes, defines the positive direction of time." This is logically untenable: (*a*) "time" occurs in the very statements of that law: there is not just an entropy increase but an entropy increase in time; (*b*) in thermodynamics, to the extent to which this science exists, the concepts of time and entropy are mutually independent (not interdefinable), and moreover whereas the former is (so far) essentially nonstochastic, the latter is statistical; (*c*) the time concept occurs in a number of theories dealing with reversible processes, hence it must be introduced independently of thermodynamics.

Third myth: "Measurements on microscopic systems define the arrow of time, as they produce irreversible changes in the measured object. Hence, no measurement, no time." This view is false on several counts: (*a*) because it presupposes the second myth; (*b*) because QM presupposes LT, without which no equation of evolution can be written; (*c*) because the orthodox account of measurement, with its irreversible reduction of the state function, is inconsistent with QM (see Ch. 5, 7); (*d*) because QM is time-symmetric and it can be applied to reversible measurements (AHARONOV, BERGMANN and LEBOWITZ, 1964); (*e*) because the view that irreversibility is not a trait of reality but is created by The Observer is sheer anthropocentrism (BUNGE, 1955 b).

The three preceding myths exhibit as many philosophical diseases that prey on physicists and nontechnical philosophers: (*a*) definitionism, a resistance to introducing new, undefined concepts; (*b*) the belief that every equation can serve as a definition — even if the equation expresses a law of nature; (*c*) the hope that the recognition of the status of primitive or basic can be made outside some definite theory. The time concepts occur, if at all, in the available physical theories. Yet they can be elucidated, namely by building and applying time theories. But it is now time to make room for space.

4. Physical Geometry

Not every physical theory concerns the motions of its objects in real space: thus thermostatics, electric circuit theory and quantum electrodynamics are unconcerned with the detailed spatial configuration of their objects. Yet every physical theory takes for granted that its objects do have some spatial configuration or at least spatial relations to other objects — i.e. that its referents "exist in space", short for "contribute to constituting real space". But no theory save GR makes it its business to investigate the structure of physical space: with this exception physical theories either assume no definite space structure at all or they do presuppose one — most often Euclidean, occasionally Riemannian, less frequently just affine. When Euclidean geometry (EG) is in fact assumed — and this happens less often than is usually believed — it is adopted without questioning: physicists would sooner sacrifice any other trait of their theories. This does not entail that physical Euclidean geometry (PEG) holds a priori, i.e. independently of experience: it means just that we judge PEG to be reasonably well corroborated at least for small regions of space.

A geometry is true a priori or by stipulation or convention as long as it is consistent and does not purport to refer to anything real. In this sense all mathematical geometries are equally true; or perhaps the truth concept is pointless in their regard. What is astonishing is that one of them, namely EG, should — when properly interpreted — hold also for the spatial relations among things, i.e. that it should have a factual truth value and indeed a very high one. It is surprising because psychology has shown that visual space — the tangle of relations satisfied by things as seen by us — is Lobachevskyan rather than Euclidean. (The builders of the Parthenon may have had an inkling of this when they corrected for the perceptual sagging of the phrieze.) Consequently EG is not summarized perception but a conceptual creation that conflicts slightly with sense experience but fits, to a much better approximation, real space in the small.

Of course it is not the mathematical (formal) theory that holds for real space but an interpretation or model of it: what is factually true is not EG but PEG. This latter discipline is the object of no special treatise because (a) the interpretations of EG in terms of solid edges and light rays are customary, and (b) the relations between mathematical theories and their physical models have been cleared up only recently. The automatic interpretation of EG in physical terms has hindered the realization that PEG is a physical not a mathematical science and moreover one that, being the concern of no single branch of physics if GR is excepted, is adequately included in protophysics or, more exactly, in non-generally covariant protophysics.

The distinction between a geometrical and a physical theory of space can be clearly seen by attaching different physical meanings to the primitives of the abstract theory D of metric spaces. The primitives of D are a set U and a nonnegative real valued function d on the set $U \times U$ of ordered pairs $\langle x, y \rangle$ of individuals of U. Axiomatic definition: U is a *metric space* iff

D 1 For any two points x, y in U, (a) $x \neq y \Rightarrow d(x, y) > 0$; (b) $d(x, y) = 0 \Leftrightarrow x = y$.

D 2 For any three points x, y, z in U, $d(x, y) + d(y, z) \geq d(x, z)$.

The symmetry condition $d(x, y) = d(y, x)$ is postulated sometimes as well but unnecessarily so because it follows from $D\,1$ and $D\,2$.

Let us now give a physical interpretation of the abstract theory D. We first try the interpretation

$$I(U) = \text{Particles}, \qquad I(d) = \text{Interparticle separation.}$$

This interpretation satisfies the triangle inequality $D\,2$ but it fails to satisfy $D\,1\,b;$ indeed any number of numerically different but otherwise identical classical point particles or of bosons can be piled up at a single point. Only fermions with certain spin relations satisfy $D\,1\,b$. Hence while the previous interpretation is false ($=$ it is not a model of D) the more restricted one

$M_1(\text{D})$: $I(U) = $ Fermions, $I(d) = $ Separation between fermions

is physically true: it constitutes a *physical model* of D. Another possible physical interpretation is in terms of point events:

$M_2(\text{D})$: $I(U) = $ Possible events, $I(d) = $ Separation between events.

In either case a *relational* as opposed to an *absolute* theory of physical space is obtained, in which geometrical predicates mirror relations among physical entities or changes thereof. In such a view space is not a preexistent passive container into which things and events fall, but it is a net of relations among possible things or events. (The furniture need not be restless: unchanging things would make up a space as well, but not time.) Hence the precise space metric need not be assigned except hypothetically: finding it out becomes a task for physicists. Yet the discovery of the structure of real space is not performed by going outdoors and measuring distances and angles with no geometry in mind: every mensuration presupposes some geometry or other. The precise structure of space is discovered in a devious way: not just by laying yardsticks one after another but by hypothesizing various connections, embedding them into physical theories (particularly GR), and finally testing the latter. Needless to say, it is not easy to disentangle the geometrical components among the logical consequences of the initial assumptions of such a theory involving conjectures concerning physical space — but no other procedure is known.

The abstract theory D is generic: it covers infinitely many possible specific metric geometries — as many as choices of the metric coefficients. The specific metric spaces employed in physics occur applied to physical space or to spacetime — the stuff of which is the set of events. The concept of spacetime is of course more inclusive than either that of space or that of time and it need not be tied to relativity although it originated historically with it. In any case, as soon as the basic set U (the space) and the basic function d (the metric) are specified and attached a physical interpretation, a specific physical metric geometry emerges. Such a physically interpreted formalism is either true or false to some degree. It is up to physics to determine what its truth value can be.

The conversion of a mathematical geometry into a physical one requires only the physical interpretation of the primitives of the former, i.e. a coordination between mathematical building stones and physical objects. But this is insufficient for the empirical test of a physical geometry as well as for its use in spatial measurements. Either requires the adoption of concrete counterparts of two nonbasic concepts: those of unit distance and coordinate system. A material model or realization of a length unit is of course any pair of relatively fixed physical points such as the endpoints of a rod or two successive maxima of a light wave. Every such *length standard* is introduced by convention and it can be said to constitute a referition (not a definition) of the unit length concept (see Ch. 1, 2.5.2). And a material realization of the geometrical concept of coordinate system is a physical thing in which permanent and preferred spatial directions are recognizable — e.g., a laboratory or a static net of light rays. In the case of a geometry of spacetime, i.e. one referring to a four-dimensional space of events, the physical system serving as a reference frame will, in addition, be ticking regularly. In any case it will constitute a more or less adequate *reference frame* which the corresponding coordinate system is supposed to map.

A *coordinate system* is not a thing but a concept, namely a mapping of a patch of a manifold into a Euclidean ball. (More precisely, a coordinate system over a region of an n-dimensional manifold M^n is a function that maps that region onto a subset of R^n, i.e. that sends every point in that region to an n-tuple of reals. If M^n is itself Euclidean then a single coordinate system will cover the whole space; otherwise the coordinate systems will be local.) The coordinates of a space point are labels that identify or name that point in a conventional way:

they are not intrinsic properties. Such geometrical coordinates of a point in space need not be physically meaningful and they must be distinguished from the *physical coordinates* of a particle or a wave front: these identify physical things and are relative to some physical frame; therefore they are not arbitrary names but represent a physical situation. We can interpret some coordinate systems as reference frames and conversely symbolize any reference frame as a coordinate system, but we should not confuse geometrical concepts with physical entities. One and the same physical frame of reference such as a Copernican frame — with center in the Sun and light rays coming from three suitably chosen fixed stars — can be associated with infinitely many mathematical coordinate systems, all of which are interconvertible.

A coordinate system is a concept and therefore it cannot move in space; accordingly a coordinate transformation need not be interpreted as involving motion. On the other hand reference frames can move relative to others. We must correspondingly distinguish *coordinate transformations* from *frame transformations*. The transformation from one set of Cartesian axes to another coordinate system, whether or not Cartesian, is of the former kind; but a Lorentz transformation, i.e. a transformation relating two Lorentz frames, is a frame transformation (see Ch. 4, 2). The distinction is important if only because none of the features of a theory which depend on the coordinate system are objective — the choice of coordinate system being free — whereas a dependence on the reference frame, as in the case of particle masses and wave frequencies, can be objective.

Besides distinguishing coordinate systems from reference frames we must distinguish the latter from *observers*, although this is not usually done either. Since every observer constitutes his own reference frame, and since frames occupy such a prominent role in experimental physics, it is customary to identify the two concepts. But most of the reference frames occurring in physics are uninhabitable and none are supposed to talk back. Regarded as a purely physical system, an observer is an example of a reference frame; but an observer which is a passive lump of matter should not be called 'observer'. Moreover, it is misleading to identify reference frames with observers, for a kinematical relation among reference frames will then be interpreted as an actual connection among observers, which could then be taken in a psychological sense. The observer has no place in physical theory — save as builder, tester and user of theories.

The functions of coordinate systems, physical frames, and observers are not the same in physical theory as they are in experimental physics, and a lot of confusion arises from exporting the experimenter's viewpoint to theoretical physics. In experimental physics one must always adopt

some reference frame (usually the laboratory) and some mode of observation (apparatus and technique), both of which must be accessible to
some operator if only vicariously. Therefore empirical data do not
refer to the objects of study alone but to physical situations as actually
observed: they involve the observer's point of view. (No wonder experimental physicists tend to be operationalists.) On the other hand the
theoretician adopts, as it were, the object's point of view. Thus he will
prefer to compute a scattering cross section in (relative to) the center
of mass system; and only in order to test his calculations is it necessary
to transform his formulas to the laboratory reference frame, as the
experimental set-up is not usually attached to the center of mass of the
things concerned. Theoretical physics raises above the level of observation
which, even when intersubjective, is unavoidably subject-centered. The
whole point of theoretical physics is to grasp the actual or possible
physical situations without the admixture of manifestly subjective
elements: the target of all science is indeed objectivity (BUNGE, 1967b).

Now in order to attain this goal two things are done. One is to
sieve data trying to disclose those which remain invariant under changes
in modes of observation and description (in particular, reference frames,
coordinate systems, and techniques). This procedure yields a few approximate invariants such as length and weight. But it is lengthy and clumsy
and it cannot lead to exact invariants for the following reasons: (a) the
means of observation cannot be arbitrarily varied and (b) most genuine
invariants, such as the spacetime interval and entropy, happen to be
unobservable — though often measurable *via* devious means employing
theoretical formulas. The second, far deeper and therefore more powerful
method, is to invent transobservational predicates and formulas involving no relation to any particular condition of observation, hence automatically invariant under changes in point of view.

Such invariant magnitudes — paradoxically called 'observables' in
GR — and covariant formulas are candidates to objectivity. (Example
of invariant: the determinant of a nonsingular matrix under a similarity
transformation; example of covariant: any tensor equation under a
coordinate transformation.) In this way we obtain statements holding
in any coordinate system and, a fortiori, possibly true for any observer.
But they are of course just candidates to objectivity: the theoretical
physicist is not interested in creating ghost-like properties and statements that would stand no matter what experience may say for being
irrelevant to experience: this is the mathematician's game. The invariants
hypothesized by theoretical physics must in principle be related to
observables — but not reducible to them — just as the covariant statements must entail low level consequences that can be checked in any
suitable observation platform. In short, invariance and covariance are

best attained by raising above experience though not in order to evade it but to make sense of it.

So much for the items — length standard and physical frame — that must be added to a physical geometry in order to test it or to use it. Now to these two radically different operations. In order to find out whether a given geometry is true of physical space in some region of the universe, we may utilize any of several techniques for distance and angle measurement, or rather comparison. Let us focus on the radar technique, which reduces distance ratios to time reckonings and, given the advanced state of chronometry, is the most exact available method. Suppose a radio signal is sent out in a certain direction, bounces off a reflector without delay (a pretence), and is recorded back, again without delay (another pretence) at the sender t seconds later. If, in accordance with SR, we assume that the signal speed c is independent of the direction of propagation (isotropy of space), then we infer that the emitter-reflector distance is $d = ct/2$. Any homogeneous relation between distances postulated in a metric geometry can then be tested *via* this simple formula without having to determine the value of c. For example, in the Pythagorean theorem or any of its extensions a common factor c^2 will occur in each term, so that it can finally be cancelled. For this reason the test of a physical metrical geometry can be performed with clocks and e.m. signals exclusively. But this is only because we are relying on the hypothesis of the constancy of c, the test of which requires two sets of independent measurements, one of distances, the other of durations.

In order to perform "absolute" distance measurements we must assume a lot more — indeed whole fragments of geometry and physics. Thus the radar technique calls for in this case a previous determination of the value of c. And this requires not just clocks but also yardsticks as well as assumptions concerning the behavior of clocks and yardsticks and the propagation of e.m. signals: all these assumptions enter the construction and the reading of clocks and graduated rules. And a more precise determination of the spacetime metric will require, in addition, some fragments of general relativity — the tool *par excellence* for the exploration of spacetime. (See LEVI CIVITA, 1929 for the theoretical presuppositions of measurements of the metric tensor components.) Therefore the proposal (WIGNER, 1957) of completely reducing length measurements to duration measurements is as impracticable as the proposed reduction of every magnitude to length (WHEELER, 1962). It is not just a technical difficulty that might conceivably be resolved in the future, but a logical impossibility. Indeed distance and duration are mutually independent (not interdefinable) concepts and moreover all four spacetime coordinates occur, if at all, as independent and equally basic (though with different meanings) in all of contemporary physics.

Now most distance measurements are indirect, and indirect measurements involve, by definition, some hypotheses of physical geometry. Thus to measure the distance from an inaccessible point to a material frame such as a rod we start by analyzing the physical triangle made up of one of the edges of the rod and two light rays passing through the point and grazing the extremities of the rod; we then measure the base length and the angles and finally compute the distance using some metric geometry. We therefore conclude (a) that while physical geometries can be tested by means of duration measurements, distance measurements are not reducible to duration ones and (b) they have hardly a test value because most of them are based on the very geometry they purport to test. Therefore any discussion of the problem of physical geometry that starts by analyzing geometrical measurements betrays neglect of the conceptual background of every measurement, whether geometrical or not (see Ch. 1, 5).

If a physical geometry cannot be tested through distance measurements alone, a fortiori it cannot emanate from them: it must first be invented, then tried out. This does not mean that it is a priori in the epistemological sense: it certainly is an empirical science yet not because it derives from experience or is concerned with it but because it can be put to the empirical test and applied — e.g. in geodetics. This is hard to realize because EG, when physically interpreted, is abnormally near the truth. For this reason when dealing with small regions we may continue to prescribe the metric as if it were independent of the stuff space is made of, although from a principled point of view this is wrong.

The fixed metric one adopts when not probing the structure of space need not be Euclidean: it can also be Riemannian as long as the metric tensor is independent of matter and fields, i.e. as long as it is assigned no independent physical meaning — otherwise we would be doing general relativity. Indeed, in many cases the properties of the physical systems of interest are insensitive to details of the space structure: in these cases the choice of a geometry for the given physical theory is up to a point a matter of convenience not of truth. But only up to a point, for not all geometries are equally rich, flexible, and natural. For instance an intrinsic geometry contains concepts such as "space curvature" which are absent from other theories; furthermore it lends itself better to the discovery and description of invariants; and it is more natural than EG because it can use one of the system's own configurations as a frame instead of introducing artificial outsiders. In short, wherever the concept of space plays an auxiliary rather than an essential role, the choice of geometry is to some extent, though not entirely, a matter of convention. This is not the case of physical geometry itself, i.e. of the set of theories whose object (referent) is physical space or spacetime.

By contrast with mathematical geometries, a physical geometry assumes that physical space is as real as the things it nets; moreover, that neither geometrical relations nor their relata exist independently from one another. This does not commit it to the substance doctrine of space (ARISTOTLE, DESCARTES, NEWTON). The reality of the things-space complex is also consistent with LEIBNIZ' relational hypothesis that space, far from being an entity — e.g., a passive container — is the set of all geometrical relations among things: no physical system (bodies or fields), no physical space. In any case, a physical geometry is a branch of physics: its framework alone is mathematical. Only, its experimental test is anything but straightforward. In particular, the adoption of a geometry by a successful physical theory does not automatically confirm that particular geometry as a faithful mapping of physical space: firstly because the experimental errors will mask differences among any number of geometries when applied, as usual, in the small; secondly because several or even all of the statements of the given physical theory may be actually independent of the structure of space. For this reason it is good policy, when building or when reconstructing a physical theory, to be as parsimonious as possible in the geometrical assumptions. It is astonishing how many theories that have heretofore been expounded presupposing EG can be formulated by assuming just a differentiable manifold with at most an affine connection. We shall try to adhere to this strategy in the following chapters.

5. General Systems Theory

A *physical system* is anything existing in spacetime and such that it either behaves or is handled as a whole in at least one respect. If endowed with mass, a physical system is called a body, a particle, or a matter quanton; otherwise, a field or a field quantum. In either case a physical system is not an arbitrary set of elements but either a simple or a complex of coexistent individuals which may or may not be physically interconnected but in any case may be regarded as a physical unit not just as a conceptual unit. A complex system must be distinguished from a set, which is a thing of reason. Thus all past, present and future neutrons, both the real and the imaginary ones, constitute a set but not a physical system; on the other hand any collection of coexistent neutrons, however feebly interacting, may be regarded as a system. Likewise a Gibbs ensemble, being conceptual, is a set but not a system like a Boltzmann ensemble of noninteracting particles. The difference will be made more precise as we advance.

We next encounter the concept of spatiotemporal *part*, which differs from the concept of set membership not only because it holds among physical systems but also because, unlike $\in$, it is transitive. In terms of the part relation, symbolized $\dot\in$, we can introduce the remaining key concepts of the systems theory, and first of all those of physical sum or juxtaposition ($\dot+$) and physical product or superposition ($\dot\times$). An aggregate of bodies may be regarded as the *physical sum* of its parts even if they are not in contact. On the other hand the system composed of an electric field superposed to a magnetic field may be regarded as the *physical product* of its parts. Loosely speaking, physical addition is external joining whereas physical product is interpenetration. More precisely: if σ_1 and σ_2 are physical systems, then: (*a*) $\sigma = \sigma_1 \dot+ \sigma_2$ iff every part of σ is either a part of σ_1 *or* a part of σ_2; (*b*) $\sigma = \sigma_1 \dot\times \sigma_2$ iff every part of σ is both a part of σ_1 *and* a part of σ_2. In particular, any given system is its own sum and its own product.

These notions enable us to express correctly a number of familiar ideas. For example, the total relative mass of the body b made up of the parts b_1 and b_2 equals the arithmetical sum of the partial masses:

$$M(b_1 \dot+ b_2) = M(b_1) + M(b_2)$$

which is an approximate truth of fact not of mathematics. The total force on b_1 equals the force exerted on b_1 by b_2 plus the share of b_1 in the external force exerted jointly over $b_1 \dot+ b_2$:

$$F(b_1) = F(b_1, b_2) + F(b_1, b_1 \dot+ b_2).$$

Finally, the strength of the electrostatic interaction between two charged particles p_1 and p_2 in their common field φ is ε times the force between the charges inside a perfect dielectric body b:

$$F(p_1, p_2 \dot\in \varphi) = \varepsilon \cdot F(p_1, p_2 \dot\in \varphi \dot\times b).$$

It is tempting to write '$b_1 \cup b_2$' instead of '$b_1 \dot+ b_2$' in the first two examples and '$\varphi \cap b$' instead of '$\varphi \dot\times b$' in the third. But bodies and fields are things not sets, and forces act on things not on ideas: otherwise physics would be a branch of mathematics. Yet in these cases the mistake has no computational consequences hence no empirical ones either and therefore it usually goes unnoticed. (Although every experiment can be criticized by some theory, not every mistake in physics can be corrected by experiment.) But elsewhere the identification of "system" and "set" does lead to empirically detectable consequences. Thus if it is assumed that a body is a Borel set with a certain structure, then the temptation is strong to regard mass as a set measure. By the additivity of mathematical measure (on Borel sets), this leads to the additivity of mass — which is slightly false for the proper mass.

Things and ideas are not identical and this is why it is worth while to try to map things on ideas. And our ideas on the ways things combine to make up wholes are not mathematical although they involve mathematics: whilst sets obey the laws of set theory, physical systems obey physical laws, among them the trivial laws that characterize the part-whole relation. But before we exhibit them we must generalize the previous definitions and introduce three other concepts.

Df. 1. A system σ is a *physical sum* or *juxtaposition* of all the individuals $\sigma_i \in \Sigma'$, with $\Sigma' \subseteq \Sigma$ and $i \in N$, iff every part of σ is a part of at least one of the members of Σ'. Symbol: $\sigma = \sum\limits_{\sigma_i \in \Sigma'} \sigma_i$.

Df. 2. A system σ is a *physical product* or *superposition* of all the individuals $\sigma_i \in \Sigma'$, with $\Sigma' \subseteq \Sigma$ and $i \in N$, iff every part of σ is a part of every member of Σ'. Symbol: $\sigma = \prod\limits_{\sigma_i \in \Sigma'} \sigma_i$.

Df. 3. A system is a *null individual* of the kind Σ iff its physical sum to an arbitrary individual of the same kind equals the latter individual. Symbol: 0_Σ. [Examples: darkness is the null light field, empty space the null body, and a vanishing electric field is the null electric field.]

Df. 4. A system is a *universe* of the kind Σ iff it is the juxtaposition of all the systems of that kind. Symbol: 1_Σ. [Examples: the aggregate of all coexistent electrons, the collection of all artificial satellites, the sum total of physical systems.]

Df. 5. An individual such that its physical sum to another individual σ of the same kind equals 1_Σ, is called the *complement* of σ in Σ. Symbol: $\bar{\sigma}$. [Examples: the environment of an atom is its complement in the cosmos, and the extragalactic universe is the complement of the Milky Way.]

The five above concepts have been defined in terms of Σ and $\in$. Let us now shift the focus to the 6-tuple $\langle \Sigma, +, \times, {}^-, 0_\Sigma, 1_\Sigma \rangle$, where Σ is a set, $+$ and $\times$ are binary operations in Σ, ${}^-$ is a unary operation in Σ, and 0_Σ and 1_Σ are fixed elements of Σ. We shall postulate that this relational system is a *Boolean algebra*. This is plausible, for (a) the operations $+$ and $\times$ are associative and commutative; (b) each of these operations distributes over the other, as exemplified by:

field $+$ (cream $\times$ coffee) $=$ (field $+$ cream) $\times$ (field $+$ coffee)

field $\times$ (air $+$ needles) $=$ (field $\times$ air) $+$ (field $\times$ needles)

body $\times$ empty space $=$ empty space,

electron $+$ all electrons $=$ all electrons;

(c) for every σ in Σ, $\sigma \dotplus 0_\Sigma = \sigma$ and $\sigma \mathbin{\dot\times} 1_\Sigma = \sigma$; and (d) every σ in Σ has a complement $\bar\sigma \in \Sigma$ such that $\sigma \dotplus \bar\sigma = 1_\Sigma$ and $\sigma \mathbin{\dot\times} \bar\sigma = 0_\Sigma$. By postulating that the systems theory S is a model of Boolean algebra, we get the uniqueness of every juxtaposition, superposition, null individual, and universe into the bargain. Moreover once the algebra of S is so stipulated, the $\dot\in$ relation can be retrieved by means of a definition, as will be seem in a moment. (For a different ontological model of Boolean algebra, see LEJEWSKI, 1960—61.)

An algebra of physical systems is necessary but insufficient: if we are to make use of analysis we also need a bridge between systems and sets. To this end we shall use the semantical concept $\triangleq$ of modelling or conceptual representation of a physical system (see Ch. 1, 1.3.2); it will occur in formulas like '$s \triangleq \sigma$', short for 's models [mirrors] σ'. We shall also rope in the concept of point set s in a differentiable manifold M, which concept will be taken over from manifold geometry. Finally, from measure theory we shall borrow the idea of Lebesgue measure μ of a region of M. The theory S shall then concern the 8-tuple $\langle \Sigma, \dotplus, \mathbin{\dot\times}, {}^-, 0_\Sigma, 1_\Sigma, M, \mu \rangle$. (The concept $\triangleq$ belongs of course to the general background of every protophysical theory.) The reference class Σ is partially ordered by the relation $\dot\in$, here newly introduced by way of the following

Df. If $\sigma_1, \sigma_2 \in \Sigma$, then $\sigma_1 \mathbin{\dot\in} \sigma_2$ iff $\sigma_1 \dotplus \sigma_2 = \sigma_2$.

The axioms of the theory S are:

S 0: *Existence.* (a) Σ is nonempty. (b) Every σ in Σ is a physical system.

S 1: *Structure.* The 6-tuple $\langle \Sigma, \dotplus, \mathbin{\dot\times}, {}^-, 0_\Sigma, 1_\Sigma \rangle$ is a Boolean algebra.

S 2: *Fabric of models.* (a) M is a 4-dimensional twice differentiable manifold. (b) M represents spacetime.

S 3: *Uniqueness of models.* (a) For every $\sigma \in \Sigma$ there exists exactly one $s \subset M$ such that $s \triangleq \sigma$. (b) For every Σ, $\emptyset \triangleq 0_\Sigma$.

S 4: *Set-theoretical model of juxtaposition and superposition.* For every $\sigma_i \in \Sigma' \subseteq \Sigma$ and every $s_i \subset M$, with $i \in N$ and $s_i \triangleq \sigma_i$, (a) If $\sigma = \overset{\cdot}{\underset{\sigma_i \in \Sigma''}{\sum}} \sigma_i$, then $\cup s_i \triangleq \sigma$; (b) If $\sigma = \underset{\sigma_i \in \Sigma'}{\prod} \sigma_i$, then $\cap s_i \triangleq \sigma$.

S 5: *Bulk-measure relation.* If $\sigma \in \Sigma$ and $s \subset M$ and $s \triangleq \sigma$, then the volume of σ equals the Lebesgue measure of s.

In addition to the theorems of Boolean algebra — now physically interpreted — we get the following logical consequences.

Thm. 1. The partial models of noninterpenetrating systems are pairwise disjoint. [If $\sigma_i, \sigma_j \in \Sigma' \subseteq \Sigma$ and $s_i \subset M$ and $s_i \triangleq \sigma_i$ and $s_j \triangleq \sigma_j$ with $i, j \in N$ and $\sigma_i \mathbin{\dot\times} \sigma_j = 0_\Sigma$ for $i \neq j$, then $\cap s_i = \emptyset$.]

Thm. 2. The model of a part is included in the model of the whole. [If $\sigma_i, \sigma_j \in \Sigma' \subseteq \Sigma$ and $s_i \subset M$ and $s_i \triangleq \sigma_i$ and $s_j \triangleq \sigma_j$ with $i, j \in N$, then: $\sigma_i \in \sigma_j$ iff $s_i \subset s_j$.]

Thm. 3. The volume of a system of noninterpenetrating parts equals the sum of the Lebesgue measures of the partial models. [Under the assumptions of Thm. 1, $V(\overset{.}{\sum}\sigma_i) = \sum \mu(s_i)$.]

We whall develop S no further. We close with a few comments. (*1*) S 2—S 4 bridge the gap between physical systems and sets. (*2*) Thm. 2 makes it possible to translate statements about part-whole relations into statements about set inclusions, and conversely. Should one mistake logical equivalence for synonymity, Thm. 2 would lead one to mistaking things for their conceptual models, as is the case when the expression 'a field is a manifold' is employed. By the same token, Thm. 2 explains why such a reification of constructs is seldom harmful. (*3*) S 5 enables us to speak of the volume of a system while letting the integration variable (not the elements of the system) range over the subset (of M) representing that system. (*4*) S is a part of *mereology*, the theory of the most general relations among objects of any kind. This theory, an essay in modern metaphysics, was originated by S. LEŚNIEWSKI in 1916 and worked out by LEJEWSKI 1955, 1960—61, R. MARTIN, 1943; SOBOCIŃSKI, 1955; TARSKI, 1956; WOODGER, 1937 and a few others. It has been neglected apparently because of the widespread belief that it is just a model of set theory. Our version of S differs markedly from the previously mentioned works, both with regard to the interpretation of the Boolean operations and with regard to the system-set bridges laid by S 2—S 5.

6. Analytical "Dynamics"

Analytical "dynamics" is a class of theories that can be used almost anywhere in physics because they are only partially interpreted. They are, indeed, general frameworks for describing the evolution of physical systems. The nature of the systems and consequently of their states are left unspecified. The fact that these theories are usually included in mechanics is due to their historical origin; nowadays the name 'analytical dynamics' is a misnomer. We shall start with a comprehensive theory G (J. L. MARTIN, 1959) from which the usual Hamiltonian theory H can be derived. The lagrangian framework L will then be deduced from G as another particular case, and its equivalence to the Hamilton-Jacobi theory HJ will be pointed out. Finally the various relations of these theories to elementary Newtonian mechanics proper will be studied. The foundations of each of the general dynamical theories will be displayed in an orderly fashion, i.e. axiomatically. Since the aim is to lay these foundations bare only a few theorems will be derived in the usual informal way.

6.1. General "Dynamics"

6.1.1. Background and Primitives. All the theories to be developed in this section are based on PC$=$, set theory, algebra, topology, analysis, and manifold geometry, as well as LT (Sec. 3) and S (Sec. 5). The concept of elementary physical system σ will be taken as primitive. The concept of complex system σ^N is derived:

Df. 1. $\sigma_i \in \Sigma \Rightarrow \sigma^N \overset{\mathrm{df}}{=} \dot{\Sigma} \sigma_i$.

The other specific building blocks of G will be those of state space Φ, dynamical variable φ, poissonian P, and hamiltonian H. More explicitly, here goes our

Primitive base

Σ Mathematical nature: set. Interpretation: the class of discrete physical systems.

T Mathematical nature: set. Interpretation: set of instants.

Φ Mathematical nature: manifold. Interpretation: a point φ in Φ represents the state of the system σ^N.

P Mathematical nature: tensor field on Φ. No physical meaning other than that P refers to σ^N.

H Mathematical nature: scalar function on Φ. No fixed interpretation save that H refers to the evolution of σ^N. In most cases, H will be interpretable as the energy of the system.

Remarks. (1) The concepts of physical space, configuration space and reference frame do not occur in our list. Likewise the concepts of position, mass and force are absent from it, which anticipates that G will be applicable to systems of any kind, even nonlocalizable ones. (2) From the absence of the concept of ordinary space we cannot infer that the concept is altogether dispensable: it lurks behind G just as behind any other physical theory since the referents of it exist in (or rather constitute) physical space. Every point φ represents, among other traits of the system, its configuration in ordinary space.

6.1.2. Axioms

G1 (*a*) Σ is a nonempty denumerable set. (*b*) Every $\sigma^N \in \Sigma$ represents a physical system consisting of N parts.

G2 For every $t, t' \in T$ and every $i \in N$, if $t' \neq t$, then: if $\sigma_i \in \Sigma$ and $\sigma_i \dot{\in} \sigma^N$ at t, then $\sigma_i \in \Sigma$ and $\sigma_i \dot{\in} \sigma^N$ at t'.

G3 (*a*) For every $\sigma^N \in \Sigma$ there exists a Φ. (*b*) Φ is a differentiable manifold of at most $6N$ dimensions, whose points φ are functions from $\Sigma \times T$ to R. (*c*) The point $\varphi(\sigma^N, t) = \langle \varphi_1(\sigma^N, t), \ldots, \varphi_{6N}(\sigma^N, t) \rangle$ represents the physical state of σ^N at time t.

G4 For every $\sigma^N, \sigma'^N \in \Sigma$, every $t \in T$ and every $i \in N$, if $\varphi(\sigma^N, t) = \varphi(\sigma'^N, t)$, then $\sigma^N = \sigma'^N$.

$G5$ (a) H is a function from Φ to R. (b) For any fixed σ^N, H is continuous and differentiable.

$G6$ (a) P is an antisymmetric tensor of the 2nd rank on Φ. (b) For any fixed σ, the components of P are real valued, continuous and differentiable. (c) For every triple $\langle i, j, k \rangle$,

$$P_{in}\frac{\partial P_{jk}}{\partial \varphi_n} + P_{jn}\frac{\partial P_{ki}}{\partial \varphi_n} + P_{kn}\frac{\partial P_{ij}}{\partial \varphi_n} = 0 \tag{2.4}$$

(summation over repeated indices).

$G7 \qquad\qquad \dot{\varphi} = P \cdot \nabla H \qquad \left[\text{I.e., } \dot{\varphi}_i = P_{ij} \cdot \frac{\partial H}{\partial \varphi_j}\right]. \tag{2.5}$

Remarks. (1) At first sight our axiom system is not p-complete since it does not specify the structure and content of T; but this is the task of the protophysical theory LT (see Sec. 3). The two conjuncts of $G1$ make up a hypothesis of physical existence; this triviality is necessary in order to prevent the whole theory from being trivial. $G2$ is an assumption of physical immutability: once an entity of the kind Σ, always such an entity. $G3$ says that the dynamical variables φ_i span the state space Φ and that every point in it represents a state of the system concerned. $G4$ states that a system is unambiguously specified by the set of values of its dynamical variables. $G5$ and $G6$ are purely mathematical assumptions necessary to write down the last axiom. $G6c$ is a condition for the Jacobi identities to hold and therefore to rope in the assistance of Lie algebra; it does not occur in Hamiltonian or in speudo-Hamiltonian dynamics where $P_{ij} = \pm 1$. The only law proper in the whole axiom system, and also its central hypothesis, is $G7$. It is a law of motion since it determines the evolution of the dynamical variables — or, in pictorial terms, it determines the trajectory of the representative point in the Φ space. (2) The kind of system is specified jointly by H and P rather than by a single source function. Since P has a maximum of $6N(6N-1)/2$ mutually independent components, we can anticipate that G can describe more complex kinds of system and motion than any of the more special dynamical theories it entails. (3) Φ has been assigned no metric. The condition that it be a differentiable manifold can be relaxed to make room for impacts.

Let us now introduce

Df. 2: Generalized Poisson bracket:

$$(F, G)_\varphi \overset{\text{df}}{=} P_{ij}\frac{\partial F}{\partial \varphi_i}\frac{\partial G}{\partial \varphi_j}. \tag{2.6}$$

With the help of this concept the law of motion $G7$ is rewritten

$$\dot{\varphi} = (\varphi, H) \tag{2.7}$$

and the components of the poissonian turn out to be

$$P_{ij} = (\varphi_i, \varphi_j).$$

Remarks. (1) It can be seen that, thanks to $G6c$, the Poisson bracket satisfies the axioms of a Lie algebra (antisymmetry, linearity, and Jacobi identity). (2) The independent introduction of the generalized Poisson brackets shows that G goes quite far in the direction of matrix QM, in which the equations of motion and the Poisson brackets are postulated independently.

6.1.3. Three Theorems

Thm. 1: *spinning system.* Let $\varphi = \langle \varphi_1, \varphi_2, \varphi_3 \rangle$ and $P_{ij} = \varepsilon_{ijk} \varphi_k$ where $\varepsilon_{ijk} = 1$ if (ijk) is an even permutation of $(1, 2, 3)$ and $\varepsilon_{ijk} = -1$ if it is an odd permutation, all other components being zero. Assume in addition that H is separable in the form $H = \omega \cdot \varphi$, ω being an arbitrary time-dependent vector in Φ. Then the equations of motion $G7$ become

$$\dot{\varphi} = \omega \times \varphi. \tag{2.8}$$

Remarks. (1) A physical interpretation of this result is: the vector φ rotates with angular velocity ω. This interpretation is not dictated by the semantical components of our axiom system but is suggested by an analogy with well known formulas of mechanics and electrodynamics. (2) The preceding interpretation is partial and purely kinematical: we have not said what is rotating and where or what makes "it" turn, as the theory makes no provision for the introduction of forces although it does not preclude it. But in this particular case no force concept can be introduced because the hypothesized hamiltonian contains only two (pseudo) vectors, none of which can be interpreted as a force. For this reason H cannot be interpreted as the mechanical energy of the system. (Besides, H is not conserved in this case; in fact, $H = $ const just in case H does not depend on t.) In other words, the microtheory of this spinning system is essentially non-Newtonian. We shall see later on that this is just a specimen of a whole class of problems admitting of no Lagrangian formulation either.

Thm. 2: *transition to Hamilton's Dynamics.* Suppose the number of dimensions of Φ is even. Call $\varphi_i = q_i$ for $1 \leqq i \leqq 3N$, and $\varphi_i = p_i$ for the remaining. Call '1_m' the $m \times m$ unit matrix, with $m = 3N$. Then if

$$P = \begin{pmatrix} 0 & 1_m \\ -1_m & 0 \end{pmatrix} \tag{2.9}$$

$G7$ becomes

$$\dot{q}_i = \frac{\partial H}{\partial p_i}, \qquad \dot{p}_i = -\frac{\partial H}{\partial q_i}. \tag{2.10}$$

Proof: straightforward matrix algebra.

Remarks. (1) Since the (2.10) are the canonical equations, HAMIL-TON's dynamics H is seen to be included in G: H<G. (Actually the inclusion relation holds for the corresponding sets of formulas.) (2) Although H is formally a subtheory of G (see Ch. 1, 4.1.1), H has a richer analytic structure (Ch. 1, 4.1.2) than G because it can be appended the mathematical theory of canonical transformations. (3) H is also semantically richer (more specific) than G. In fact, it is only in the case covered by Thm. 2 that the fundamental variables can be partitioned into conjugate pairs. Consequently, in the particular case in which σ is a mechanical system, the q_i are interpretable as generalized position coordinates and the p_i as the conjugate momenta, since in such a case the second half of the canonical equations (2.10) can be identified with the mechanical equations of motion on condition of introducing

Df. 3. Generalized force: $F_i \overset{dt}{=} - \dfrac{\partial H}{\partial q_i}$.

Thm. 3: *transition to pseudo-Hamiltonian dynamics.* If the number of dimensions of Φ is even and the first half of the dynamical variables are called q_i while the second half are called p_i, and

$$P = \begin{pmatrix} 0 & 1_m \\ \varepsilon 1_m & 0 \end{pmatrix} \quad \text{with} \quad \varepsilon = \text{const,}$$

then G 7 becomes

$$\dot{q}_i = \frac{\partial H}{\partial p_i}, \quad \dot{p}_i = \varepsilon \frac{\partial H}{\partial q_i} \tag{2.11}$$

which, for $\varepsilon = -1$, reduce to the canonical equations.

Remark. It can be shown (DUFFIN, 1962) that if $\varepsilon = +1$ and if H is quadratic in the dynamical variables, the pseudohamiltonian equations describe a class of dissipative systems. Consequently G applies to nonconservative systems. As usual, by modifying the equations of motion even if slightly, the reference class Σ can be altered.

6.2. Excursus: Independent Axiomatization of Hamilton's "Dynamics"

Primitives: Σ, T, Φ, and H.

The first two axioms of H are identical with their homologues in G; the others are adaptations to the new situation in which there are two sets of fundamental dynamical variables.

$H1 \Leftrightarrow G1$

$H2 \Leftrightarrow G2$

$H3$ (*a*) For every $\sigma^N \in \Sigma$ there exists a Φ. (*b*) $\Phi = Q \times P$, where Q and P are differentiable manifolds of $3N$ dimensions each. (*c*) The points $q \in Q$ and $p \in P$ are real valued functions on $\Sigma \times T$. (*d*) There is a $1:1$ correspondence between the points $\langle q, p \rangle \in \Phi$ and the states of σ^N at different times.

$H\,4$ For every $\sigma^N, \sigma'^N \in \Sigma$ and every $t \in T$ and every $i \in N$, if $q_i(\sigma^N, t) = q_i(\sigma'^N, t)$ and $p_i(\sigma^N, t) = p_i(\sigma'^N, t)$, then $\sigma^N = \sigma'^N$.

$H\,5 \Leftrightarrow G\,5$.

$H\,6$ For every $i \in N$, $\dot{q}_i = \dfrac{\partial H}{\partial p_i}$, $\dot{p}_i = -\dfrac{\partial H}{\partial q_i}$ \hfill (2.10)

Remarks. **(1)** So far H does not contain the concept of coordinate system. Consequently considerations of GALILEI and LORENTZ invariance do not come up in H; they emerge only when the q_i are interpreted as spatial coordinates. **(2)** In the particular case in which a lagrangian can be introduced and the q_i are interpreted as spatial coordinates, H can be interpreted as the energy of σ^N. But such an interpretation is not mandatory: even requiring that H be interpreted in Newtonian terms, H can be taken as an arbitrary function of the total energy rather than as the energy itself (KENNEDY and KERNER, 1965). **(3)** When a lagrangian can be introduced, only the 2nd half of the canonical equations remain as laws, the first half being definitions (see 6.4, Thm. 4).

Df. 1. The transformations $q_i' = f(q_i, p_i, t)$, $p_i' = g(q_i, p_i, t)$ constitute a *canonical transformation* iff they preserve the law of motion H 6.

Df. 2. Poisson bracket of F and G:

$$(F, G)_{qp} = \frac{\partial F}{\partial q_i}\frac{\partial G}{\partial p_i} - \frac{\partial F}{\partial p_i}\frac{\partial G}{\partial q_i}.$$

Thm. 1. The Poisson brackets of q and H, and p and H, determine the evolution of the system:

$$\dot{q}_i = (q_i, H), \quad \dot{p}_i = (p_i, H).$$

Thm. 3. The Poisson brackets are invariant under a canonical transformation:

$$(F, G)_{qp} = (F, G)_{q'p'}$$

Df. 3. Generalized force:

$$F_i = -\frac{\partial H}{\partial q_i}$$

Remark. This Df. allows us to recast the 2nd half of the canonical equations in a Newtonian form. But this does not entail that the meaning is the same. For one thing, not every force expressible in algebraic terms is derivable from a single function as required by Df. 3; in particular, nonconservative forces are not so derivable.

Thm. 4. If the new variables

$$\tau = it, \quad r_k = 2^{-\frac{1}{2}}(q_k + ip_k), \quad \text{with} \quad i^2 \overset{\mathrm{df}}{=} -1 \tag{2.12}$$

are introduced, the canonical equations can be synthesized into

$$\frac{\mathrm{d}r_k}{\mathrm{d}\tau} = -\frac{\partial H}{\partial r_k^*}. \tag{2.13}$$

Remark. This is just a reminder (*a*) that there is no 1:1 correspondence between objective laws and equations, and (*b*) that complex variables can be introduced in physics whenever certain predicates come in pairs. Besides, (2.13) can be generalized to QM (STROCCHI, 1966).

6.3. Transition from G to Lagrange's "Dynamics"

We saw in 6.1.3 that G includes H. We shall now prove that G also includes LAGRANGE's dynamics L, and that the conditions for the existence of a lagrangian are as exacting as those warranting the existence of the ordinary canonical equations. To this end we return to G and prove first the following

Lemma. If P has an inverse then there exists a vector $\psi \in \Phi$ such that its curl equals P^{-1}. *Proof.* See J. L. MARTIN (1959).

We now exploit this lemma, using ψ to introduce

Df. 4. Lagrangian: $L \overset{\mathrm{df}}{=} \psi \cdot \dot{\varphi} - H$ (2.14)

Thm. 4. If $\sigma \in \Sigma$ and $t \in [t_1, t_2] \subseteq T$ and $i \in N$ and $\delta\varphi(t_1) = \delta\varphi(t_2) = 0$, then

$$\delta \int_{t_1}^{t_2} \mathrm{d}t\, L(\sigma, \varphi, \dot{\varphi}, t) = 0. \tag{2.15}$$

Proof sketch. Upon performing the variations $\delta\varphi_i$, integrating the 2nd term by parts and using the above Lemma, we are left with

$$\int_{t_1}^{t_2} \mathrm{d}t\, \Sigma_i\, \delta\varphi_i \left[P_{ij}^{-1} \cdot \dot{\varphi}_j - \frac{\partial H}{\partial\varphi_i} \right]$$

which in turn vanishes by (2.5), provided P is nonsingular.

Remarks. (1) There exists a lagrangian satisfying a variational principle — and therefore leading to LAGRANGE's equations — provided there exist both a hamiltonian and a nonsingular poissonian. For the latter it is necessary that the number of φ_i's be even, since an odd antisymmetric P is singular. In short, the conditions for the existence of a lagrangian are more exacting than those warranting the existence of a hamiltonian. The spinning system of Thm. 1, characterized by an odd number of dynamical variables, was a case in which there existed a hamiltonian but not a lagrangian. (2) The condition that Φ be even-dimensional ensures both the existence of a lagrangian and the usual canonical equations (recall Thm. 2). Hence although G is more comprehensive than L, H and L are equivalent. In other words, hamiltonians are more common than lagrangians but the restricted hamiltonian theory H covers the same ground as L. The advantage of H over L is that it can be generalized in a way which L cannot — namely to G. Also, H benefits from the mathematical theory of canonical transformations, which just do not make sense in L. But lagrangians can be Lorentz invariant, which hamiltonians cannot. In any case H and L are inequivalent in several respects and therefore utilizable in different circumstances.

6.4. Excursus: Independent Axiomatization of Lagrange's "Dynamics"

Primitives: Σ, T, Q, and L. Q is the configuration space and L the lagrangian.

Axioms

$L\,1 \Leftrightarrow G\,1$

$L\,2 \Leftrightarrow G\,2$

$L\,3$ (a) For every $\sigma^N \in \Sigma$ there exists a Q. (b) Q is a differentiable manifold of at most $3\,N$ dimensions, whose points q are functions from $\Sigma \times T$ to R. (c) The set of all couples $\langle q_i(\sigma^N, t), \dot{q}_i(\sigma^N, t) \rangle$ represents the state of σ^N at time t.

$L\,4$ For every $\sigma^N, \sigma'^N \in \Sigma$, every $t \in T$ and every $i \in N$, if $q_i(\sigma^N, t) = q_i(\sigma'^N, t)$ and $\dot{q}_i(\sigma^N, t) = \dot{q}_i(\sigma'^N, t)$, then $\sigma^N = \sigma'^N$.

$L\,5$ (a) L is a function from $Q \times T$ to R. (b) For any fixed σ^N, L is continuous and differentiable.

$L\,6$ If $\sigma^N \in \Sigma$ and $t \in [t_1, t_2] \subseteq T$, and $\delta q_i(t_1) = \delta q_i(t_2) = 0$ for every $i \in N$, then

$$I \equiv \int\limits_{t_1}^{t_2} \mathrm{d}t\, L(\sigma^N, q_i, \dot{q}_i, t)$$

is invariant under the infinitesimal one-parameter group of displacements $t \to t' = t + \delta t$, where $\delta t = \varepsilon \cdot \xi(t)$ with ε infinitesimal and ξ arbitrary.

Remarks. (1) No metric has been specified for Q — nor, for that matter, has a metric been adopted in G and H. But if desired Q can be regarded as a Riemannian space provided (a) L refers to a system of particles and (b) L is separable in the form $L = T - V$, where V depends on q, while $T = \frac{1}{2} g_{ik} \dot{q}_i \cdot \dot{q}_k$. In this case, the metric becomes $\mathrm{d}s^2 = 2\,T\,\mathrm{d}t^2 = g_{ik}\,\mathrm{d}q_i\,\mathrm{d}q_k$. If this is done, the problem of minimizing I may be treated as the geometrical problem of finding the geodesics in Q (LANCZOS, 1949). But even with the above restrictions, the introduction of a definite metric is optional. (2) Lagrangians depending on higher-order derivatives of the generalized coordinates are occasionally used but, since they cannot occur in Maxwellian electrodynamics (POST, 1962), their interest seems to be restricted.

Thm. 1: differential laws of motion.

$$L\,6 \Rightarrow \frac{\mathrm{d}}{\mathrm{d}t}\frac{\partial L}{\partial \dot{q}_i} - \frac{\partial L}{\partial q_i} = 0 \qquad \text{for all } i \in N. \tag{2.16}$$

Df. 1. *Generalized momentum:* $p_i \overset{\mathrm{df}}{=} \dfrac{\partial L}{\partial \dot{q}_i}$.

Thm. 2. *Generalized Newton-Euler eq.;* If L is separable in the form

$$L(\sigma, q, \dot{q}, t) = T(\sigma, \dot{q}, t) - V(\sigma, q, \dot{q}, t)$$

then

$$\dot{p}_i = F_i$$

where

Df. 3. *Generalized force:* $F_i \overset{\mathrm{df}}{=} \dfrac{\mathrm{d}}{\mathrm{d}t}\dfrac{\partial V}{\partial \dot{q}_i} - \dfrac{\partial V}{\partial q_i}$.

Remarks. (1) Thm. 2 is a generalization of the central axiom of CM: it holds now irrespective of the kind of system, provided only that the above separation can be made. (2) Momenta and forces are now defined, and moreover in kinematical terms. L is therefore a purely kinematical theory. Mass values will occur in L only if it refers to mechanical systems, but even so as numerical coefficients with no physical meaning, since so far L provides no interpretation for them.

Thm. 3. *Conservation laws.* L 6 and Thm. 1 imply

$$\frac{\mathrm{d}}{\mathrm{d}t}\left[L\,\delta t + \frac{\partial L}{\partial \dot{q}_i}(\delta q_i - \dot{q}_i \delta t)\right]. \tag{2.17}$$

Proof. Substitute Thm. 1 into NOETHER's first theorem (NOETHER, 1918).

Corollary 1. If $\delta t \neq 0$ and $\delta q_i = 0$ for every $i \in N$ (time translation), then

$$\frac{\mathrm{d}}{\mathrm{d}t}\left[L\,\delta t - \frac{\partial L}{\partial \dot{q}_i}\dot{q}_i \delta t\right] \vdash H \overset{\mathrm{df}}{=} p_i \dot{q}_i - L = \text{const.} \tag{2.18}$$

Corollary 2. If $\delta t = 0$ and $\delta q_j \neq 0$ ("spatial" translation along the jth coordinate), then

$$\frac{\mathrm{d}}{\mathrm{d}t}\left(\frac{\partial L}{\partial \dot{q}_j}\cdot \delta q_j\right) = 0 \vdash p_j = \text{const.} \tag{2.19}$$

Remark. These corollaries are consequences of any integral variational principle, whether or not the variables occurring in it are physically meaningful. The association between symmetry operations and conserved quantities was a mathematical discovery (NOETHER, 1918). When the fundamental variables q_i, t are assigned a physical meaning, the H figuring in Cor. 1 can be interpreted as the energy of the closed system σ^N. When the q_i are interpreted as curvilinear position coordinates, the conserved p_j are the momenta. In these cases Corollaries 1 and 2 are conservation laws proper ("strong conservation laws": see Ch. 1, 3.4).

Thm. 4. If $L \neq 0$ and the hessian Det $|\partial^2 L / \partial \dot{q}_i \partial \dot{q}_j| \neq 0$, then Df. 1 can be inverted to yield

$$\dot{q}_i = \frac{\partial H}{\partial p_i} \quad \text{with} \quad H = p_i \dot{q}_i - L.$$

Proof. See LEVI-CIVITA and AMALDI 1927.

Remark. In other words, under the above conditions the first half of HAMILTON's equations are not physical laws but conventions. This is a semantical theorem: it does not speak about nature but about the status of certain symbols.

6.5. Transition from G to Hamilton-Jacobi's "Dynamics"

Thm. 5. If there exists a lagrangian $L \overset{\mathrm{df}}{=} p_i \dot{q}_i - H$, then there exists a function(al)

$$S[\sigma, q, t] \overset{\mathrm{df}}{=} \int_{t_0}^{t} dt L(\sigma, q, \dot{q}, t) \tag{2.20}$$

such that

$$p_i = \frac{\partial S}{\partial q_i}, \quad \frac{\partial S}{\partial t} + H\left(q_i, \frac{\partial S}{\partial q_i}, t\right) = 0. \tag{2.21}$$

Proof. Differentiate w.r.t t and compare

$$L = \frac{dS}{dt} = \frac{\partial S}{\partial t} + \frac{\partial S}{\partial q_i} \cdot \dot{q}_i$$

with its definition.

The theory HJ centered in eqs. (2.21) is a sort of synthesis of H and L. This can best be seen by axiomatizing it independently. To this end we adopt the following

Primitives: Σ, T, Q, H, S.

and introduce

Df. 1. $p \overset{\mathrm{df}}{=} \nabla S$.

$HJ\ 1. \Leftrightarrow L\ 1$ $HJ\ 2 \Leftrightarrow L\ 2$

$HJ\ 3. \Leftrightarrow L\ 3$ $HJ\ 4 \Leftrightarrow L\ 4$

$HJ\ 5.$ H and S are functions from $Q \times T$ to R.

$HJ\ 6.$ If $\sigma^N \in \Sigma$ and $t \in T$, then

$$\frac{\partial S}{\partial t} + H(\sigma, q, \nabla S, t) = 0. \tag{2.22}$$

Remarks. (1) Of all the general dynamical theories examined so far this is the only one in which the law of motion is expressed by a single equation. This apparent advantage is overcompensated by the nonlinear character of (2.22). (2) In most presentations HJ appears as a result of the mathematical theory of canonical transformations rather than as a theory of its own. In that way the heuristic (theory construction) aspect is not distinguished from the formal one.

Thm. 1. If $H = E = \text{const}$, then $S(q, t) = -Et + W(q, \alpha)$, with $\alpha = \text{const}$.

Remarks. (1) Whereas in G, H and L the state of the many-component system is represented as a point in a certain space (Φ or Q), in HJ it is represented as a surface $W = \text{const}$. that propagates in the $3N$-dimensional configuration space Q. An intuitive but purely analogical picture emerges if W is regarded as a shock wave in Q, and the lines orthogonal to it as rays. But these rays come in infinite number and therefore do not represent particle trajectories. (2) From a purely mathematical point of view HJ can be regarded as a field theory, whether or not it refers to particles. This way of looking at HJ was heuristically instrumental in the invention of wave mechanics and it may not have exhausted its genesic power (DIRAC, 1951). (3) Unlike the Hamilton and the Lagrange equations, the Hamilton-Jacobi eq. of motion has complex as well as real solutions. The former are discarded as unphysical in PM but are retained in QM. Thus $S = ikq - Et$, with $k, q, E \in R$, corresponds to an imaginary momentum $p = \dfrac{\partial S}{\partial q} = ik$ which characterizes a stationary state in QM and has no classical analog.

6.6. Extensions to Continuous Systems

The preceding theories concern discrete systems σ^N composed of N separate individuals each. Any of them can be reformulated so as to refer to a continuous system such as a drum or a field. Mathematically any such continuum theory subsumes the corresponding discrete systems theory. For example, the extension of L to continuous systems requires essentially the following modifications: (a) introduce a differentiable manifold M representing ordinary space (or spacetime as the case may be); (b) assume that the N fundamental field variables q_i are smooth functions on $M \times T$; (c) introduce a lagrangian density $\mathscr{L}$ as a primitive; (d) postulate that there is a region $V \subseteq M$ such that

$$I \equiv \int\limits_{t_1}^{t_2} dt \int\limits_V d^3x \, \mathscr{L}(\sigma, q_i, q_{i,\mu}, t), \qquad q_{i,\mu} \stackrel{\text{df}}{=} \frac{\partial q_i}{\partial x^\mu}$$

is invariant under a 4-parameter group of infinitesimal transformations $x^\mu \to x'^\mu = x^\mu + \delta x^\mu$ with $\mu = 0, 1, 2, 3$. The field equations are then derived, and NOETHER's theorem warrants the existence of a 2nd rank tensor T such that Div $T = 0$. If the q_i are interpreted as genuine field variables then T can be interpreted as the energy-momentum-stress tensor of the field and consequently 'Div $T = 0$' as four conservation laws. But so far nothing of the sort has been specified: T is an empty form and so is the whole L for continuous systems as long as these are nondescript. Whether for continuous or for discrete systems, L is a protophysical theory.

6.7. Intertheory Relations

In order to facilitate the comparison between the four general "dynamical" theories G, H, L and HJ, as well as between them and Newtonian particle mechanics PM, it will be convenient to display their essential concepts and central postulates:

General	Hamiltonian	Lagrangian	Jacobian	Newtonian
Σ Physical system	Σ Physical system	Σ Physical system	Σ Physical system	Σ System of particles
T Time	T Time	T Time	T Time	T Time
Φ Phase space	Φ Phase space	Q Configuration space	Q Configuration space	E^3 Ordinary space
H Hamiltonian	H Hamiltonian	L Lagrangian	H Hamiltonian	K Reference frame
P Poissonian			S Action	X Material coordinate M Mass F^e Applied force F^i Interparticle force
$\dot{\Phi} = P \cdot \nabla H$	$\dot{q} = \dfrac{\partial H}{\partial p}$, $\dot{p} = -\dfrac{\partial H}{\partial q}$	$\dfrac{d}{dt}\left(\dfrac{\partial L}{\partial \dot{q}}\right)$ $-\dfrac{\partial L}{\partial q} = 0$	$\dfrac{\partial S}{\partial t} + H = 0$	$\dot{p} = F^e + F^i$

The sole basic concept all five theories share is that of time; accordingly their axioms are definitely different. The most remarkable difference is the one between the set of all four protophysical theories, on the one hand, and Newtonian PM on the other. To begin with, the latter is the only specific theory amongst the five, in the sense that its referents are systems of particles rather than systems of nondescript physical entities. In the second place, Newtonian mechanics is, of all five theories, the only one dealing with the configuration and the motion of systems in a fairly direct fashion: it places them in ordinary space rather than representing their state of motion in a mathematical space —

the phase space or the configuration space. In the third place, in its usual 3-dimensional formulation PM (and also CM) attributes space a definite metric, whereas no such assumption is made in the other theories, although it can be made in some cases in L (see 6.4). Fourthly, the concept of reference frame plays no role in these protophysical theories, which for this reason are vacuously invariant with respect to changes of reference frame. Fifthly, in Newtonian mechanics the causes of the changes of dynamical states are handled in a rather direct way, by representing them by the force concept, whereas in the protophysical theories causes are either absent or represented in a roundabout way; they are phenomenological theories, the more useful the less one knows about the mechanisms responsible for change.

For all these reasons PM alone, of all five theories, is a specifically mechanical theory. The other four are partially interpreted theories (semisemantical systems), which can in principle be employed to state and solve physical, biological, or social problems of any kind (KERNER, 1962, 1964). In other words, G, H, L and HJ are *general frameworks for physical theories* rather than physical theories (BUNGE, 1957). Only when the source functions H, P, L, and S are restricted to refer to entities both localized and endowed with mass (i.e., to bodies) do the general dynamical theories acquire a mechanical interpretation. This point is often overlooked because, historically, those theories were introduced as mere reformulations of PM. We now realize that their merit is not so much that of being general mechanical theories — which they turn into provided the suitable semantical assumptions are added — but that of being general schemata for physical theories, frameworks to be filled by specifying the physical meaning of the reference class Σ and the fundamental variables, as well as the mathematical form of the source functions. (For example, the theory of e.m. radiation can be given a hamiltonian formulation, in which

$$H = \sum_k H_k, \quad \text{with} \quad H_k = \tfrac{1}{2}(p_k^2 + \omega_k^2 q_k^2)$$

where q_k is the time-dependent part of the kth fourier component of the vector potential. Since H_k has the same form as the hamiltonian for a mechanical harmonic oscillator, it is often said that the fourier analysis of the field consists in decomposing it into a system of oscillators. But this is a misleading metaphor, for the e.m. field is not a mechanical system. In short, H is not restricted to mechanical systems.)

How are the five theories related? We can distinguish three kinds of *theory inter-relation*: logical, semantical, and methodological. The logical relation we are interested in is the one of inclusion of the respective sets of formulas; the semantical relation is the one of supplying meaning;

the methodological relation is the one of affording indirect empirical tests. In a nutshell the situation is this: logically the more abstract dynamical theories include the less abstract ones, which in turn include PM; semantically the relation is reversed: the more specific theories help supply the meaning of the symbols of the more general theories, which would otherwise remain uninterpreted; methodologically something similar is the case: the more specific theories, and in the first place the properly physical ones (e.g., PM), are the more fully testable ones. Let us take a closer look at these relations.

We have seen that, of the four protophysical theories, G is the more inclusive. H follows from G on specifying P and restricting the number of dimensions of Φ to be even. H and L are the same as regards their *sets of formulas*, all of which are mutually translatable *via* the code: $p = \partial L/\partial \dot{q}$, $\dot{q} = \partial H/\partial p$, $L = p\dot{q} - H$. As to HJ, it may also be regarded as equivalent to L since its two source functions, H and S, are defined explicitly in terms of L. In short, as regards the total sets of formulas, HJ $=$ L, L $=$ H, and H $<$ G. But as regards their *structure* these are definitely different theories rather than variations of one and the same theory: they differ in their primitive base, in their initial assumptions, and in their transformation properties. Were it not for these differences they would not be used concurrently.

What is the logical relation between these general frameworks and PM? For a long time it was believed that the general "dynamical" theories are restricted to problems involving monogenic and more particularly conservative forces, and constraints expressible in a closed algebraic form (holonomic auxiliary conditions). In other words, it was thought that there remains an adamant Newtonian core that could not be translated into either of the general formalisms. It was gradually realized, from HELMHOLTZ onwards, that the appropriate translation of Newtonian problems into the Lagrangian language was a matter of mathematical ingenuity (BIRKHOFF, 1927 and HAVAS, 1957). In short, the logical relation of the protophysical theories of evolution and CM is one of inclusion: CM $<$ L and similarly for the others. (Again, the inclusion relation holds between the sets of formulas.) This much for the logic of inter-theory relation. Let us now consider the semantics of this relation.

Such as they stand, our four "dynamical" theories are semantically indeterminate to a large extent, in the sense that the very objects or referents of the theories, and a fortiori the functions that represent their state, are left unspecified for the sake of generality. In particular the generalized coordinates need not represent position components and consequently need not have the dimensions of a length; consequently the generalized momenta, when introduced at all, need not be quantities of motion and therefore H and L need not have the dimensions of an energy. No such theory postulates a definite model for its referent save that it consists of a collection of a finite number of entities. A fortiori, the theory specifies no mechanism concerning the mode of action of such entities. In short, G, H, L and HJ are amechanical and acausal, although they can be narrowed down to mechanical and causal theories by the addition of suitable interpretation assumptions. In this process of semantical specification, comparisons with simpler and more specific theories such as PM and ray optics play an important role.

As to empirical testability, none of the general "dynamical" theories can be taken to the laboratory as long as it is given no definite interpretation in addition to the one supplied by the semantical members of its postulate set. In fact the experimenter must know what are the symbols supposed to stand for before he can devise means for testing the accuracy of such a reference, i.e. the degree of truth of the theory. And, as we saw a while ago, it is of the essence of G, H, L and HJ not to commit themselves as regards meanings. If the reference is to mechanical systems such as particles, a complete specification of any of the general "dynamical" theories will require borrowing from elementary mechanics. Consequently CM, by virtue of being a more fully interpreted theory, is a tool for the test of general dynamical theories once these have been interpreted in mechanical terms. Similarly if the referent of any such theories is, say, a set of light rays, then elementary optics will be employed as a tool for attaching the general theory a precise meaning as well as for making it testable. Futhermore, since actual motion takes place in physical space, in order to test the consequences of a mechanically interpreted general dynamical theory we shall have to resort to elementary (vectorial) mechanics, which presupposes a definite space-time structure. We shall then pay the price of losing the flexibility of invariance under rather general transformations such as the canonical ones: indeed when testing we fix the reference frame. In short, elementary mechanics is no less needed for methodological than for semantical reasons even after the general dynamical theories have been built. The same holds for other elementary theories in relation to the general dynamical ones. One more reason for not neglecting the clean formulation of elementary theories.

Every one of the five theories we are examining has its peculiar virtues and shortcomings. H, L and HJ are best suited for problem-solving, on account of their transformation properties. G, H and HJ can best be generalized to QM. As to L, it is unsurpassed as a tool for the preliminary exploration of new ground and for relativistic extensions. Finally, CM remains the standard framework for posing mechanical problems and hypothesizing laws of force. Once such statements are gotten, more general and computationally more convenient forms can be adopted by going over to either of the general theories. But CM remains the main source of inspiration in the particular case of mechanics, if only because it works with ordinary space, just as H and L are the standard, yet not the sole possible, forms in which nonmechanical theories are cast. The value of these more inclusive frameworks does not lie in that they give us a deeper insight into specific physical processes but (a) in that they constitute comprehensive frameworks and therefore tools for investigating new domains without being hampered by the limitations of a mechanical picture, and (b) in that they evolve powerful mathematical methods for the solution of problems (e.g., the case of HJ in celestial mechanics). But they are not the sole possible comprehensive frameworks and progress may come from inventing new, even wider protophysical theories (DIRAC, 1950 and 1951).

We may draw three general conclusions from the preceding discussion: (a) not every physical theory is a fully interpreted system; (b) the more general theories do not suppress the need for all the less general ones, as these may supply part of the meaning and the test of the former; and (c) a physical theory often results from the precise interpretation of a ready-made protophysical framework.

Let us now examine the foundations of a few important specific theories, some of which employ the general "dynamics" formalisms.

Chapter 3

Classical Mechanics

Introduction

Until the turn of the century mechanics held the place we now assign protophysics: it was regarded as the basis of physics. The field revolution initiated by FARADAY and brought to its apex by EINSTEIN showed that classical physics consists of two equally basic sets of theories: mechanics and field physics. (The quantum theories have begun to synthesize these two without quite merging them.) The only reasons for beginning a physics training with mechanics are tradition — which is no reason at all — and man's peculiar biopsychical constitution. The latter ground is sometimes reinforced with the epistemological argument that mechanical experiences are in the foreground and anyhow we have no direct experience of fields save in the optical range. But this argument holds no water, as physical science is not derivable from sense experience and is not restricted to checking our ideas about the world but is concerned with all the basic traits of the world, none of which are perceptible. From a purely logical point of view mechanics and classical field theory are on an equal footing.

Classical mechanics is the union of three theories: continuum mechanics (CM), its chapter particle mechanics (PM), and statistical mechanics (SM). The first two are still called *rational mechanics,* a name which is deserved but slightly misleading as it suggests that it is a purely rational, i.e. a priori discipline, the other branches of physics not being quite so rational. The three disciplines have remote origins and they have constituted the kernels of as many cosmologies: SM is found *in statu nascendi* in the void-atoms-chance cosmology of EPICURUS and LUCRETIUS; CM in the plenistic cosmologies of ARISTOTLE and DESCARTES; and PM is the nucleus of the brand of mechanism dominant in the 18th and 19th centuries. All three cosmologies are mechanistic, although the one centered in PM is usually evoked when mechanism is mentioned: indeed, for all three the world is a system of bodies, i.e. of entities endowed with both localization and mass. Mechanism dominated even optics and electromagnetism until EINSTEIN showed that the ether, if it exists at all, cannot be assigned mechanical properties. The field concept was then regarded as the suitable candidate for building a monistic cosmology. But all attempts to reduce classical mechanics to field theories have failed as thoroughly as the converse attempts: it would seem that the variety of the cosmos calls for a pluralistic cosmology.

Having attained the adult state before other branches of physics, classical mechanics has been the subject of most of FR in physics. For this reason we shall devote to it less attention than to the more neglected domains, particularly field physics. Yet we will have to deal with it not only for the sake of completeness but also because most of the foundational work in classical mechanics is either inadequate or unknown to physicists or both. When the foundations of mechanics are mentioned most physicists think of MACH, a few of HERTZ, practically none of the modern work by McKINSEY and SUPPES, NOLL, and TRUESDELL. Yet MACH's (and KIRCHOFF's) work in FR was inadequate though effective, and HERTZ' was ineffective. Some of MACH's criticisms of CM (MACH, 1883) mainly those of absolute space, time, and motion, were correct though wrongly inspired; but his attempt to replace the whole of mechanics by an experimental proposition and a definition was no less than an attempt to destroy the theory rather than reinterpret it. In particular, the famous elimination of the mass concept by defining it as an acceleration ratio was a blunder: (*a*) far from being theory-independent it presupposes NEWTON's second and third law; (*b*) the formula breaks down when referred to an accelerated frame: unlike the mass, the acceleration ratio is not invariant; (*c*) the whole idea rests on a confusion between a definition (logical identity) and a numerical equation. In short MACH's attempted renewal of CM was a flop: it was neither purely empirical as alleged, nor general, nor logically correct (BUNGE, 1966). The same holds for some recent work on PM that dispenses with nonobservational concepts and — following D'ALEMBERT, 1743; KIRCHHOFF, 1883 and MACH, 1883 — tries to build dynamics out of kinematics. It could not be otherwise: an empiricist theory is a contradiction in terms, and in particular the reduction of mechanics to obervables ("operationally defined quantities") is hopeless: experience is not that generous, and the aim of science is not to summarize sense data but to explain reality (BOLTZMANN, 1905, PLANCK, 1909), which task requires the invention of transempirical concepts (see Ch. 1, 4.3).

But why should we care for the foundations of CM and in general of classical physics since it is supposed to be false? Even assuming that it is false (as if factual truth were not always partial), there are at least three reasons for pursuing studies of this kind: (a) they may teach us something concerning the structure and content of physical theories in general; (b) the foundations of classical physics are far from being cut and dried: in fact there is far more obscurity than clarity in their regard; (c) classical theories are the correspondence limits of nonclassical ones and should therefore be in good shape to serve as guides and checks. A body of knowledge that is regularly used and is assigned a role of control of practically every new idea yet at the same time is despised as basically false and is not regarded as worth being repaired, is like a concubine who is worth being kept — only miserably and indoors. Such a haughty attitude towards classical physics is both silly and immoral. Since we use it let us do it decently, by overhauling it so as to free it from inconsistencies and deadwood, renewing and strengthening its formalisms, completing it as far as possible, and clarifying its meaning. This is one of the goals of the young and vigorous "natural philosophy" movement (TRUESDELL, 1966) which, in addition to revamping classical physics, is making new discoveries in it. As usual, love and research are more fruitful than scorn and ignorance.

1. Particle Mechanics

In this section we shall investigate the structure and content of classical particle mechanics (PM) in its elementary (vectorial or Newtonian) form. This investigation is interesting for the following reasons: (a) PM still passes for the paradigm of scientific theory and is accordingly the object of many, perhaps most, discussions on the nature of scientific theory in general; (b) PM is employed, either heuristically or constitutively, in several fields, including elementary QM, which preserves the "c-numbers" of PM and their classical relations; (c) PM is still a source of inspiration for the physical interpretation of more general mechanical theories and, unfortunately, for many which, like QM, are not fully mechanical; (d) up to now no generally accepted axiomatization of elementary mechanics seems to exist, and even the best among them (McKinsey et al., 1953, 1955) is largely formal, i.e. concerned with the mathematics rather than with the physical meaning of the theory — and anyway it has unjustifiably failed to attract the attention of physicists.

1.1. Background and Primitives

Formal background: (a) elementary logic (PC=); (b) elementary set theory; (c) elementary topology; (d) vector spaces theory; (e) analysis and the algebraic and number-theoretical fragments underlying it.

Protophysical background: (a) chronology, in particular the theory of universal time (UT); (b) physical Euclidean geometry; (c) the general systems theory S (see Ch. 2). These theories are employed in building PM; consequently any failure of PM to match facts may be traced back either to its specific assumptions or to some of the factual components of its background — e.g., its theory of time. It is characteristic of PM that it presupposes no specific physical theory: it is, in this sense, a fundamental theory — and so is of course CM.

Primitive base: E^3 (space), T (time), Σ (point particle), K (reference frame), $\{X\}$ (material coordinate), $\{M\}$ (mass), $\{F^e\}$ (external force), and $\{F^i\}$ (inter-particle force). E^3 and T are the chronogeometrical primitives; Σ, K and X the kinematical ones; and M, F^e and F^i the dynamical primitives of PM. The first four concepts are sets, the last four are families of functions. E^3 comes of course with any number of global bases or systems of axes (linearly independent unit vectors) $e = \langle e_1, e_2, e_3 \rangle$. Relative to any given basis e, every point $x \in E^3$ is uniquely characterized by the triple $x = \langle x_1, x_2, x_3 \rangle$. These coordinates must not be mistaken for the physical coordinates X_i $(i = 1, 2, 3)$ of a material point: these are time-dependent and relative to a $k \in K$, which is a physical frame; e is a conceptual image of k. Since protophysics takes care of E^3, T, K, and X, our axiom system will not say much about them; the chief concern of PM is the set of its specific primitives Σ, M, F^e and F^i. The reason for introducing two logically independent force concepts is that they are structurally different: whereas F^e is a real valued vector on $\Sigma \times K \times E^3 \times T$, F^i is a real valued vector on $\Sigma \times \Sigma \times K \times E^3 \times T$. Moreover F^e can often be assigned to an external field, although in PM one is not interested in the origin of the external force. On the other hand F^i represents one side of the interaction between the particles σ and σ', the other side being the reaction. In some cases it is possible to determine F^e in terms of F^i, namely when it can be postulated that every force is an inter-particle action — i.e. $F^e(\sigma) = \Sigma_{\sigma'} F^i(\sigma, \sigma')$. But this is a hypothesis, not a definition.

By means of the preceding primitives and some logical and mathematical concepts we can define all the other concepts of PM, in particular

Df. 1. *Instantaneous velocity* of $\sigma \in \Sigma$, at $t \in T$, relative to $k \in K$:

$$v(\sigma, k, t) \overset{\text{df}}{=} \mathrm{d}X(\sigma, k, t)/\mathrm{d}t \equiv \dot{X}(\sigma, k, t).$$

Df. 2. *Linear momentum* of $\sigma \in \Sigma$, at $t \in T$, relative to $k \in K$:

$$p(\sigma, k, t) \overset{\text{df}}{=} M(\sigma) \cdot v(\sigma, k, t).$$

These two concepts occur in our axioms for PM. All other derived concepts occur in the theorems: in particular, the important concepts of angular momentum, torque, and energy are derived not primary (primitive) concepts. Some of them will be introduced as needed. In limiting the number of concepts we abide by the rule that commands introducing no new concept unless it occurs in a law statement.

1.2. Axioms

Chronogeometrical axioms

PM 1.1. (*a*) T is an interval of the real line. (*b*) Every $t \in T$ represents (refers to) an instant of time. (*c*) The relation $\leq$ that orders T means "earlier than or simultaneous with".

Remarks. (1) By virtue of this axiom time is endowed with all the formal properties of the real number continuum. In other words, the theory of universal time underlying PM is a model or interpretation of the mathematical theory of real numbers. But we might as well have postulated the much deeper theory UT (see Ch. 2, 3). (2) The members of T are not clock readings: the latter constitute a finite, hence mathematically insignificant, subset of T. In an operationalist formulation T would have to be such a finite set — but then the calculus could not be used. In other words, the time concept is not operationally characterized in PM. This does not mean that 't' and 'T' lack a physical meaning: it is the job of the clauses (*b*) and (*c*) in *PM* 1.1 to suggest such a meaning even if sketchily. (3) *PM* 1.1 consists of one statement concerning the structure and two concerning the meaning of the symbol concerned. We shall employ the same strategy throughout, as ours are supposed to be physical not just mathematical theories; but even so we won't be able to do more than sketch the semantical profile of our theories (see Ch. 1, 4.2.5 and 4.2.6).

PM 1.2. (*a*) E^3 is a three-dimensional Euclidean space. (*b*) E^3 maps (represents) ordinary space.

Comment. This axiom could be rephrased: "Physical space has a Euclidean structure." But it is advantageous to make a clear-cut separation between the mathematical assumptions and their physical interpretation (see Ch. 2, 4).

Kinematical axioms

PM 2.1. (*a*) Σ is a nonempty denumerable set. (*b*) Every $\sigma \in \Sigma$ represents a corpuscle (particle).

Remarks. (1) In brief: there are corpuscles. And the dynamical axioms will make it clear that PM models them as material points, i.e. as structureless entities. In other words, the corpuscle is the mediate, the point mass the immediate referent of PM (see Ch. 1, 1.3.5). (2) The set Σ should not be taken for a system of coexisting particles: PM is about individuals and aggregates of individuals of the kind Σ.

PM 2.2. For any two different $t, t' \in T$, if $\sigma \in \Sigma$ at t, then $\sigma \in \Sigma$ at t'.

Remarks. (1) In words: once a particle always a particle. This is not true in microphysics. (2) This postulate is needed in order to warrant that predictions and retrodictions calculated for a σ on laws of motion and initial conditions will still refer to σ. If particles came into being or ceased to be no such unlimited extrapolations could be made beyond the life span of such entities.

PM 2.3. (*a*) K is a nonempty denumerable set. (*b*) Every $k \in K$ is a rigid system of corpuscles at least four of which lie on the vertices of a regular trihedral. (*c*) For every $k \in K$ there exists a Cartesian system of axes $e = \langle e_1, e_2, e_3 \rangle$ such that $e \triangleq k$. (*d*) No $k \in K$ interacts with any $\sigma \in \Sigma$ that is not a part of k.

Remarks. (1) This axiom was unnecessary in NEWTON's original formulation, in which positions were absolute. (2) The last component of this postulate is an indispensable pretence.

PM 2.4. (*a*) For every $\sigma \in \Sigma$ and every $k \in K$, every $X \in \{X\}$ is a continuous and real valued function from T to E^3. (*b*) For every $\sigma \in \Sigma$ and every $k \in K$, $\dot{X}$ is a piece-wise continuous and real valued function from T to E^3. (*c*) $X(\sigma, k, t)$ represents the position of σ, relative to the frame k, at the instant t. (*d*) Every quintuple $\langle \sigma, X_1, X_2, X_3, t \rangle$ represents an event. (*e*) For any given $\sigma \in \Sigma$ and any given $k \in K$, the set $\{X(\sigma, k, t) \mid t \in T\}$ represents a trajectory (motion) of σ.

Remarks. (1) In order to make room for impacts, no stronger continuity properties are imposed on X and $\dot{X}$. (2) *PM* 2.4 makes it possible to add position vectors and velocities. Moreover, it ensures the existence and uniqueness of every linear combination of any number of X's and their time derivatives, which will be needed to define the total linear momentum and the center of mass of a system of particles. (3) The continuity of motion, as expressed by (*a*) and (*b*), overreaches experimental data (see Ch. 2, 1). (4) The components (*c*), (*d*) and (*e*) are not definitions (although they are taken as such in alternative formulations) but full-fledged semantical hypotheses that relate mathematical concepts to physical ones. (5) The relative position coordinate X — the physical not the geometrical coordinate — takes here the place of NEWTON's absolute space (ZANSTRA, 1924). We thereby get rid of some of the ambiguous and therefore hardly testable statements in NEWTON's original formulation.

PM 2.5. For any two $\sigma, \sigma' \in \Sigma$ and every $k \in K$ and every $t \in T$, if $X(\sigma, t, k) = X(\sigma', t, k)$ and $\dot{X}(\sigma, t, k) = \dot{X}(\sigma', t, k)$, then $\sigma = \sigma'$.

Remark. In words: the position and velocity values uniquely identify (characterize) a particle. Coordinates alone are insufficient for this: thus two different particles may be at the same place at the instant of collision but then they will (if *PM* 2.5 is true) have different (vector) velocities.

Dynamical axioms

PM 3.1. (*a*) M is a function from Σ to the set R^+ of nonnegative reals. (*b*) The value of M at $\sigma \in \Sigma$ represents the mass (inertia) of σ.

Remarks. (1) Mass is the only intrinsic property PM attributes particles; this ceases to hold in relativistic theories. (2) As a consequence of the intrinsic (nonrelational) character of mass in PM, no field can possibly alter it; therefore it is mistaken to speak of electromagnetic mass just because the e.m. field can contribute to the particle momentum. It is equally mistaken to interpret 'hv/c^2' as the photon mass just because the photon "momentum" is hv/c. No function can validly be interpreted as a classical mass unless it satisfies all the dynamical axioms of PM or CM. (3) *PM* 3.1 is neither obvious nor conventional: if there were particles with negative mass, the laws of motion would be different. (4) The relation $M(\sigma) = m \in R^+$ established by *PM* 3.1 between the two primitives M and Σ does not amount to a definition although in the incorrect jargon one says that M is defined (given) on the set Σ. That relation does not entitle us to define, say, the concept of particle in terms of the mass concept, by inverting M, if only because M is not a bijective map (1:1 and onto). In fact, M sends members of Σ to points in R^+ but not conversely, since there are different particles with the same mass and there are presumably real numbers corresponding to no mass values. We can certainly characterize σ as the independent variable of the function M or, equivalently, Σ as the set of first elements of the couples $\langle \sigma, m \rangle$ that constitute the extension of M. But a characterization is not a definition, i.e. the expression of an identity the stipulation of which enables us to eliminate one concept in favor of others.

PM 3.2. If $\sigma_i \in \Sigma$ for every finite $i \in N$, then $M(\sum_i \sigma_i) = \sum_i M(\sigma_i)$.

Remarks. (1) In words: the mass of a system of point particles equals the sum of the partial masses. (For the operation $\dotplus$, see Ch. 2, 5). (2) The mathematical counterpart of this axiom is: mass is a fully additive measure. This additivity is not conventional: it can be falsified by experiment. (3) We shall make no use of this axiom but it is indispensable to derive theorems concerning particle systems. (4) The additivity of mass is inconsistent with the so-called *Mach principle*, according to which the mass of a body is a measure of the sum total of its interactions with the remaining bodies in the universe. Indeed, if this were true then the mass of the universe would be zero (even though every one of its components bodies would have a nonvanishing mass) because the universe has nothing to interact with.

PM 3.3. (a) Every $F^e \in \{F^e\}$ is a real valued vector on $\Sigma \times K \times E^3 \times T$. (b) The value of F^e at any given 6-ple $\langle \sigma, k, x, t \rangle$ represents the external force acting on σ, relative to the frame k, at the place x and the instant t.

PM 3.4. (a) Every $F^i \in \{F^i\}$ is a real valued vector on $\Sigma \times \Sigma \times K \times E^3 \times T$. (b) The value of F^i at any given 7-ple $\langle \sigma, \sigma', k, x, t \rangle$ represents the force exerted by σ' on σ relative to k at x and t.

Remarks. (1) The distinction between two force concepts does not occur in field theories, where every kind of force is assigned a field. (2) In the light of the semantical components (b), these axioms are seen to have a physical meaning. In particular, the vector representation of forces is correct only if the forces in fact add linearly. (3) These postulates allow one to deduce the law of the parallelogram of forces for a finite number of forces without having to postulate explicitly that forces add linearly, associatively, and commutatively. (4) Empiricists have always distrusted the force concept, because forces are said to be imperceptible, and because the concept of force constituted, together with the one of substance, the core of the coarse materialistic philosophy popular in the 18th and 19th centuries. But the concept has survived in many physical theories. True, it can often be eliminated in favor of the even less empirical concept of potential, but this only shifts the logical status of the force concept, which is turned from a primitive into a defined concept. Such definitions ($F^e \overset{\mathrm{df}}{=} -\nabla V^e$, $F^i(\sigma, \sigma') \overset{\mathrm{df}}{=} -\nabla_\sigma V^i$) are necessary for the mathematical expansion of mechanics and for its connection with field theories, but (a) they are restricted to conservative forces and (b) they do not render the force concept dispensable. Even if theoretically eliminable, the force concept is needed for test purposes, i.e. as a bridge between theoretical and experimental physics. (5) As long as the force functions are given (hypothesized) PM remais a semiphenomenological theory. And even the derivation of forces from potentials makes no difference in the character of the theory as long as the potentials are assigned no physical interpretation but are regarded as auxiliary symbols or intervening variables. Only when they are assumed to represent a field existing between the particles does the theory become thoroughly representational (nonphenomenological) — but then it ceases to be a theory of particles alone. The study of the origin of forces does not belong to PM but to CM and field theories.

PM 3.5. For every $\sigma, \sigma' \in \Sigma$, every $t \in T$, every $F^e \in \{F^e\}$ and every $F^i \in \{F^i\}$, there is at least one $k \in K$ such that

(a) $$\dot{p}(\sigma, t, k) = F^e(\sigma, t, k) + \sum_{\sigma' \neq \sigma} F^i(\sigma, \sigma', t, k) \qquad (3.1)$$

and

(b) $$F^i(\sigma, \sigma', t, k) = -F^i(\sigma', \sigma, t, k). \qquad (3.2)$$

Remarks. (1) These are of course the basic laws of PM in the elementary (vectorial) version of it (see Ch. 2, 6). (2) If some of the forces are reactions of the constraints then they can be taken out of (3.1) and replaced by constraint equations. (3) Since no restriction has been placed either on the magnitude of the forces or the continuity of the acceleration, the given formulation covers the case of impulsive forces: there is no need for a special theory of impacts (see 1.3, Thm. 8). (4) By calling $-\dot{p}$ the force of inertia, D'ALEMBERT rewrote the *lex secunda* in the form of a balance of forces and deluded himself into believing that he had reduced dynamics to statics. (5) For $F^i = F^e = 0$, (3.1) entails the "principle" of inertia. This corollary is sometimes regarded as a definition of the concept of inertial frame not as a physical law. But a logical consequence of a law cannot be a convention: it is a subordinate law, hence a refutable statement. (6) From MACH onwards, many authors have regarded (3.1) as a law and at the same time as a definition of "force". A famous author even derived the law from a definition, namely in the following delightful way. Let a_i be the accelerations induced in a system — but do not say by what. Call $a = \Sigma a_i$ the total acceleration. Now multiply both sides by m and call F the right hand side. Lo and behold: you have NEWTON's law. (7) Should a particle fail to obey (3.1), it would be possible to introduce the *ad hoc* hypothesis that some force other than the known ones was acting. This trick will be justified if the new force can be determined when the given forces are switched off. Otherwise the hypothesis remains a mere conjecture or the case blackballs *PM*. (8) The law of action and reaction holds only for pairwise interactions among particles, and provided this interaction is not effected through an e.m. field. (9) The *lex tertia* is sometimes dispensed with but, within the domain in which it holds, it is indispensable: it occurs not only in important mechanical problems (e.g., rocket propulsion) but it has the virtue of killing the traditional concept of matter as a passive substance. (10) The possibility of building a theory in which (3.2) is missing (McKINSEY et al., 1953) shows that it is an independent postulate. This independence can also be shown directly. If $F^e = 0$, then by (3.1) $\dot{p}(\sigma) = F^i(\sigma, \sigma')$. Exchanging σ and σ', $\dot{p}(\sigma') = F^i(\sigma', \sigma)$. But these formulas alone do not give us the relation between the action $F^i(\sigma, \sigma')$ of σ' on σ to the reaction $F^i(\sigma', \sigma)$ of σ on σ'. To this end we need a separate postulate; the one that happens to hold for static forces is (3.2).

This completes our axiom system for PM. We can now give axiomatic definitions of two important concepts.

Df. 1. A reference frame $k \in K$ is called *galilean*, or *inertial*, iff it satisfies the preceding axioms. Notation: $g \in G \subset K$.

Remarks. (1) The concept of galilean frame is here taken to be a dynamical not a kinematical concept, in the sense that its is defined in terms of a dynamical theory. (2) The collection G of galilean frames is an equivalence class. This concept had its ancestor in C. NEUMANN's alpha body (1870). It plays in contemporary classical mechanics the role absolute space played in the older theory: indeed PM and CM are supposed to hold relative to some member g of G. On the other hand PM and CM retain the concept of universal (not necessarily absolute) time. (3) It is often stated that G is an infinite set. This is not necessary: to test the theory a single accessible galilean frame would suffice. Moreover we do not know how many G's there are in nature; we only know that the c.m. of our solar system is reasonably inertial, and hope to find some other systems that qualify.

Df. 2. A system $\sigma^N = \sum \sigma_i$ of N particles $\sigma_i \in \Sigma$ is called a *classical mechanical system of point particles* iff every part σ_i of σ^N satisfies the preceding axioms.

The theory of point particle systems is developed exclusively on the previous foundations with the help of a few definitions such as those of c.m. and reduced mass and certain simplifying assumptions. But the development of many-body theories requires the adoption of richer formal structures, chiefly Hamiltonian dynamics (see Ch. 2, 6). We shall now derive a few theorems with the sole purpose of illustrating the role of our axioms.

1.3. Sample of Theorems

Thm. 1: *absoluteness of simultaneity.* If $k, k' \in K$ and two events are simultaneous in k then they are simultaneous in $k' \neq k$. *Proof:* From the frame-independence of time asserted by *PM* 1 and the characterization of "event" supplied by *PM* 2.4*d*.

Thm. 2: *independence of motions and forces.* Velocities and forces compound and decompose in accordance with the vector law of the parallelogram. *Proof* for velocities: from *PM* 2.4*b* and *Df.* 1; for forces: from *PM* 3.3 and *PM* 3.4.

Remark. The law of composition of forces is sometimes regarded as a mathematical theorem, at other times as a rule (the "parallelogram rule"). Every one of these interpretations contains a grain of truth: (*a*) once forces are represented by vectors in a linear vector space they automatically obey the laws of such mathematical spaces, which enables us to derive the parallelogram law in a purely formal fashion — but once the derivation is completed we restore the physical meanings to the terms involved, e.g. we speak of forces not of uninterpreted vectors, thereby gaining a physical statement; (*b*) any statement, whether physical or not, can be used as a rule — e.g., a rule for computing a resultant force; but this does not prove that the statement itself is a rule, just as the fact that books can be used to kill mosquitoes does not prove that they are nothing but mosquito killers.

Thm. 3: *impenetrability.* If two particles are different then they have different positions or different velocities or both. *Proof:* from *PM*2.5 and the law of contraposition (see Ch. 1, 1.2.4).

Thm. 4: *constancy of mass.* (*a*) For every $\sigma \in \Sigma$ and any two different $k, k' \in K$, $M(\sigma, k) = M(\sigma, k')$ *(invariance).* (*b*) For every $\sigma \in \Sigma$, every $k \in K$ and every $t \in T$, $dM(\sigma, k, t)/dt = 0$ *(conservation).* *Proof:* from *PM* 3.1.

Thm. 5: *invariance under time reversal.* If, for all $t \in T$, $X(-t) = X(t)$, and if the forces are time-symmetrical, then the equation of motion (3.1) is invariant under the reversal $t \rightarrow -t$. *Proof:* by performing the indicated operation on (3.1).

Remark. This, being a theorem about a theoretical statement, is a metatheorem and more particularly a metalaw statement. It is usually mistaken for the hypothesis of the reversibility of mechanical processes.

Df. 5. Total linear momentum of a system of point particles:

$$\sigma^N = \sum \sigma_i \Rightarrow p(\sigma^N) \overset{\text{df}}{=} \sum p(\sigma_i).$$

Remark. This Df. presupposes Df. 1 and *PM* 2.4*b*, which ensures the vector character of the partial linear momenta $p(\sigma_i)$ and consequently their additivity.

Thm. 6: *conservation of total linear momentum.* If no external forces act on a system of particles, then the total linear momentum of the system is conserved in every $g \in G$: $(\text{d}/\text{d}t)\sum p(\sigma_i, g, t) = 0$. *Proof:* by *PM* 3.5.

Thm. 7: *no self-force.* For every $\sigma \in \Sigma$, every $k \in K$ and every $t \in T$, $F^i(\sigma, \sigma, k, t) = 0$. *Proof:* from *PM* 3.5 *b*.

Remark. This statement does not hold in electrodynamics.

Thm. 8: *impulsive force.* Let $F^e = p\,\delta(t)$, $F^i = 0$, where $p \in R$ represents the impulse on the particle and $\delta(t)$ is DIRAC's generalized function. Then (*a*) the trajectory equation is

$$x(t) = \frac{p}{m}\frac{1}{2}t\,\varepsilon(t), \quad \text{with} \quad \varepsilon(t) = \begin{cases} -1, & t < 0 \\ 0, & t = 0 \\ +1, & t > 0 \end{cases} \qquad (3.3)$$

and (*b*) the jump across the time origin is

$$[p] \overset{\mathrm{dt}}{=} p(t+0) - p(t-0) = [m\dot{x}] = [\tfrac{1}{2}p\,\varepsilon(t)] = p.$$

Proof: (*a*) follows upon fourier-analyzing the equation of motion; (*b*) results from performing the indicated computation with the help of (3.3) (BUNGE, 1960 for the corresponding integral).

Remark. Upon incorporating the theory of generalized functions (originated in QM) in the formal background of PM and CM, the theory of impacts becomes an integral part of them. This is an instance of progress achieved by adopting tools wrought in different fields.

Thm. 9: *torque law.* If $g \in G$, then

$$\frac{\mathrm{d}}{\mathrm{d}t}\sum_i X(\sigma_i, g, t) \times p(\sigma_i, g, t) = \sum_i X(\sigma_i, g, t) \times F^e(\sigma_i, g, t)$$

$$+ \frac{1}{2}\sum_{i,j}[X(\sigma_i, g, t) - X(\sigma_j, g, t)] \times F^i(\sigma_i, \sigma_j, g, t).$$

Proof: by *PM* 3.5.

Df. 6. Total angular momentum:

$$\sigma^N = \sum_i^\cdot \sigma_i \Rightarrow L(\sigma^N) \overset{\mathrm{dt}}{=} \sum_i X(\sigma_i) \times p(\sigma_i).$$

Df. 7. Total torque (static momentum) of the applied force:

$$\sigma^N = \sum_i^\cdot \sigma_i \Rightarrow N(\sigma^N) \overset{\mathrm{dt}}{=} \sum_i X(\sigma_i, k, t) \times F^e(\sigma_i, k, t).$$

Using Dfs. 6 and 7, Thm. 9 can be rewritten in a more compact way:

Thm. 9. If $g \in G$, then

$$\dot{L} = N + \tfrac{1}{2}\sum_{i,j}[X(\sigma_i, g, t) - X(\sigma_j, g, t)] \times F^i(\sigma_i, \sigma_j, g, t).$$

Remark. Nothing has been added either in the way of form or in the way of meaning by recasting Thm. 9 in terms of the defined concepts of angular momentum L and torque N. The gain has been inkwise and psychological — the latter to the extent to which we have got used to handling the complex concepts L and N as units.

Corollary 1. If $g \in G$ and $F^i(\sigma_i, \sigma_j, g, t) \,\|\, X(\sigma_i, g, t) - X(\sigma_j, g, t)$, then $\dot{L} = N$. *Proof*: by vector algebra.

Remark. As it is impossible to deduce this corollary in CM, it will have to be postulated in that theory.

Corollary 2. If $g \in G$ and both the internal and the external forces are central, then the total angular momentum is conserved: $\dot{L} = 0$. *Proof*: by vector algebra on Corollary 1 and Df. 7.

All the above theorems except Thm. 8 are comprehensive in the sense that they do not involve a special choice of forces and constraints. These are not determined by the postulates of PM but must be conjectured *ab extrinseco*. For example, it is not the business of PM to find out the gravitational force — a problem for gravitation theory. Strictly speaking in PM there is no other mechanical force than the one occurring in impacts. CM does introduce further forces, e.g. the elastic ones, but it takes the whole of physics to cover all kinds of force. And all of them are received by PM.

This completes our version of the foundations of PM. Let us now turn to an analysis of it.

1.4. Analysis

A detailed analysis of all the aspects of PM would take a whole volume. Let us here just sketch a few possible lines of analysis.

To begin with, an imprecise characterization of the formal background of PM was given in 1.1. Thus it was stated that analysis underlies it. But exactly what chapters of analysis? This should not be specified: one should be able to expand the background as needed. (Recall that in Thm. 8 we made use of DIRAC's delta, which was not legitimated until the 1950's.) As to the formal structure of PM, and more particularly the mathematical one, it is not particularly interesting both because the central equations of PM are ordinary differential equations and because their transformation properties are rather poor. In fact the law of motion $PM\ 3.5\,a$ is covariant only under the very restricted group of Galilei transformations. It is not even generally covariant under time and parity reversals (see, e.g., Thm. 5). This mathematical rigidity imposes severe limitations on the applicability of PM both as regards problem-solving and as regards the referents of PM; thus, strictly speaking PM does not hold on Earth because this is not an inertial frame. Yet it is possible to expand the group under which the equations of motion remain unaltered: like nearly every other theory, PM can be reformulated in a generally covariant way (see Ch. 4, 3.1.3).

An examination of the logical properties of PM is more rewarding but it will not be undertaken here as it demands the highly specialized machinery of metamathematics (see Ch. 1, 4.1.6). Thus we shall intuitively assume rather than prove that our axiom set is consistent. We shall likewise suppose that it is independent but shall not subject it to an independence test. Either test would take more time than reconstructing a new theory. The p-completeness of our system can be ascertained by inspection: indeed, there is at least one postulate per primitive. And the axiom system is very likely d-complete as well, i.e. capable of entailing all the usual theorems. On the other hand we know PM is not expressively complete: i.e. not every formula written in the symbolism of PM holds in PM; for example, certain relativistic formulas can be written in the language of PM but are false in it. Nor is PM decidable: there is no algorithm (mechanical method) to decide whether any given formula written in the language of PM is true in it; in fact, the formal background of PM includes undecidable theories such as PEANO's arithmetic (TARSKI et al., 1953).

It seems that the basic concepts of PM are mutually independent. Take e.g. Σ and X. They are mutually independent in PM because two different particles may occupy the same position in space as long as their velocities differ (recall PM 2.5). Conversely, one and the same particle may occupy (successively) different positions. That is, given the particle its position is not uniquely determined and vice versa. Nor is X definable by M or conversely. In fact, let $F^i = 0$ and $F^e = ma$ with $a(k,t) = g = $ const, whereupon (3.1) becomes $\ddot{X} = g$ with the help of Thm. 4. By integration GALILEI's law is obtained, in which the mass concept does not occur. That is, the trajectory of a particle under a constant force is independent of the mass of the particle. Equivalently: one and the same X function is compatible with infinitely many numerical interpretations of the symbol 'M', whence the two are mutually independent. Consider finally the concepts of mass and force: although they are mathematically related they are not uniquely related and they are prescribed independently. Given a particle with a certain mass, it may be subjected to any of infinitely many forces; conversely, any given force may be applied to particles of any given mass values. In particular, the whole universe may be assigned a mass value (finite or infinite according to the cosmological theory) whereas the very concepts of force and acceleration are inapplicable to it. Consequently the widespread opinion that NEWTON's law of motion supplies a definition of "force" in terms of "mass" (or conversely) is technically false. Proceeding in a similar vein we should test for the mutual independence of the eight basic concepts of CM; we leave the $8.7/2 = 28$ tests to the interested reader. (For an analysis in a similar case see McKINSEY et al., 1953.)

Let us now take a look at the semantics of PM. As a purely mathematical theory it is of course uninterpreted because the reference class Σ has no natural interpretation within mathematics although it might be assigned any number of *ad hoc* interpretations, say as a number set. Yet our axiom system provides not only for the mathematical structure but also for the content of PM — e.g., by adding the hypothesis that every member of Σ represents a particle. Still, our semantic assumptions supply only the core meaning of PM. The full determination of the physical meaning of PM would necessitate the derivation of all its infinitely many theorems and its application to all possible cases — which is of course impossible. In fact, meanings are in part stipulated at the outset and partly discovered as the theory is worked out and applied (see Ch. 1, 4.2.6). Which, along with the changes in mathematical structure and the revolutions in other fields, explains the differences in content assigned to PM in the course of nearly three centuries.

The meaning of PM was outlined by stating correspondences between primitive symbols and certain physical concepts, i.e. concepts that are supposed to stand for physical objects — things like those represented by Σ, properties like those mapped by M, relations like those mirrored by F. No reference to empirical tests was made in the process because the reconstruction of a theory was not mistaken for its empirical validation. In particular, no restriction concerning the observability of the referents of the various primitives was placed, since the aim was not to summarize experience with pebbles but to explain an aspect of reality. Moreover none of the primitives of PM is observational, since they are either sets or mappings among sets, neither of which belongs in the real world although both can be assumed to be images of real objects. It would be impossible to restrict all the sets occurring in PM to those whose members are perceptible for this would leave us with no theory at all: just think of replacing, say, the continuous function X by the finite collection of *values* of X determined by *measuring* the positions of an *observable* particle, relative to a given *observer*, when the latter *reads* successive time values on a clock. In the first place we would have to include the concepts of observer, observability, clock reading, and the like among the primitives of our theory. In the second place no set of empirical estimates of the values of a function can mimic the function. This does not entail that X and the other functions occurring in PM have no physical meaning: it does show that X has no empirical meaning and that operationalism confuses, among other things, functions with their values.

Finally one word concerning the testability of PM. Being a fairly fully interpreted theory it is highly testable. Moreover in a few cases its central axiom, *PM* 3.5, can be subjected to fairly direct tests, i.e.

tests not involving the prior transformations of the axioms or the intervention of other theories (in addition to those already used to design and read the instruments). Thus ATWOOD's machine provides a fairly direct test of the law of motion while any experiment in statics puts the *lex tertia* to the test. Yet by far the more numerous and important tests of PM are the indirect ones, i.e. those contrasting the solutions of the equations of motion with empirical data concerning real things. But every such test requires the additional specification of the force functions; in many cases some of the mass values must also be assigned hypothetically. There is nothing wrong with this procedure as long as the logical consequences can be checked. And this in turn is possible whenever the referents of the theory are taken to be observable bodies. If they are not, the test will be more complex: it will call for additional theories supplying bridges between unobservable mechanical objects and observable effects. One of these bridges is statistical thermodynamics, which brings together nonobservational concepts such as the one of molecular motion and observational concepts such as the one of macroscopic pressure. Needless to say, a statement in PM does not lose its meaning when it is assumed to refer to an unobservable entity: it only ceases to be directly testable. Were this not the case how could we have been concluded that PM is in fact far from true in reference to microphysical systems?

In any case, before PM can be tested it must be worked out, and this elaboration requires the introduction of additional hypotheses (e.g., concerning the forces). The actual tests of PM show the limits of scientific prediction. Since the initial conditions and the other relevant data are known only approximately, the calculation of future and past states of a system is subject to error, and over a long period the error can be so large as to ruin the forecast altogether. This has been taken as a proof that even classical mechanics is indeterministic (BORN, 1958). Actually this inference rests on two misunderstandings. First, it is not the business of PM to compute the propagation of errors: this is a job for the theory of errors on the basis of the supposedly exact formulas of PM. This, like all other physical theories, handles theoretical not empirical predicates (see Ch. 1, 2.1.3). Second, any limits to exact predictability are not indications of the objective indeterminateness or fuzziness of nature but a reflection on our own limited ability to determine the present: such limits restrict the haughty thesis of epistemological determinism but do not affect ontological determinism (BUNGE, 1959a). Classical determinism must be given up for quite different reasons: not because predictability is as fallible as any other human ability but because randomness is an objective mode of being (EXNER, 1922 and BUNGE, 1961c).

2. Continuum Mechanics

The matter we experience with is bulk matter, the motions of which are studied by continuum mechanics (CM). Hence CM is the chapter of mechanics closest to experience and the one needed to explain and predict the behavior of perceptible bodies — in particular those constituting the laboratory paraphernalia, which for some queer reason are often said to obey the laws of microphysics. Therefore *methodologically* CM is prior to PM and all other physical theories: the design and interpretation of experimental tests, in particular of those of PM, electrodynamics, and quantum mechanics, presupposes CM. The latter is also logically prior to PM in the sense that, when suitably weakened, CM yields PM (see 2.4). Either reason would suffice to justify the treatment of CM before that of PM. In any treatment without pedagogical concessions one should start with CM.

That does not entail that CM is an ultimate or irreducible theory but only that, within classical physics, it is as basic as electromagnetism. From the point of view of DEMOCRITUS' programme of explaining the large in terms of the small, the perceptible in terms of the unobservable, CM is a nearly black box theory, a science of incompletely analyzed systems of particles and average motions — a discipline whose laws should eventually be obtained as the statistical outcome of the largely random motions of myriads of atoms. In other words, it is desirable to explain CM in terms of microphysics — of QM not of PM, since after all the point particle is an exaggerated idealization. The failure to perform such a logical (not ontological) reduction of CM to microphysics in an exhaustive way does not count against the programme itself, the more so since some progress has been achieved (GRAD, 1967).

But the success of the reduction programme depends both on our mastery of microphysics and on a correct formulation of CM: otherwise one would not know what should microphysics average to. (The same holds for thermodynamics, electromagnetism, and every other theory of bulk matter.) In short CM must be studied by itself, on its own level, as well as in the light of microphysics if we want to understand it in statistical (but not purely corpuscular) terms. In other words, in addition to striving for a representational theory of matter we must secure a semiphenomenological theory of it. (Semiphenomenological rather than phenomenological, for the stress tensor is anything but an observable.) And this both for the logical reason just mentioned and because the laws of bulk matter are different from those of its constituents and are therefore entitled to a separate study, much as the laws of societies differ from those of their individual components. Take, e.g., any of the laws in which the stress tensor occurs: since this property is not

hereditary (it does not hold for atoms), such laws are characteristic of
bulk matter even though we hope to be able to show that they emerge
from microlaws: they express patterns that do not exist at the microlevel, just as the latter is characterized by laws of its own. In other
words macrolaws, though in principle deducible from microlaws, do not
reduce ontologically to them but constitute an emergent novelty —
only, hopefully an understandable one.

In the general foundations of CM no details are assumed about the
constitution, structure and state of the bodies dealt with: the internal
properties are represented in bulk (though locally) by the stress tensor
and the stress vector. This generic stage is then superseded by adding
certain special assumptions leading to a bifurcation of CM into elasto-
mechanics and hydrodynamics — not to mention exogamous unions
such as thermomechanics and magnetohydrodynamics. Every one of
those main branches ramifies in turn upon the adjunction of further
assumptions aiming at characterizing species of materials. Of all the
subtheories thus built by the addition of constitutive equations, rigid
body mechanics is the simplest and therefore the least realistic and least
interesting; moreover, it is so special that it leads nowhere. On the other
hand hydrodynamics is a more pliable subject that lends itself to be
studied within the frames of relativity and quantum theory — both
literally as the study of quantum liquids and by way of analogy. But
we shall deal only with the comprehensive foundations of CM, i.e. with
the principles common to rigid body mechanics, elasticity, rheology,
and hydrodynamics.

Unbelievable as it may seem, the foundations of CM were not seriously investigated until the 1950's. The whole interest in CM had died
with the birth of relativity and the quantum. The subject was buried
alive and has ever since been officially regarded as dead, to the point
that no Ph. D.'s in physics are normally awarded in the field. But in
the mid 1940's it was realized that not enough was known of CM even
for engineering purposes: that not enough theorems were available,
and that the existing theory did not account for newly discovered
effects such as the swelling of the fluid emerging from a pipe and the
climbing of a fluid up the walls of a rotating cylinder — not to speak
of superfluids. A quick expansion followed and soon the subject began
to be investigated in depth, until the foundations of CM were subjected
to detailed revision by the "natural philosophy" group. The upshot
was a gain in generality, mathematical rigor, and logical organization;
in particular, the bondage with respect to PM was shown to be fictitious.
The gain was such that CM is now the best example of the benefits of
FR for "hard" physics. (The fact that the subject is sometimes classified
as a mathematical science and at other times as a branch of engineering

only goes to show how thin our metascience can be.) In the following we shall borrow freely from that pioneer work (TRUESDELL, 1952, 1953; NOLL, 1959; TRUESDELL and TOUPIN, 1960; TRUESDELL and NOLL, 1965). Our treatment will we more schematic and mathematically less sophisticated, but on the other hand we shall add some assumptions necessary to make its physical content more apparent.

2.1. Background and Building Blocks

The formal background of CM is constituted by PC=, semantics, algebra, topology, analysis, and manifold geometry as well as by all the theories presupposed by these mathematical disciplines. The material background of CM is protophysics (Ch. 2). As far as its material background goes CM is one of the few self-contained physical theories: it presupposes no specific physical theory. In particular it does not presuppose PM although some treatments of it use corpuscular hypotheses in a heuristic fashion. Moreover by dropping certain continuity assumptions PM could be deduced from CM while the converse deduction is impossible: one does not get a continuum by increasing the cardinality of a denumerable collection. But this is physically awkward since extended bodies are made of microsystems that look like particles — not bare corpuscles but "particles" glued by their accompanying fields, and not classical particles but quantons (see Ch. 5). The continuity of extended bodies is real not illusory, hence it is wrong to oppose the discontinuous table of classical atomic theory to the continuous table of common sense: the real table is not just a swarm of classical particles but a compositum of quantons and fields. Yet this is not the continuity of CM, which assumes that bodies are *mechanical* continua, in the sense that there exist mass and velocity fields over them. In other words, bodies are in fact continuous media and CM builds a simplified yet quite complex model or representation of them as mechanical continua. The task of CM is to specify such a model as a conceptual image of the body as a whole, whereas the task of many-body physics is to perform an analysis of it — not however to deny its existence or to say that "at bottom" of "ultimately" there are only particles, which is as false as saying that in the last instance there are no men but only atoms.

Primitive base

Σ: the set of bodies

$\{\mathscr{B}\}$:   a set of manifolds $\mathscr{B}$ representing a body σ each

E^3:   the Euclidean 3-space representing physical space

T: time

K: the set of physical reference frames k

$\{X\}$: a set of maps whose values represent body places

$\{\varrho\}$: a family of scalar fields representing the body density

$\{f_b\}$: a family of vector fields representing the body force density

$\{\mathfrak{T}\}$: a family of tensor fields representing the inner stress.

Remarks. (1) Σ is the reference class of CM: whatever is said in CM is said about members of Σ. But it will usually be said in a roundabout way, namely by speaking of the images or models $\mathscr{b} \triangleq \sigma$ of the real bodies $\sigma \in \Sigma$. In short, while $\mathscr{B} \triangleq \Sigma$ is the immediate referent of CM, Σ is its mediate referent (see Ch. 1, 1.3.5). This distinction should be made everywhere in physics; it was not made in PM because the referents of it were supposed to be structureless. (2) As far as their form is concerned all of the above concepts are studied by mathematics. This makes CM a mathematical science yet not a branch of mathematics, as every one of its primitives has some physical referent even if indirect. (3) E^3 and T are borrowed respectively from physical geometry and chronology; we shall leave their detailed study to protophysics (Ch. 2). (4) From a microphysical point of view ϱ is a local average, i.e. one taken over a volume element. As smaller and smaller bodies are taken the values of ϱ fluctuate more wildly until the very concept of density becomes pointless and correspondingly the formulas in which it occurs break down. It is not within the power of CM to indicate what its extension is: this is a task for many-body physics and for experiment. (5) The stress tensor $\mathfrak{T}$ is the typical magnitude of CM. It describes in a global way the inner mechanical state of a body, which determines its external behavior. It is clearly a nonobservational concept. Only the stress vector $t = \mathfrak{T}n$ is accessible on the body boundary. Yet $\mathfrak{T}$ is perfectly meaningful and hypotheses concerning special forms of $\mathfrak{T}$ are indirectly testable, as they will show up in the mechanical behavior of the body.

With the help of the preceding primitives all the other concepts of CM can be built; we shall list only a few. Where the clause "$\mathscr{b} \triangleq \sigma$" occurs it shall mean that the theoretical model $\mathscr{b} \in \mathscr{B}$ represents the real body $\sigma \in \Sigma$.

Df. 1. *Particle* $\pi \overset{\mathrm{df}}{=}$ element of $\sigma \in \Sigma$. Notation: $\pi \overset{.}{\in} \sigma$.

Df. 2. *Material point* $\beta \overset{\mathrm{df}}{=}$ element of $\mathscr{b} \in \mathscr{B}$. Notation: $\beta \triangleq \pi$.

Df. 3. σ_1 is a *proper part* of $\sigma_2 \overset{\mathrm{df}}{=} \mathscr{b}_1 \triangleq \sigma_1 \wedge \mathscr{b}_2 \triangleq \sigma_2 \Rightarrow \mathscr{b}_1 \subset \mathscr{b}_2$.

Df. 4. $\beta \triangleq \pi \Rightarrow$ *Location* of $\pi \overset{.}{\in} \sigma$ at t relative to $k \in K \overset{\mathrm{df}}{=} X(\beta, k, t)$.

Df. 5. *Configuration* of σ at $t \overset{\mathrm{df}}{=}$ image of $\mathscr{b}$, at t, under the function X. Notation: $\mathscr{b} \triangleq \sigma \Rightarrow C(\sigma, t) = X(\mathscr{b}, t)$.

Df. 6. *Reference configuration* of σ: $\mathscr{b} \triangleq \sigma \Rightarrow C_0(\sigma) \overset{\mathrm{df}}{=} X(\mathscr{b}, t_0)$.

Df. 7. If $\mathscr{b} \triangleq \sigma$, then: *motion* of $\sigma \overset{\mathrm{df}}{=} \{X(\mathscr{b}, t) \,|\, -\infty < t < \infty\}$.

Df. 8. *Volume* of σ at t:

$$\mathcal{b} \triangleq \sigma \Rightarrow V(\sigma, t) \overset{\mathrm{df}}{=} \int\limits_{X(\mathcal{b}, t)} \mathrm{d}^3 x \equiv \mu(\mathcal{b}, t).|$$

Df. 9. σ_1 is *distinct* from $\sigma_2 \overset{\mathrm{df}}{=} \mathcal{b}_1 \triangleq \sigma_1 \wedge \mathcal{b}_2 \triangleq \sigma_2 \Rightarrow \mu(\mathcal{b}_1 \cap \mathcal{b}_2) = 0$.

Df. 10. *Mass* of body:

$$\mathcal{b} \triangleq \sigma \Rightarrow M(\sigma, t) \overset{\mathrm{df}}{=} \int\limits_{X(\mathcal{b}, t)} \mathrm{d}^3 x \varrho(X).$$

Df. 11. *Material derivative* of A (scalar, vector or tensor):

$$\dot{A} \overset{\mathrm{df}}{=} \frac{\partial A}{\partial t}\Big|_{\beta=\mathrm{const}} = \frac{\partial A}{\partial t}\Big|_{X=\mathrm{const}} + \dot{X} \cdot \nabla A.$$

Df. 12. *Linear velocity* of particle

$$\beta \triangleq \pi \Rightarrow v(\pi) \overset{\mathrm{df}}{=} \dot{X}(\beta).$$

Df. 13. *Angular velocity* of particle

$$\beta \triangleq \pi \Rightarrow \Omega(\pi) = \tfrac{1}{2}[\nabla \times \dot{X}(\beta)].$$

Df. 14. *Linear momentum* of body

$$\mathcal{b} \triangleq \sigma \Rightarrow p(\sigma) \overset{\mathrm{df}}{=} \int\limits_{X(\mathcal{b}, t)} \mathrm{d}^3 x \varrho(x) \dot{X}(\beta, x).$$

Df. 15. *Moment of momentum* of body about a fixed point X_0

$$\mathcal{b} \triangleq \sigma \Rightarrow L(\sigma) \overset{\mathrm{df}}{=} \int\limits_{X(\mathcal{b}, t)} \mathrm{d}^3 x \varrho(x) X_0 \times \dot{X}(\beta, x).$$

Df. 16. *Stress vector* (force per unit surface exerted at boundary ∂X with outer normal n on the material enclosed in it):

$$t \overset{\mathrm{df}}{=} \mathfrak{T}n.$$

Df. 17. *Resultant contact force*

$$\mathcal{b} \triangleq \sigma \Rightarrow F_c(\sigma) \overset{\mathrm{df}}{=} \oint\limits_{\partial X(\mathcal{b}, t)} \mathrm{d}^2 x \, t(x).$$

Df. 18. *Resultant body force*

$$\mathcal{b} \triangleq \sigma \Rightarrow F_b(\sigma) \overset{\mathrm{df}}{=} \int\limits_{X(\mathcal{b}, t)} \mathrm{d}^3 x \varrho(x) f_b(x).$$

Df. 19. *Resultant force* on body

$$\mathcal{b} \triangleq \sigma \Rightarrow F(\sigma) \overset{\mathrm{df}}{=} F_c(\sigma) + F_b(\sigma).$$

Df. 20. *Torque* on body about a fixed point X_0:

$$\mathcal{b} \triangleq \sigma \Rightarrow N(\sigma) \overset{\mathrm{df}}{=} \oint\limits_{\partial X(\mathcal{b}, t)} \mathrm{d}^2 x X_0 \times t(x) + \int\limits_{X(\mathcal{b}, t)} \mathrm{d}^3 x \varrho(x) X_0 \times f_b(x).$$

Remarks. (1) Whereas particles are regarded as parts of real bodies, material points are conceived as the elements of manifolds, hence as concepts. (2) The concept of configuration constitutes a refinement or elucidation (explication) of the concept of shape. Though at first blush an empirical concept, "configuration" is transempirical since the empirical determination of the instantaneous configuration of a body would require infinitely many measurements performed at a given instant and lasting no time. Hence an operationalist should conclude that bodies in motion have no definite configuration. Which would render CM impossible. The same holds for the concept of motion (see Df. 7). (3) In this axiomatization the contact force, which in CM is more important than the body force, is defined in terms of the concept of stress vector, which is in turn defined in terms of the deeper concept of stress tensor; on the other hand the body force density is taken as a primitive. Obviously, this choice of primitives is not mandatory. (4) Some concepts are local whereas others are global. The integrations are to be performed over the configuration $X(\mathscr{B}, t)$ of a body representative and/or the boundary $\partial X(\mathscr{B}, t)$ of the previous set: mathematically it would make no sense to let the integration variables scan the body itself, which is not a mathematical object. (5) In theoretical work the integrands, in particular the densities, are prescribed (hypothesized) and the overall behavior of the body is then calculated. The experimenter, on the other hand, has access to the integrated quantities. Since one and the same global quantity, such as the resultant contact force, can be obtained from infinitely many expressions for the corresponding densities, the inference from experimental data to the true densities is impossible. The densities must be conjectured and experiments can check such hypotheses but cannot dictate them. (6) On applying Df. 11 to Df. 12 we get: $\ddot{X} = \partial \dot{X} / \partial t + \dot{X} \cdot \nabla \dot{X}$, which shows why CM is essentially a nonlinear theory, i.e. one with a much more complex formalism than PM and consequently much harder to work out and test, since in many cases only approximate solutions of the basic equations can be obtained. (7) The expression for the resultant force, here regarded as a definition, is sometimes regarded as a principle. (8) The definitions in which integrals are involved make sense only if the integrands are smooth enough; since a continuity condition is a hypothesis, the preceding definitions actually presuppose some of the axioms to follow. (9) Several other concepts can be introduced upon working out the postulates but we shall not even use all of the above defined concepts; we have introduced them just to show how certain concepts can be given an exact elucidation.

2.2. Axioms

The axioms for E^3 and T are the same as those for PM (see 1.2). But it is possible to formulate CM in a curved space, thereby acquiring the freedom allowed by curvilinear coordinates as well as rendering a general covariant formulation possible. This is indeed necessary in order to bring CM and general relativity together, but otherwise unnecessary. The importance of alternative choices of geometry lies in showing that CM is not indissolubly married to any of them and that therefore the experimental confirmation or refutation of CM has no bearing on the physical truth or falsity of Euclidean geometry. Let us now characterize the remaining seven primitives of CM.

Axiom group 1: Body

CM 1.1 (a) Σ is a nonempty denumerable set. (b) Every $\sigma \in \Sigma$ is a body.

CM 1.2 (a) $\{B\}$ is a nonempty family of point sets. (b) Every $\mathscr{b} \in \mathscr{B}$ is a 3-dimensional differentiable manifold. (c) For every $\sigma \in \Sigma$ there exists a $\mathscr{b} \in \mathscr{B}$ such that $\mathscr{b} \triangleq \sigma$ pointwise.

CM 1.3 (a) X is continuous and bounded for every $X \in \{X\}$, every $\beta \in \mathscr{b}$, every $k \in K$, and every $t \in T$. (b) $X(\beta, k, t)$ represents the place (location) of the particle π of the body σ relative to the frame k, at the instant t.

CM 1.4 (a) For every $X \in \{X\}$, every $\mathscr{b} \in \mathscr{B}$ and every $t \in T$, the map $X : \mathscr{b} \to C$ is 1:1, onto and continuous. (b) For every $X \in \{X\}$, every $\mathscr{b} \in \mathscr{B}$ and every $t \in T$, the image $C = X(\mathscr{b}, t)$ of $\mathscr{b}$ under X is a compact set with a piecewise continuous boundary $\partial X(\mathscr{b}, t)$. (c) For every $\sigma \in \Sigma$ and any $t \in T$ there exists an $X \in \{X\}$ such that the image $X(\mathscr{b}, t)$ of $\mathscr{b} \in \mathscr{B}$ under X represents the place occupied by the body σ represented by $\mathscr{b}$.

CM 1.5. (a) Every $\varrho \in \{\varrho\}$ is a bounded and continuous function from $\mathscr{B}$ to the set of nonnegative reals. (b) If $\mathscr{b} \triangleq \sigma$, then $\varrho(\mathscr{b}, x, t)$ represents the mass density of σ at x, t.

CM 1.6 (a) For every $\sigma \in \Sigma$, if $\mathscr{b} \triangleq \sigma$ and C is the image of $\mathscr{b}$ under X, then M is a continuous function of the volume (Lebesgue measure) $\mu(\mathscr{b}, t)$ of $\mathscr{b}$ at t. (b) For any two distinct $\sigma, \sigma' \in \Sigma$ (see Df. 9) if $\mathscr{b} \triangleq \sigma$ and $\mathscr{b}' \triangleq \sigma'$ and $\mathscr{b} \cup \mathscr{b}' \triangleq \sigma \dotplus \sigma'$, then $M(\mathscr{b} \cup \mathscr{b}') = M(\mathscr{b}) + M(\mathscr{b}')$. (c) If $b \triangleq c$, then $M(\mathscr{b})$ represents the mass of σ.

CM 1.7 For every $\sigma \in \Sigma$, if $\mathscr{b} \triangleq \sigma$ and $C = X(\mathscr{b})$ and $C' = X'(\mathscr{b})$ are two different configurations of σ at a given time t, then the corresponding masses are the same:

$$M(\sigma) = \int\limits_{X(\mathscr{b})} d^3 x \, \varrho_X = \int\limits_{X'(\mathscr{b})} d^3 x \, \varrho_{X'}.$$

Remarks. (1) By *CM* 1.2, the concrete thing called σ is mapped into the conceptual object $\mathscr{b}$. In symbols, $\mathscr{B} \triangleq \Sigma$, where '$\triangleq$' designates the semantical relation of modelling (see Ch. 1, 1.3.2). Once we have lifted

the discourse to the mathematical level we remain in it until the calculations are over: we then reinterpret the resulting statements in physical
terms. In other words, CM 1.2 says what the theoretical model $\mathscr{E}$ of
a concrete body looks like. (2) By CM 1.3, that theoretical model is
isomorphic to a region of space. More precisely, X maps the manifold $\mathscr{E}$
(a conceptual object) onto a chunk of space, namely the configuration
$X(\mathscr{E})$. We have then the map composition:

$\triangleq \circ X : \Sigma \overset{\triangleq}{\Rightarrow} \mathscr{B} \overset{x}{\rightarrow} C$. Whereas the members of Σ name concrete
bodies in motion, those of $\mathscr{B}$ and C are concepts: in reality there are
neither bodies in abstraction from configurations nor conversely. The
analysis of a real system into substance $(\mathscr{B})$ and form (C) is a conceptual
not an empirical analysis. (3) We give axioms for ϱ because this concept
is here taken as a primitive. In alternative formulations postulates are
given for the total mass; then, assuming that M is smooth on $X(\mathscr{E})$,
the Radon-Nikodym theorem warrants the existence of ϱ — not however
its uniqueness. (4) The additivity of total mass (CM 1.6b) and its
conservation (CM 1.7) are here taken as basic physical law statements,
hence as hypotheses that are in principle refutable. In alternative formulations (e.g., NOLL, 1959) the total mass is defined as an invariant
measure and consequently its additivity and conservation are obtained
as straightforward measure-theoretical consequences. This procedure is
not adopted here because (a) the phrase 'mass of the set $\mathscr{E}$' makes no
sense: sets are not physical entities, (b) mass conservation is a law not
a convention, and (c) we know from relativity that the total relative
mass is subadditive.

Axiom group 2: Kinematics

CM 2.1   (a) K is a nonempty denumerable set included in Σ. (b) The
configuration of every $k \in K$ is time-independent. (c) For every $k \in K$
there exists a Cartesian system of axes $e = \langle e_1, e_2, e_3 \rangle$ such that $e \triangleq k$.
(d) No $k \in K$ interacts with any $\sigma \in \Sigma$ that is not a part of k.

CM 2.2 Let $X' = a + QX$, with a and Q functions of time, represent
a change of frame. Then (a) If V is a vector referring to $\pi \overset{\cdot}{\in} \sigma \in \Sigma$,
which upon a change of frame becomes $V' = QV$, then V' and V refer
to the same physical situation, and (b) if S is a 2nd rank tensor referring
to $\pi \overset{\cdot}{\in} \sigma \in \Sigma$, which upon a change of frame becomes $S' = QS\tilde{Q}$, where $\tilde{Q}$
is the transposed of Q, then S' and S refer to the same physical situation.

CM 2.3   (a) The derivatives $\dot{X}$ and $\ddot{X}$ exist and are bounded for
every $X \in \{X\}$, every $\beta \in \mathscr{E}$, every $k \in K$ and every $t \in T$. (b) If $\beta \triangleq \pi$
and $\pi \overset{\cdot}{\in} \sigma$, then $\dot{X}(\beta, k, t)$ and $\ddot{X}(\beta, k, t)$ represent the velocity and the
acceleration respectively of the particle π in the body σ relative to the
frame k at time t.

Remarks. (1) By *CM* 2.2, one and the same physical situation can be represented in infinitely many ways. These representations constitute an equivalence class every member of which can be gotten from any other one through a change of frame. The transformation concerned should not be mistaken for a coordinate transformation, under which vectors and tensors transform in a homogeneous way; in other words, *CM* 2.2 is not concerned with the arbitrary choice of label for the points in space but with displacements and rotations of material frames. (2) The continuity of the trajectory is asserted, not so the continuity of the velocity and the acceleration, which can jump across certain boundaries — e.g., in (idealized) shock waves. In other words, positions are changed but not lost altogether, as they would if the constituents of bodies ceased to exist.

Axiom group 3: Dynamics

CM 3.1 (*a*) Every $f_b \in \{f_b\}$ is a real and bounded vector field over $\mathscr{B} \times K \times E^3 \times T$. (*b*) If $\mathscr{B} \triangleq \sigma$, then $f_b(\mathscr{B}, k, x, t)$ represents the force per unit volume on the body σ.

CM 3.2 (*a*) Every $\mathfrak{T} \in \{\mathfrak{T}\}$ is a real and bounded tensor field of valence (2,0) over $\mathscr{B} \times K \times E^3 \times T$. (*b*) If $\mathscr{B} \triangleq \sigma$ and $\beta \triangleq \pi$, then $\mathfrak{T}(\beta, k, x, t)$ represents the body stress at the particle $\pi \overset{.}{\in} \sigma$.

CM 3.3 *Law of balance of momentum:* For every $\sigma \in \Sigma$, every $t \in T$, every $\varrho \in \{\varrho\}$, every $f_b \in \{f_b\}$ and every $\mathfrak{T} \in \{\mathfrak{T}\}$, there exists a $k \in K$ such that
$$\dot{p} = F$$
(see Dfs. 14 and 19).

CM 3.4 *Law of moment of momentum:* For every $\sigma \in \Sigma$, every $t \in T$, every $\varrho \in \{\varrho\}$, every $f_b \in \{f_b\}$ and every $\mathfrak{T} \in \{\mathfrak{T}\}$, there exists a $k \in K$ such that
$$\dot{L} = N$$
(see Dfs. 15 and 20).

This completes our axiom system. It allows us to introduce the following axiomatic definition:

Df. 21. A reference frame $k \in K$ relative to which all the preceding axioms hold is called *inertial* or *galilean.*

Remarks. (1) The leading hypotheses of CM, those which cannot be changed unless the whole theory is to be altered, are *CM* 1.7 (conservation of mass), *CM* 3.3 (Newton-Euler law of motion), and *CM* 3.4 (moment of momentum). They are integral or global statements but they will be reformulated in differential (local) form in 2.3. (2) Contrarily to what happens in PM (see 1.3, Thm. 9), the torque law (*CM* 3.4) cannot here be derived from the other axioms of CM. The text-book "deductions" from the homologous theorem in PM are phony. (3) In actual theoretical work ϱ, f_b and $\mathfrak{X}$ are prescribed (assigned hypothetically) and the corresponding integral quantities, as well as the solutions of the basic equations, are computed. This does not mean that the total mass, the resultant force and torque, and other logically derived quantities are known a priori: being physical ideas we have no a priori information about their referents and values. Only, by hypothesizing them or by proposing them as data to the mathematician, we act momentarily *as if* we did have such an information. (4) Having included the principle of material objectivity in protophysics (see Ch. 2, 1), we need not state it here. This policy is further justified by recalling that the principle operates in the whole of macrophysics not just in CM, and that it is not a physical law but a metalaw statement. (5) The preceding axiomatization is one among many possible organizations and elucidations of the foundations of CM. Thus NOLL (1963) replaced the two balance postulates by a single axiom restricting the possible configurations and forces to those preserving the mechanical power under changes of frame.

2.3. Typical Consequences

The preceding axiom system is p-complete in the sense that it characterizes all the specific basic concepts of CM; and it is d-complete as it suffices to derive all the known comprehensive theorems of CM — those which do not depend on special assumptions about the nature of the body. Let us recall a few such consequences.

Firstly the law of conservation of mass CM 1.7, in conjunction with the hypothesis CM 1.5 that ϱ is a scalar density, entails that, for any two configurations $X(\mathscr{E})$ and $X'(\mathscr{E})$ of a body σ represented by $\mathscr{E}$, $\varrho_X = J \cdot \varrho_{X'}$, where $J \stackrel{\mathrm{df}}{=} \mathrm{Det}\,|\partial X'/\partial X|$ is the Jacobian of the transformation. Taking the material derivative $(Df.\ 11)$ of this equation we get

Thm. 1: *conservation of mass.* For every $\sigma \in \Sigma$, every $\varrho \in \{\varrho\}$ and every $t \in T$, if $\mathscr{E} \triangleq \sigma$, then: $\dot{\varrho} + \varrho \cdot \nabla \dot{X} = 0$.

Comment. This statement is not exactly equivalent to the corresponding integral statement, for it says that mass is conserved not only globally but also locally. This shows that the integral principles of CM are not properly interpretable as statements about averages.

On introducing the Df. 16 of the stress vector in the law of momentum balance and using the divergence theorem we get

Thm. 2: *Cauchy's first law of motion.* For every $\sigma \in \Sigma$, every $k \in K$, every $t \in T$, every $\varrho \in \{\varrho\}$, every $f_b \in \{f_b\}$ and every $\mathfrak{X} \in \{\mathfrak{X}\}$, if $\mathscr{E} \triangleq \sigma$, then: $\varrho \ddot{X} = \mathrm{div}\,\mathfrak{X} + \varrho f_b$.

Remarks. (1) The linear momentum is balanced not only globally but also locally. (2) If f_b, ϱ and $\mathfrak{X}$ are prescribed, the motion is uniquely determined (by integration). But if f_b, ϱ and $\ddot{X}$ are given, the stress state is not uniquely determined: only div $\mathfrak{T}$ acquires a unique value. Consequently, though empirically we must remain on the surface of the body (restricting ourselves to the stress vector t), theoretically we must start from the inside, hypothesizing (prescribing) $\mathfrak{T}$. The hypothetical knowledge of the 9 components of $\mathfrak{T}$ suffices to predict what will happen on the boundary but not conversely. In short, external appearances are determined by unobservables that must be hypothesized.

Proceeding in a similar fashion with the torque law, we derive

Thm. 3: *Cauchy's second law of motion.* For every $\sigma \in \Sigma$, every $k \in K$, every $t \in T$, every $\varrho \in \{\varrho\}$, every $f_b \in \{f_b\}$ and every $\mathfrak{T} \in \{\mathfrak{T}\}$, if $\mathscr{E} \triangleq \sigma$, then: $\mathfrak{T}$ is symmetric.

Remarks. (1) It seems queer to call this a law of motion since apparently it refers to no changes; it even seems odd to call it a law as it would seem to express just a mathematical property of $\mathfrak{T}$. Yet it is equivalent to the differential statement of the conservation of the moment of momentum. One finds similarly puzzling law statements in elementary particle physics, where the unitarity of a certain matrix, say, is equivalent to the conservation of the total number of "particles". (2) The mention of ϱ and f_b in the antecedent of this theorem would seem to be superfluous since those functions do not occur in the equation; but they do figure in the Df. of torque that appears in one of the premises. (3) The two Cauchy laws, which derive from the integral principles of CM, are field equations in the mathematical sense. From a physical point of view they are not, as they refer to a (fictitious) substance smeared out over an expanse of space: by definition, genuine fields are, among other things, not bodies. More on this in Ch. 4. (4) Thms. 2 and 3 constitute 6 equations for the 9 components of the stress: the system of general equations for $\mathfrak{T}$ is indeterminate. Hence the equations cannot be solved exactly and explicitly unless additional assumptions are added which remove that indeterminacy. This is characteristic of the foundations level, where the source functions — such as $\mathfrak{T}$ and the lagrangians — can be neither computed nor extracted from experience but must be assumed to see what happens with them. Yet a qualitative study of the equations of motion can be made even if they are indeterminate: in this way the infinite class of permissible stress fields and the corresponding infinite class of solutions can be determined. The theoretical importance of such a study lies in that it shows precisely in what way do the postulates restrict the class of motions. Its practical importance lies in that it introduces a number of arbitrary functions, every one of which calls for the introduction of a special hypothesis representing in a global way either the constitution and structure of the body or its environment. In this way a mathematical analysis of the theory points the way to physical problems (see TRUESDELL, 1961).

Up to now we have stated only laws holding within domains of continuity. But CM contains important statements about what can happen at surfaces of discontinuity, i.e. where certain magnitudes, such as the velocity or the stress, suffer a jump: otherwise CM would deal only with infinitely large bodies. As a specimen of the discontinuity laws of CM let us mention the theorem on the jump of the stress vector across a singular surface; it is proved by applying the law of balance of momentum to an imaginary cube containing a portion of a singular surface s across which the stress vector jumps from t to t'. It reads

Thm. 4. For every $\sigma, \sigma' \in \Sigma$, if $\mathscr{B} \triangleq \sigma$ and $\mathscr{B}' \triangleq \sigma'$ have a common boundary s such that t is continuous on $\mathscr{B}$ and t' is continuous on $\mathscr{B}'$ and if they approach unique limits t_s and t'_s respectively as s is approached along a path lying wholly on one side of s, then: $t_s + t'_s = 0$.

Remark. This is one member of the set of laws known as law of action and reaction. In CM there is at least one more member of this set: the corresponding law for the resultant force, which can be derived from Thm. 4 in conjunction with the additional postulate that the *lex tertia* holds for the body force density.

This completes our review of the foundations of CM — i.e. the principles assumed to hold for all bodies. The set of basic assumptions we have laid down is mathematically indeterminate (recall Remark 4 on Thm. 3). As a consequence it is insufficient in order to state and solve any problem in a complete way — whence it cannot be put to the empirical test without further ado. Before this general framework can be applied to special problems, and consequently put to the test, further assumptions must be added, namely (*a*) boundary conditions specifying the exterior shape of the body, (*b*) constraints representing in a global way the action of the environment, and (*c*) constitutive equations expressing the response of the body to a given kind of environmental stimulus. Whereas (*a*) and (*b*) may be regarded as data (often hypothetical ones), the constitutive equations are full-fledged hypotheses.

The best known class of constitutive equations — i.e. conjectures about the specific constitution of materials — is the set of equations that relate the stress tensor to the deformation gradient tensor $D = \|\partial X^i / \partial X^j_0\|$, where X_0 is the position of the representative particle when the body is in the relaxed state (reference configuration). Every relation of the form $\mathfrak{T} = G(X, D)$ is a constitutive equation characterizing a species of deformable materials. Just as in the case of the equations of motion, it is convenient to make a general study of the constitutive equations: this will determine the taxonomy of materials in CM (NOLL, 1958). But as soon as specific realistic constitutive equations are looked for, even at the general level, a number of nonmechanical variables — such as temperature and dielectric power — pop up. Something similar happens also with regard to the conservation equations. In either case CM must be broadened to make room for thermomechanical and ponderomotive effects (COLEMAN, 1964). In short, like every other theory CM is factually incomplete in the sense that it cannot fully account for any single macroevent, because macroevents are many-faced.

At the opposite extreme stands PM, which can be built on CM by leaving aside all the pointless formulas — essentially those involving $\mathfrak{X}$ — and adding two hypotheses: (a) that the equality of action and reaction holds for interparticle forces and (b) that the mass vanishes everywhere except at the particles, i.e.

$$\varrho = \sum_{i=1}^{N} m_i \delta[x - X_i(t)],$$

where δ is DIRAC's delta. In short, PM is built by adding certain restrictive hypotheses to CM. (This shows that PM is not a subtheory of CM in the technical sense discussed in Ch. 1, 4.1.1: there is *construction*, not *reduction* in this case.) The converse construction of CM on the basis of PM is mathematically as impossible as the construction of reals out of integers alone. CM cannot even be stated unless its object is modelled as a continuous set — e.g., a differentiable manifold. This does not exclude the possibility of obtaining the integral laws of CM as statistical averages of molecular motions — but this by far transcends the boundaries of PM, for molecules are not point particles and anyway they do not abide by classical mechanics.

2.4. Tests

As if to refute positivism, classical CM, such as it grew during the 18th and 19th centuries, was believed on reason rather than on experiment. In fact only a few exact solutions of the basic equations were known, chiefly because of the nonlinearities — a good example of poor testability and even poorer actual tests due to the underdeveloped state of the theory (due partly to the nonexistence of a general theory of nonlinear equations) rather than to the lack of experimental facilities (see Ch. 1, 5). Then, even when exact solutions were available, the corresponding experimental data were often irrelevant to them: they concerned, say, the turbulent flow of a viscous liquid (usually water) endowed with some memory, whereas the formulas concerned the lamellar flow of a nonviscous and forgetful liquid (TRUESDELL, 1961

and TRUESDELL and NOLL, 1965 — and recall that the problem of the
rotation sense of a liquid going down a drain was not solved until the
mid 1960's). No wonder that CM was insufficiently confirmed by experiment. Finally the available measuring instruments were and still are
too few and too coarse: the field did not attract experimental physicists.
Yet people believed CM. Why?

CM was believed to be true not only because some of its theorems
had been confirmed but mainly because it was assumed (*a*) that CM was
just an application of PM — a theory in turn believed to be completely
true at least in the classical approximation, because it was endorsed
by astronomy, and/or (*b*) that, being thoroughly mathematical in form,
CM was analytic — rational — and consequently in no need of experimental test. We now know that the first belief is false; and a semantical
analysis of CM shows that CM has an intended physical referent — an
arbitrary body — whence it must have some factual truth value, even
if not adequately known, in addition to being mathematically correct.

The testability of CM has improved since the mid 1940's with its
own rapid maturation — growth of the body of theorems and deepening
of the foundations — as well as with its connection with other fields
of physics (e.g., thermodynamics) and with progress in experimental
techniques. Yet the situation is still unsatisfactory not only because of
the mathematical complexities of CM and the complexities of real
situations — most of which must be ironed out in a theoretical treatment — but also because of some persistent misunderstandings concerning the functions of CM *vis à vis* experiments. The experimentalist —
who in this field is often a research engineer — often thinks that CM
is too far from reality for it does not cover all the details of the behavior
of his pet material: he sometimes complains that CM is not specific
enough and at other times that CM employs too many experimentally
inaccessible ("operationally meaningless") functions. Yet a similar gap
between basic theory and experiment pervades all advanced science
and it is unavoidable: (*a*) no fundamental theory can be expected to
cover all the details of a real situation; (*b*) no experiment can determine
all the components of a theory — in fact the most important ones must
be assigned hypothetically, for they refer to unobservable (though
supposedly real) traits. These two statements belong to the philosophy
of science, whence the philosopher — often within the physicist's skin —
is entitled to enter this dispute between the theoretician and the experimentalist of CM.

In recent years CM has grown both theoretically and experimentally
to a point where its testability has become adequate (see, e.g., COLEMAN,
MARKOVITZ and NOLL, 1966). Yet it will probably turn out that its
extension or domain of truth will be less than the one it claims. In

fact CM does not seem to apply to an arbitrary body in an arbitrary state: for example it does not hold for superfluids such as liquid helium, which exhibits nonclassical properties — e.g. a quantized vorticity. (This motivates and justifies introducing the concept of *classical body* as that body which satisfies CM within experimental error.) Quantum fluids seem to require a quantum continuum mechanics and both this theory and CM can in principle, though not yet *de facto*, be based on atomic-statistical theories. Yet such a further foundation will not decrease the value of CM as a theory of bulk matter: CM will stay because it is largely insensitive to the particular atomic structure hypothesized to underlie it. Its weakness is its strength.

Chapter 4

Classical Field Theories

Introduction

Any theory containing the mathematical concept of field (scalar, vector, tensor or spinor) may be called a *field theory*. In particular, any theory based on an action principle involving a lagrangian density is, mathematically, a field theory. (The entire class of such field theories can be handled in a unified and axiomatic way: see EDELEN, 1962.) But for a theory to represent a real field something more than such a mathematical formalism is needed: it must refer to some extended imponderable that cannot be transformed away. Consider the Coriolis acceleration at every point inside a rotating body: from a mathematical point of view this is a vector field. But it is not regarded as a real field because it can be transformed away by choosing a reference frame rotating with the body. On the other hand a magnetic field, even if it vanishes in a given frame, will reappear in other frames because of its association with electric fields. Therefore a magnetic field is a physical field although its strength is relative — just as that of an electric field. We shall then agree that a field is *real* (or physical) iff it can be transformed away only locally but not globally. On the other hand a "field" that can be transformed away everywhere is not a physical field — this being why no tensor field equations can be written for the Coriolis acceleration.

A physical field theory, then, (*a*) employs mathematical fields, (*b*) postulates certain laws for them which have definite invariance properties, and (*c*) assumes that the field quantities and the relations they enter refer (apply) to an arbitrary member of a class of extended but massless substances. Since such a theory requires no material substratum, it will be called a *pure field theory*. Thermodynamics and other theories of continua can be called *material field theories*, to emphasize that they refer to extended material systems (bodies). The difference between the two kinds of theories, the pure and the material field theories, is apparent in most of their concepts. Thus in a material field theory a velocity field represents the velocity of a material particle π at a place x and an instant t. In a pure field theory π does not occur and a velocity field represents the velocity of an extended field disturbance. Moreover in a material field theory the field concept is ultimately dispensable: whatever is said about what happens *at* a point in space can be translated into a statement concerning what happens *to* a particle. (For the translatability of the spatial into the material description, see TRUESDELL, 1961.) CM and the 19th century aether theories are material field theories for they refer to the state of motion of particles of matter or aether. After special relativity wiped out the aether from classical electromagnetism (CEM), the latter became a pure field theory. But macroelectromagnetism, which deals with fields in electrically and magnetically polarizable materials, is a *mixed field theory* (POST, 1962 for the structural or substantial and the functional tensor fields in this theory). Consequently the pure/material distinction does not amount to a dichotomy.

Every field theory poses a number of philosophical questions. The most important of them is whether the very concept of field is justified, since after all fields are unobservable — and what is unobservable is metaphysical and therefore damnable, at least according to the dusty philosophy still popular among physicists. Indeed, nearly every freshman is taught that a field is a region of space where, if we placed a test body, this would exhibit an acceleration. Inconsistently enough he is immediately warned that a test body is a fiction since it is supposed not to distort the given field — which of course presupposes that the field has an autonomous existence and that we believe the field equations even if they do not refer to test bodies. Anyhow the freshman gets the impression that fields depend on test procedures and that a field intensity is a specific force acting on a test particle, so that a field theory links measurement results. This view is not only inconsistent but is also a remnant of the stage when fields were not regarded as substances existing independently of bodies — supposedly the only real thing. It is also a piece of subjectivistic epistemology. The truth is that field strengths (or their potentials) are not defined, let alone in mechanical terms: they are introduced as primitives. What happens is that they are measured indirectly, through their ponderomotive effects. By introducing the operationalist confusion between reference and test, the student's attention is diverted from fields to bodies and so he is spared the healthy shock of facing an unexperienced kind of thing which, precisely for being beyond ordinary experience, cannot be accounted for by empiricism.

In this respect, the usual elementary teaching of CEM has not yet felt the impact field theories had on our world view (= philosophy of nature = philosophical cosmology), by rendering mechanism obsolete, and showing that nature is infinitely more complex than had been assumed before. (Recall that an arbitrary field can be analyzed as a superposition of infinitely many elementary waves, each characterized by a given value of the propagation 4-vector and perhaps also a given polarization direction, in addition to its amplitude.) This does not mean that mechanism is now dead: most people still imagine in terms of swarms of particles; and mechanism of the corpuscular brand, allied to operationalism, has stimulated the building of action at a distance theories of the e.m. and the gravitational fields. Mechanism is still alive both because it is simple and because field theories are ridden with grave mathematical difficulties — mainly the divergences. The pure field theoretical programme advocated by MIE, EINSTEIN and WEYL, has not triumphed: even in general relativity fields and particles, though interacting, are separate things, mutually irreducible substances. A unitary theory, whether corpuscular or field-like, though attractive, is still a programme (see however WHEELER, 1962). At the classical level we have a matter-field duality and at the quantum level we have a promise of synthesis: although in quantum field theory the field is the prime matter and matter proper (the quantons) comes out of the background field, neither this nor the quantization process is fully understood in physical terms.

We shall restrict our treatment to classical fields, and more particularly to MAXWELL's CEM, its offspring special relativity (SR), and EINSTEIN's gravitation theory (GR). The first two are among the few theories which have not retained important traces of the process of their construction: their leading axioms were laid bare from the start and the heuristic auxiliaries, such as force tubes, were eventually recognized to belong to the scaffolding. On the other hand GR still bears outstanding birth marks. For this reason special care will be exerted in distinguishing the heuristic guides employed in constructing GR from its actual constitutive principles. And all three theories will be analyzed in some detail because of their boldness and beauty.

1. Classical Electromagnetism

Classical electromagnetism (CEM) is a set of theories. The basic theory is the microscopic or Maxwell-Lorentz theory of the e.m. field in empty space. The referents of this theory are fields smeared over

expanses of empty space, and charged particles. Polarizable media —
dielectrics and magnetizable bodies — are introduced on a second
level — macroscopic electromagnetism. The latter can be approached
in two different ways: directly (phenomenologically) or on the basis
of microscopic CEM. In the former approach the electric displacement,
the magnetic intensity and the medium parameters are introduced as
new primitive concepts, MAXWELL's equations are written out for them,
and finally the general constitutive equations that relate the fields to
the inductions are postulated. In the second, more profound approach
these additional concepts and relations characterizing the electromagnetic
properties of matter in bulk are built on the basis of the microscopic
Maxwell-Lorentz theory and specific assumptions concerning the structure of matter — essentially, the charge and current distributions. At
this point the road forks into a classical and a quantum mechanical
theory. In the following we shall sketch the foundations of the three
above mentioned classical theories.

The presuppositions of all these theories are the same and will
therefore be mentioned only once. *Formal background:* PC=, semantics,
algebra, topology, analysis, manifold geometry and whatever these theories
presuppose. The foundations of CEM are both affinity-free and metric-
free (CARTAN, 1924). It is only for the development and application of CEM
to special cases that a definite space metric must be assumed. The fact that
Euclidean geometry is usually chosen for these purposes does not commit
CEM to the Euclidean theory of space, which is on the other hand
presupposed in every electromagnetic measurement.

Material background: protophysics — chronology, the general systems theory, and so forth, including the fragment of physical geometry
that characterizes the concept of material reference frame (see Ch. 2).
The strict field theory, referring to fields alone, presupposes no specific
physical theory and is therefore fundamental in the strict sense. But
this theory is severely limited: it cannot account for the interactions
between fields and matter. Moreover it is for this reason untestable,
as only the effects of fields on charged bodies can be observed — and
even this usually in an indirect fashion. A complete and testable theory
of electromagnetism must include equations of motion of charged bodies
(not point particles), and these make no sense unless the concepts
occurring in it are specified both formally and semantically — a specification incumbent partly on CEM, partly on CM. Therefore the construction of a realistic CEM presupposes and includes a substantial portion
of dynamics. To put it into other words, the basic equations of CEM
fall into two sets: the field equations and the equations of motion,
which are in turn specifications of the general equations of motion of
dynamics. This division of the basic equations should not occur in a

unitary theory, be it a pure field theory or an action at a distance theory. In any case CEM is a dualistic theory, in which bodies and fields are equally fundamental and mutually irreducible. We shall assign CM the job of characterizing the dynamical properties of bodies. This is of course a makeshift, for CM is inconsistent with CEM. A more rigorous treatment of classical electrodynamics must include fragments of relativistic continuum mechanics from the start. Hence our treatment will be exact for the pure e.m. field only.

1.1. Microelectromagnetism

1.1.1. Building Blocks. The primitive base of microelectromagnetism is this:

M^3 Manifold representing physical space (elements: $x \in M^3$)

T: Segment of the real line, representing duration

Φ: The set of electromagnetic fields φ

Σ: The set of charged bodies σ

$\mathscr{B}$: The set of manifolds representing charged bodies $\mathscr{b}$

K: The set of physical (reference) frames k

$\{X\}$: Set of vector valued functions representing the position of a particle of a charged body

$\{\varrho\}$: Set of scalar functions representing the charge density

$\{E\}$: Set of vector fields representing the electric field intensity

$\{B\}$: Set of pseudovector fields representing the magnetic field intensity

$\{f\}$: Set of vector fields representing the force density

c: A positive real number

Remark. CEM is a dualistic theory: it has two disjoint reference classes, Φ and Σ — or even three if K is counted. Furthermore the field concepts do not mix with the matter concepts — a situation that will change in macroelectromagnetism, where the inductions refer to fields within material media.

1.1.2. Defined Concepts. With the help of the foregoing primitives we introduce

Df. 1. *Total charge* on a body $\sigma \in \Sigma$:

$$\beta \in \mathscr{b} \wedge \mathscr{b} \triangleq \sigma \Rightarrow Q(\sigma, t) \overset{\text{df}}{=} \int_{\partial X(\mathscr{b}, t)} \mathrm{d}^3 x \varrho(\beta, x, t).$$

Df. 2. *Current density:*

$$\beta \in \mathscr{b} \wedge \pi \overset{\cdot}{\in} \sigma \wedge \beta \triangleq \pi \Rightarrow j(\pi, t) \overset{\text{df}}{=} \varrho(\beta, x, t) \frac{\mathrm{d}X(\beta, t)}{\mathrm{d}t}.$$

Df. 3. *Current* across a surface s in σ:

$$\partial X \triangleq s \wedge \mathcal{S} \triangleq \sigma \Rightarrow I(\sigma, t) \overset{\mathrm{df}}{=} \int_{X(\mathcal{S},\,t)} \mathrm{d}^2 x j(x, t).$$

Df. 4. *Electric flux* of φ across a surface s in σ with outer normal n:

$$\partial X(\mathcal{S}) \triangleq s \Rightarrow \varphi_e(\varphi) \overset{\mathrm{df}}{=} \oint_{\partial X(\mathcal{S},\,t)} \mathrm{d}^2 x E \cdot n.$$

Df. 5. *Magnetic flux* of φ across a surface s with outer normal n:

$$\partial X(\mathcal{S}) \triangleq s \Rightarrow \varphi_m(\varphi) \overset{\mathrm{df}}{=} \oint_{\partial X(\mathcal{S},\,t)} \mathrm{d}^2 x B \cdot n.$$

Df. 6. *Poynting vector* of φ:

$$S(\varphi) \overset{\mathrm{df}}{=} (c/4\pi)(E \times B).$$

Df. 7. *Energy density* of φ:

$$U(\varphi) \overset{\mathrm{df}}{=} (1/8\pi)(E^2 + B^2).$$

Df. 8. Field *momentum-energy-stress* tensor

$$T^{ij}(\varphi) \overset{\mathrm{df}}{=} \delta^{ij} U - (1/4\pi)(E^i E^j + B^i B^j).$$

Remarks. (**1**) As usual in physics, these definitions are neither capricious nor logically justified but they are physically motivated. That is, they are in no way necessary and they are conventional, but they are convenient because the defined concepts occur as units in several important law statements, they are assigned definite referents, and they are sometimes measurable — e.g., total charges are measurable, not so charge densities. (**2**) The total charge is here defined. Sometimes the attempt is made to parallel MACH's unsuccessful elimination of the mass concept, namely by means of the following pseudodefinition: If a test body 1, when at a point x, is acted on by a force $F_1(x)$ and another test body 2, when placed at the same point x instead of the first body, is acted on by a force $F_2(x)$, then the charge Q_{12} of 1 relative to 2 equals, by definition, the ratio of the forces: $Q_{12} = F_1(x)/F_2(x)$. The defects of this procedure are obvious: (*a*) the statement presupposes the concept of electric charge as occurring in the formula "$F = QE$", actually employed in it; (*b*) it cannot be extended to the case in which a single charged body is studied — which case is the generator of elementary electrostatics; (*c*) like every other "operational definition" it yields some values of a function not the function itself; (*d*) it obscures the fact that the concept of electric charge is newly introduced by CEM, not constructible in mechanical terms.

1.1.3. Axioms. We assign protophysics the task of characterizing T and K, and CM the task of characterizing X and f. Let us then proceed to determine the structure and meaning of the remaining primitives.

Axiom group 1: Characterization of specific primitives

CEM 1.1 (a) Φ is a nonempty set. (b) Every $\varphi \in \Phi$ is an e.m. field.

CEM 1.2 (a) Σ is a nonempty set. (b) Every $\sigma \in \Sigma$ is a charged body.

CEM 1.3 (a) $\{\mathscr{B}\}$ is a nonempty family of point sets. (b) Every $\mathscr{B} \in \mathscr{B}$ is a 3-dimensional differentiable manifold. (c) For every $\sigma \in \Sigma$ there exists a $\mathscr{B} \in \mathscr{B}$ such that $\mathscr{B} \triangleq \sigma$ pointwise.

CEM 1.4 (a) M^3 is a threedimensional differentiable manifold. (b) M^3 represents physical space.

CEM 1.5 (a) $\{\varrho\}$ is a nonempty family of functions. (b) Every $\varrho \in \{\varrho\}$ is a scalar valued function from $\mathscr{B} \times M^3 \times T$ to R. (c) If $\mathscr{B} \triangleq \sigma$, then $\varrho(\mathscr{B}, x, t)$ represents the charge density of σ at x, t.

CEM 1.6 (a) $\{E\}$ is a nonempty family of fields. (b) Every $E \in \{E\}$ is a real and vector valued function on $\Phi \times M^3 \times T$. (c) $E(\varphi, x, t)$ represents the intensity of the electric component of φ at x, t.

CEM 1.7 (a) $\{B\}$ is a nonempty family of fields. (b) Every $B \in \{B\}$ is a real pseudovector valued function on $\Phi \times M^3 \times T$. (c) $B(\varphi, x, t)$ represents the intensity of the magnetic component of φ at x, t.

CEM 1.8 E and B drop off with distance at least as fast as $1/r$.

CEM 1.9 c is a positive real number such that $[c] = LT^{-1}$.

Remarks. (1) No space structure has been assumed. Nor is it necessary to introduce the field coordinates x_i as separate primitives, for they come with M^3. In fact by definition of "differentiable manifold" M^3 is locally Euclidean and therefore any neighborhood U of M^3 is provided with coordinate functions $\{u_i\}$, i.e. functions such that, if $x \in U$, then $u_i(x) = x_i$. This does not hold for the particle coordinates X_i, which are different functions (with domain $\Sigma \times K \times T$). (2) From an operationalist point of view all the preceding axioms are meaningless, and particularly so the ones referring to the field: indeed, fields are unobservable. A strict operationalist will (a) reject field variables altogether or (b) regard them as auxiliary devices (intervening variables) with no physical meaning or, finally, (c) try to eliminate the reference class Φ, having Σ absorb as it were all the field effects — i.e., operate with particle coordinates alone (see, e.g., WHEELER and FEYNMAN, 1949). (3) The semantical components of the preceding axioms evoke rather than fully determine the physical meaning of the corresponding symbols. This cannot be helped: meanings grow and become more precise together with the growth of the theory (see Ch. 1, 1.3.7 and 4.2.6). (4) The preceding axioms are the minimum assumptions necessary to write down the central axioms of CEM and to assign them a seed of meaning that is to germinate in the process of deriving theorems and applying them to real situations. The central axioms are the following.

11*

Axiom group 2: Field equations

CEM 2. At any point $x \in M^3$ and any instant $t \in T$ for every field $\varphi \in \Phi$ and every charged body $\sigma \in \Sigma$, there exists a $k \in K$ such that

$$V \times E = -\frac{1}{c} \frac{\partial B}{\partial t}, \quad V \cdot B = 0, \tag{4.1}$$

$$V \times B = \frac{1}{c} \frac{\partial E}{\partial t} + \frac{4\pi}{c} j, \quad V \cdot E = 4\pi \varrho \tag{4.2}$$

Df. 9. A frame $k \in K$ such that the axioms *CEM* 1 to 2 above hold relative to it is called an *inertial frame*. Notation: $\iota \in I \subset K$.

Remarks. (1) These eight partial differential equations — MAXWELL's two sets — constitute the kernel of CEM. The two triplets specify the evolution of the e.m. field while the divergence equations may be regarded as constraints. Taken jointly, in view of the first axiom group the field equations describe the e.m. field and the charges and currents associated with it. (2) In special problems ϱ and j are usually assigned (hypothesized), wherefore the belief that they are the field sources; but nothing of the sort is said by the preceding equations, so much so that the field need not vanish as a consequence of the nonexistence of "sources" in a region. (3) It is often stated that the last equation defines ϱ in terms of E. But this is a law of nature not a convention. Moreover ϱ and E are logically independent (not interdefinable) though mathematically related. In fact, given ϱ, div E but not E itself remains uniquely determined; in particular, $\varrho = 0$ does not entail the vanishing of the electric component of the field. Moreover, that equation remains invariant under the change $E \rightarrow E + V \times A$ with A a vector subject to the sole condition $V^2 A = 0$. In short, ϱ and E are logically independent (not interdefinable) though mathematically related (in a one-many way). (4) The field equations are not easy to interpret; their interpretation is best made upon rewriting them in integral form (see 1.2.1). Even so, at the level of axioms the interpretation is always incomplete. (5) If the field values are referred to a noninertial frame moving with velocity u with respect to an inertial frame, then the term $V \times (B \times u/c)$ must be added to the right handside of MAXWELL's first triplet, and $V \times (E \times u/c) + (V \cdot E) \frac{u}{c}$ to the right handside of the second triplet. (6) Notice that CEM employs its own notion of inertial frame; it will be adopted by special relativity.

Axiom 3: Law of force

CEM 3. The force density exerted by an external field $\langle E, B \rangle$ $\triangleq \varphi \in \Phi$ on $\sigma \in \Sigma$ is

$$f = \varrho \left(E + \frac{v}{c} \times B \right) \tag{4.3}$$

Remarks. (1) Without this axiom CEM would be untestable, because the fields show through their ponderomotive effects alone. This does not mean that *CEM* 3 supplies the meaning of CEM — e.g., that E is just the force per unit charge. Surely for $B = 0$ the value of E equals the value of the force on a test body with unit charge as measured by an observer attached to the test body. But E is not a specific force and it is assumed to have a meaning and a value even when there are neither observers nor test particles around. In fact the value of E does not depend on the body upon which E may happen to act, although the corresponding ponderomotive force does so depend. The difference between E (or B) and a force function is clear from their respective structures: while the domain of a field strength is $\Phi \times M^3 \times T$, the domain of a ponderomotive force is $\Sigma \times \Phi \times M^3 \times T$. The operationalist misinterpretation of CEM involves therefore a confusion between a function and its range — or rather a sample of it. In short, CEM is chiefly about e.m. fields not about the bodies employed in exploring them. (2) Axiom *CEM* 3 is the Achilles heel of CEM: even if completed with the addition of div $\mathfrak{T}$, where $\mathfrak{T}$ is the mechanical stress tensor, the equation holds only for external fields and electric monopoles. It does not take the reaction of the proper field on the body into account, and it does not hold for a spinning body. If the spin S of the charged body is included, *CEM* 3 must be modified and a postulate referring to the spin precession must be added — e.g., $dS/dt = (Q/mc)(S \times B)$.

The preceding axiom system characterizes all the specific primitives of CEM (it is p-complete) except for c, which so far remains uninterpreted. And it suffices to derive all the known formulas of the theory, both the true and the false ones: it is also d-complete. The 2nd half of (4.1) disowns the hypothesis of magnetic monopoles — which was originally introduced (a) to build magnetostatics in analogy with electrostatics, and (b) to manufacture an operational pseudodefinition of the magnetic field intensity (see however Ch. 1, 5.1.2). Nor do tubes of force occur in our system: this too used to discharge a heuristic function in MAXWELL's time and, for being a mechanical concept, it helped promote the unmechanical concept of e.m. field; but by the same token it retarded the realization that this field is a thoroughly nonmechanical substance whose existence must be hypothesized without any mechanical supports. As to the Coulomb field, which occupies such a prominent role in the elementary expositions of CEM, it need not be mentioned in the foundations of CEM, as it is just a very special solution of the second half of (4.2). It will also be noticed that the Gaussian system of units has been employed throughout without apology to the engineer; the alternative systems make no sense in microphysics, which is after all the basis of all physics (see Ch. 1, 2.5.3). Specific references to condensers, cables and meters are missing from the previous axiomatic foundations. They can occur only if the suitable special hypotheses (theoretical models of experimental arrangements) are introduced at the application stage. Finally it would seem that our axiom system allows for an axiomatic definition of the concept of e.m. field, namely thus: The pair $\langle E, B \rangle$ is an e.m. field iff it satisfies nontrivially the axioms $1-3$ above. But this procedure is dangerous: we know that e.m. fields do not satisfy MAXWELL's equations in the small. The concept of e.m. field must be left undefined; like every other key concept of physics, it is variously characterized by a number of theories, none of which is perfect. The best we can do is to hope that, for every $\varphi \in \Phi$ and every $\iota \in I$, there exists a pair $\langle E, B \rangle$ satisfying MAXWELL's theory and such that $\langle E, B \rangle \triangleq \varphi$ in ι.

1.2. Alternative Formulations of Microelectromagnetism

The same theory can be formulated in several alternative ways, every one of which has its peculiar advantages and shortcomings. We shall sketch a few such alternative formulations.

1.2.1. Integral Formulation. This presentation retains axiom groups 1 and 3 and replaces axiom group 2 by

CEM 2'. In any region $V \subseteq M^3$, at any time $t \in T$, relative to any frame $\iota \in I$, for every field $\varphi \in \Phi$ and every body $\sigma \in \Sigma$, if ∂V is the boundary of V, n its outer normal and C a closed contour on ∂V, then

$(a) \quad \oint_C E \, ds = -\frac{1}{c} \frac{d}{dt} \oint_{\partial V} d^2 x \, (B \cdot n),$ \hfill (4.4)

$(b) \quad \oint_{\partial V} d^2 x \, (B \cdot n) = 0,$ \hfill (4.5)

$(c) \quad \oint_C B \, ds = \frac{1}{c} \frac{d}{dt} \oint_{\partial V} d^2 x \, (E \cdot n) + \frac{4\pi}{c} \int_{\partial V} d^2 x \, (j \cdot n),$ \hfill (4.6)

$(d) \quad \oint_{\partial V} d^2 x \, (E \cdot n) = 4\pi Q.$ \hfill (4.7)

The intuitive concepts of electric and magnetic flux (Dfs. 4 and 5) and of total current (Df. 3) can be used with heuristic advantage, as they suggest fluids traversing the diaphragm ∂V. Furthermore the line integrals occurring above can be taken in block as defining the electromotive and the magnetomotive "forces" — in analogy with mechanics. The familiar elementary versions of (a) and (c) are then obtained:

$$e.m.f. = -\frac{1}{c} \frac{d\varphi_m}{dt}, \quad m.m.f. = \frac{1}{c} \frac{d\varphi_e}{dt} + \frac{4\pi}{c} I. \tag{4.8}$$

The differential formulation of 1.1 can be obtained from the preceding integral formulation if suitable continuity and differentiability assumptions are added. Which shows that the integral formulation is less demanding (more general) than the differential one: the latter holds only within smooth regions. In addition to being logically stronger than the differential formulation, the integral one is important for semantical and methodological reasons. Indeed, it contributes to determining the meaning of the primitives and it is susceptible to fairly general and direct tests, whereas the differential field equations must be integrated in special cases before they can be put to the test. And such an integration requires the choice of a special metric.

1.2.2. Second Order Formulation. The preceding formulations are not quite suitable for the detailed working out of the theory: they are much too rigid and they are not suitable for making contact with the lagrangian and hamiltonian formalisms, which employ potentials not field strengths. To either end the most adequate formulation of CEM is in terms of the 4-potential A_0, A. These field variables are introduced as new primitives, with a scanty physical meaning, on top of the preceding primitives. The new axiom basis is the following.

Axiom group 1''

Axioms CEM 1.1 to 1.5 are kept.

CEM 1.6'' $\langle A_0, A \rangle$ is a real valued 4-vector field on $\Phi \times M^3 \times T$.

CEM 1.7'' A_0 and A approach constant values at spatial infinity.

CEM 1.8'' Gauge condition:

$$\partial_\mu A_\mu = 0, \quad \mu = 0, 1, 2, 3, \quad x_0 \equiv ct, \quad \partial_\mu \equiv \frac{\partial}{\partial x^\mu}. \tag{4.9}$$

CEM 1.9'' (a) $E = -\nabla A_0 - \frac{1}{c}\frac{\partial A}{\partial t}$. (b) $B = \nabla \times A$.

CEM 1.10'' (a) $E(\varphi, x, t)$ represents the intensity of the electric component of $\varphi \in \Phi$ at $x \in M^3$, $t \in T$. (b) $B(\varphi, x, t)$ represents the intensity of the magnetic component of $\varphi \in \Phi$ at $x \in M^3$, $t \in T$.

CEM 1.11'' c is a positive real number.

Axiom group 2''

CEM 2''. At any given point $x \in M^3$ and any instant $t \in T$, relative to any inertial frame $\iota \in I$, for every field $\varphi \in \Phi$ and every charged body $\sigma \in \Sigma$,

$$\Box A_\mu = \frac{4\pi}{c}\, j_\mu \quad \text{with} \quad \Box \stackrel{\mathrm{df}}{=} \frac{1}{c^2}\frac{\partial^2}{\partial t^2} - \nabla^2,$$

$$j_\mu = \varrho\,\frac{dX_\mu}{dt}, \quad \mu = 0, 1, 2, 3 \tag{4.10}$$

Axiom 3 is retained.

A typical trait of this formulation is

Thm. 1. If Λ is an arbitrary scalar field, then two solutions A_μ and A'_μ of the field equations, related by the gauge transformation of the 2nd kind

$$A'_0 = A_0 + \frac{\partial \Lambda}{\partial x^0}, \qquad A' = A - \nabla \Lambda \tag{4.11}$$

represent (refer to, describe) the same e.m. field $\varphi \in \Phi$. *Proof:* Introduce (4.11) into axiom *CEM* 1.9″ and recall that, by the next axiom, an e.m. field is characterized by $\langle E, B \rangle$.

Remarks. (**1**) Our axioms assign no physical meaning to the A_μ: these are source functions from which the field components derive. Hence the potentials may be said to have (through axiom *CEM* 1.9″) a *derivative* or indirect physical meaning (see Ch. 1, 1.3.8). As a consequence of their lack of direct physical meaning, the A_μ are not measurable. One can certainly read the generalization (4.10) of Poisson's equation as stating the proportionality of charge and current concentration, on the one hand, and the spatiotemporal bump or local deviation from the field average, but since $\langle A_0, A \rangle$ is a mathematical field, not much is gained. (**2**) Since the e.m. potentials are not directly meaningful in this theory, the gauge condition (4.9) is not a physical law but a purely mathematical restriction on the otherwise quite free potentials: it is a necessary condition for rewritting Maxwell's equations as second order differential equations. This is why it has been placed in Axiom group 1, although in actual work it will always be kept beside the field equations. In alternative theories, particularly in hamiltonian electrodynamics, $(e/c) A_\mu$ is the additional four-momentum imparted by a field to a charged particle. (**3**) At first sight it would seem that the formulas *CEM* 1.9″ relating the fields to the potentials are definitions. But Thm. 1 proves that this is not the case, for one and the same field (physical entity) may be described by infinitely many potential functions. In other words, Thm. 1 proves that A_μ, E and B are logically independent, i.e. not interdefinable, and justifies thereby our introducing the A_μ as new primitives in addition to E and B. If the primitive base were shrunk by dropping E and B, a physically void structure would remain. The genuine field variables E and B must be kept as primitives beside the A_μ in order to get a physical theory even though, given the A_μ, the e.m. field components remain uniquely determined. (**4**) The relations *CEM* 1.9″ between the intervening variables A_μ and the hypothetical constructs E, B are gauge invariant, i.e. they are preserved under the transformations (4.11). On the other hand the gauge condition *CEM* 1.8″ is not gauge invariant and need not be so since it is not a physical law. Consequently the gauge transformations have no physical meaning: they show that the choice of potentials is, just as the choice of coordinates, conventional within bounds. (5) The limits to this arbitrariness are set by the requirement that the field equations (4.10) be invariant under the gauge transformations. And this is the case on condition that the latter are subjected to the restriction $\partial_\mu \Box \Lambda = 0$. If this usually forgotten restriction is dropped, the 4-current is not invariant under (4.11). (6) By virtue of the gauge invariance of the fields, it is always possible to choose a time-independent scalar potential (Coulomb gauge). Moreover, it is possible to do with a single (super)potential Π (HERTZ' vector) such that $A_0 = -\mathbf{V} \cdot \Pi$ and $A = -\dfrac{1}{c}\dot{\Pi}$. Several other superpotentials can be introduced (PHILLIPS, 1962; POST, 1962). (7) The present formulation, being one more step removed from experience, allows one to work in a manner typical of contemporary theoretical physics: in addition to starting off with simple known fields such as those produced by an electromagnet of variously shaped polar pieces, and trying to describe them mathematically, one may also proceed inversely, i.e. toying with potentials satisfying the field equations and the gauge condition, and finally looking around in search for possible real referents. (8) The mathematical advantages of this formulation over MAXWELL's is obvious: given (i.e., hypothesizing) the 4-current, an infinite set of potentials remains determined by only four, and often three, rather than eight field equations — which refutes the claim that a simplification of the axiom basis is always attained by shrinking the primitive base. But what is gained in computational ease is lost in uniqueness of reference and in intuitability. (9) In addition to mathematical convenience (elasticity), the formulation of CEM with the help of the 4-potential has three advantages: (a) it makes the use of hamiltonian formalisms possible, upon replacing the mechanical 4-momentum p_μ by the kinetic 4-momentum $\pi_\mu \overset{\mathrm{df}}{=} p_\mu - \dfrac{e}{c} A_\mu$; (b) in quantum mechanics the e.m. potentials acquire a physical meaning (AHARONOV and BOHM, 1959, 1961); (c) on looking at (4.10) for $j_\mu = 0$ one notices its similarity with the equation for elastic waves and one conjectures (but does not conclude) that there are e.m. waves that propagate in vacuum with velocity c. (This hypothesis depends then, heuristically, on the pre-existing false elastic theory of light.) This endows c with a physical meaning.

1.2.3. Covariant Formulation. A fourth useful formulation is obtained by synthesizing E and B into a single tensor field F. It is further assumed that the components of F are related to E and B by

$$E_i = F_{0i}, \qquad B_i = \varepsilon_{ijk} F^{jk} \tag{4.12}$$

where ε_{ijk} is the Ricci symbol, and to the 4-potential by

$$F_{\mu\nu} = \partial_\nu A_\mu - \partial_\mu A_\nu. \tag{4.13}$$

These relations may but need not be regarded as definitions of F. It is more natural, though perhaps undidactic, to postulate that the e.m. field is represented by a tensor field, particularly since a Galilei transformation, and a fortiori a Lorentz transformation, show that the division of F into an electric and a magnetic component is frame-dependent. If F is accordingly handled as a primitive concept, then (4.12) and (4.13) are not definitions but bridges linking the present formulation to the more familiar ones; moreover, the (4.13) become redundant as the A_μ need not be introduced except to use results gotten in the A_μ-formulation. When this is done MAXWELL's equations collapse automatically to two sets that are *manifestly Lorentz covariant:* If $F \triangleq \varphi$, then

$$\partial_{[\lambda} F_{\mu\nu]} \equiv \frac{1}{3!} [\partial_\lambda F_{\mu\nu} + \partial_\mu F_{\nu\lambda} + \partial_\nu F_{\lambda\mu}] = 0, \qquad \partial_\nu F^{\mu\nu} = \frac{4\pi}{c} j^\mu. \tag{4.14}$$

If now M^3 is endowed with a metric or just an affinity, the ordinary differential operator ∂_λ can be replaced by the covariant V_λ, and MAXWELL's equations acquire a *general covariant form:*

$$V_{[\lambda} F_{\mu\nu]} = 0, \qquad V_\nu F^{\mu\nu} = \frac{4\pi}{c} j^\mu. \tag{4.15}$$

This form is particularly convenient to investigate the relations between the e.m. and the gravitational field, but it need not be related to general relativity: the latter does not make its entrance unless there is a gravitational field. The fact that CEM can be formulated in a generally covariant way — and even must be so formulated (POST, 1967) — shows once again that it does not concern observers. It is only for solving the basic equations and testing them that special kinds of coordinate systems and particular reference frames will be chosen — but even then the frames need not carry observers. To introduce observers and measurement operations in CEM at the foundations level is to ignore the natural covariance of the theory.

Finally, an action principle involving F and A_μ can be set up which leads to MAXWELL's equations. The main advantages of the variational formulation of CEM are: (a) the consistency of the field equations is automatically warranted; (b) conservation theorems are obtained methodically rather than by chance or insight; (c) the connection with dynamics and other theories is facilitated; and (d) the canonically conjugate field variables are spotted and consequently the second quantization of the theory is made possible. Apart from this, the lagrangian and hamiltonian formulations of CEM throw no light on its physical meaning.

1.2.4. Subtheories. Once the comprehensive foundations of CEM are laid down they can be worked out in any particular direction, by focusing on certain postulates and adding special assumptions. Among others two subtheories result in this fashion. One is electrodynamics, which focuses on the dynamical aspects of the interactions between charges and currents (see 1.4). Another conspicuous subtheory is optics, which is concerned with the propagation of e.m. waves. For the special case of empty space, this theory can be derived, e.g., from (4.14) by making the following simplifications: (*a*) set the 4-current equal to zero; (*b*) set $F = Re(Re^{i\psi})$, and assume that the field amplitudes $R_{\mu\nu}$ are slowly varying functions of x and t, while the phase ψ is a rapidly fluctuating function of the same arguments. Optics provides a clear illustration of theory reduction, i.e. of the deduction of a theory from another to which certain subsidiary assumptions have been adjoined. This does not mean that the whole physics of light has become a chapter of CEM. No theory covers all the facts in a given domain, much less the experimental facts, which require for their explanation a number of different theories. Theories do not cover all the sides of their own experimental tests (see Ch. 1, 5).

1.3. Some Typical Theorems

1.3.1. Charge and Force. The main theorems of a field theory can be classed into three categories: theorems concerning symmetry properties (in particular conservation statements), solutions of the equations of motion, and solutions of the field equations. Let us take a hurried look at each class.

The field equations of CEM entail the so-called continuity equation $\partial_\mu j^\mu = 0$, a name borrowed from CM. This name is misleading in the new context both because the electric charge is not a substance but a state of certain bodies, and because the theorem holds even if the charge is quantized, i.e. if $Q = ne$ with $n \in N$ and $e \in R^+$. At any rate the theorem can hardly be interpreted as it stands: in order to find its meaning apart from mechanical analogies one must restrict it to a system contained in a volume V across whose boundary, ∂V, j vanishes. For in this case, by GAUSS' theorem the total charge is conserved:

$$\frac{\partial}{\partial t} \int_V d^3x \varrho = 0 \vdash Q = \text{const.} \tag{4.16}$$

This is not a principle but a theorem of CEM, but so important a statement that at the present time we would pay no attention to a theory that did not respect (4.16), one of the best confirmed of all physics. And it is one more example of a law statement that must be subjected to mathematical transformations before it can be interpreted. The newly discovered conservation theorems (LIPKIN, 1964; MORGAN, 1964) are still waiting for an adequate interpretation.

Concerning the force law, only two points will be mentioned. One is that the field occurring in it is the external (incident) field not the total field (self plus external). In other words, the law of motion holds for fictitious test bodies. The introduction of the self field originates difficult problems about whose solution there is so far no consensus — an interesting example of a classical theory which, far from being dead, still poses challenging problems. Another interesting point is that, when conjoined with the relativistic law of motion, the force law *CEM* 3 entails trajectories that have been confirmed to an astonishing accuracy, particularly with high energy accelerators. This is surprising in view of their acknowledged incompleteness and of the current belief that quantum mechanics alone can account for the behavior of elementary particles.

1.3.2 Field solutions. As to the solutions of the field equations — typical of a field theory — let us remark, firstly, that few exact solutions are known, which shows that CEM is far from being a highly developed theory. Thus exact solutions for simple antennas, joining smoothly to free space solutions, were not obtained until recently (SCHELKUNOFF, 1941), and there are no exact solutions to seemingly simple problems such as the diffraction of e.m. waves by a finite slit of arbitrary width (BOUWKAMP, 1954). Secondly, the task of finding such solutions is considerably simplified if either there are no charged bodies in the region under study or they are schematized as discrete collections of identical point charges — though not necessarily as point masses. (Such a separation is indeed possible in the relativistic theory of electrons: see BUNGE, 1955a.) For a collection of point charges, the current, an intensive magnitude, can be treated as an extensive magnitude thanks to DIRAC's delta:

$$\sigma_i \stackrel{.}{\in} \sigma^N \Rightarrow j^\mu(\sigma^N) = \Sigma j^\mu(\sigma_i),$$

with
$$j^\mu(\sigma_i) = \varrho(\sigma_i)\, dX^\mu(\sigma_i)/dt = e\,\delta\,[x - X(\sigma_i,t)] \cdot \frac{dX^\mu(\sigma_i)}{dt}. \qquad (4.17)$$

That this hypothesis is absurd, since it entails that the current diverges along the particle trajectories, is clear; yet it yields good approximations.

We finally come to the central equations (4.10), interpreted as wave equations from an analogy with elastic waves. The most important solution of these equations and the gauge condition (4.9) is

$$A_\mu(x,t) = k_\mu^- A_\mu^-(x,t) + k_\mu^+ A_\mu^+(x,t) \qquad (4.18)$$

(no summation) with

$$A_\mu^\pm (x, t) = \int \frac{d^3 x' j_\mu (x', t_\pm)}{|x' - x|} , \qquad t_\pm \overset{\text{df}}{=} t_-^+ \frac{|x' - x|}{c} . \tag{4.19}$$

Consequently the field tensor becomes: $F(x, t) = k^- F^- + k^+ F^+$. Usually the advanced field F^+ is discarded by setting $k^+ = 0$, on the ground that F^+ violates the principle of antecedence (misnamed causality). But in certain theories the whole F is employed. In particular, the combination $k^+ = k^- = 1/2$ is sometimes (e.g., SCHÖNBERG, 1946) regarded as representing the field attached to the body. Interestingly enough, the corresponding self-energy is finite, whereas the self-energy calculated with the retarded field diverges.

1.3.3. Field Sources? From a mathematical point of view the elimination of the advanced fields is arbitrary: if one accepts MAXWELL's second order equations then one must put up with the advanced solutions; otherwise one should abstain from introducing potentials altogether. In order to keep MAXWELL's second order equations and at the same time discard its advanced solutions in a consistent way one must add the hypothesis that the charged bodies are the *sources* of the e.m. field — a hypothesis that is taken so much for granted that it is hardly stated explicitly. For in such a case it is clear, on the principle of antecedence (see Ch. 2, 1), that the future vicissitudes of the body cannot influence the present state of the surrounding field. But the source hypothesis is an extra assumption: as our axiom systems show, fields and currents are conjoined but not causally associated: only field changes are causally associated with charged bodies in case there are any in the region considered.

The persistent belief that MAXWELL's theory does involve the source hypothesis seems to have several sources. One is that the Coulomb field, which is a very particular solution of the field equations, is too often taken as a paradigm. Another is that, in order to produce fields, we have to manipulate charged bodies — but then we have to do the same in order to detect "source"-free fields. A third source of the belief is focusing on equation (4.19), which says "No 4-current, no 4-potential — hence no field". But it should not be forgotten that this is not the most general solution of the field equations and cannot therefore be regarded as summarizing MAXWELL's theory: indeed, if a general solution of the homogeneous ("source"-free) equations is added to (4.18), the inhomogenous equations are satisfied. In order for the field to vanish everywhere and always it is not enough that the current vanishes, as shown by the case of e.m. waves. The necessary and sufficient condition for the vanishing of F everywhere and always is that it vanishes both asymptotically and initially: if it vanished only initially there still could exist an incoming field, and if it vanished only asymptotically there still could be outgoing waves. There are similar situations in all field theories, particularly in EINSTEIN's GR. Both in CEM and in GR the "source"-free field equations have infinitely many nontrivial solutions, in particular waves. In short, the source hypothesis is an extra assumption that may but need not be made — not any more than its converse, the sink hypothesis. Consequently there is no reason to discard the "advanced" F's, particularly as they need not be interpreted as advanced fields.

For example, $j(x', t_+)$ might be interpreted as the current at x', t_+ associated with the field F^+; i.e., F^+ might be regarded as the part of the field that reacts on the current and influences its value at a later instant. Consequently the total solution (4.18) could be interpreted as describing both the action of the currents on the field and the latter's reaction on the former. This interpretation does not violate the principle of antecedence. Whether or not this interpretation holds water is here immaterial: the point is that some of the difficulties of CEM are of a semantic nature and some of them are caused by tacit and controvertible assumptions such as that the charges produce fields but not conversely — a hypothesis that is clearly absent from the axiom basis of CEM. This particular problem does not arise in macroelectromagnetism, where the 4-current is prescribed (hypothesized). But as a matter of principle one might as well prescribe the fields and compute the corresponding 4-currents. And in the case of the theories of elementary particles, whether classical or quantal, neither can be prescribed.

This closes our sketchy discussion of microelectromagnetism.

1.4. Classical Electrodynamics

In order to study the effects of e.m. fields on the motion of bodies and conversely, CEM must be adjoined a dynamical theory — not just force hypotheses (e.g., *CEM* 3) but laws of motion. The resulting theory will be some member of the set of theories known as classical electrodynamics — CED for short. The construction of a theory of CED requires making some assumptions about the constitution of bodies — e.g., that they consist of charged particles, or that they are continuous fluids. For then we shall be able to adopt a dynamical theory — PM, CM, or any of its subtheories. The next step is to effectively bridge the two theories, i.e. to plug the force law *CEM* 3 or some generalization of it into a set of laws of motion. This requires changing *CEM* 3 so as to cover both the external field acting on the body, the body self-field, and the contact forces deriving from its mechanical stress. The simplest assumption is of course that $CEM\ 3$ holds for the total field — but this may be wrong. In any case the construction of CED requires some change in CEM: it is not just a question of adding CM to CEM. And CM itself should be changed into a theory compatible with MAXWELL's equations — i.e. into relativistic continuum mechanics (RCM). After a somewhat different electromagnetic theory CEM' has been formulated, it can be smoothly joined to, say, RCM to constitute CED = CEM' U RCM, which will encompass the electrodynamics of particles, magnetohydrodynamics, etc. If on the other hand quantum mechanics is adopted as the dynamical theory, then the elementary q.m. theory of radiation is obtained, in which the field is not quantized.

All of the above theories have been highly successful except in accounting for the very existence of charged bodies. Granted, a macrobody is made up of smaller parts; but what model of these units shall we make? If it is to be a classical model then either it has to be a point particle or it must have some definite shape, though not necessarily a constant one. If the former model is adopted, then all questions about particle structure are dodged, but the field at the particle becomes infinite and consequently the particle self-energy diverges. If the particle is assigned a definite shape then the electrostatic repulsion among its parts makes it explode (on paper): its self-stress diverges unless *ad hoc* adhesive forces are added. The former model leaves us with no fields to speak of, the latter with no particles. If these fundamental difficulties are ignored, CED can still work miracles in predicting particle trajectories — e.g., for cosmic ray research and for the construction of accelerators — and in explaining the emission and absorption of light and many electromagnetic and optical properties of bulk matter. Which is almost understandable: real electrons do not worry about the infinities associated with our wrong theoretical models, and the theory gets into no trouble as long as it asks no troublesome questions and as long as we make no use of the bad parts of the egg. Nor do the experimental physicist and the engineer who use CED as a tool need to worry.

But the thoughtful theoretical physicist does worry about the state of CED: even if he pays lip servise to a pragmatist philosophy, he does not live by it. He wants to understand what makes the existence of charged particles possible despite CED — or, if he postulates their existence, then he wants to build an adequate theory of CED. And he wants to solve these problems at the classical level because a quantum theory patterned after a classical one will inherit most of its absurdities. Two paths have been tried to get out of the *impasse:* one is to tamper with the mathematics, the other is to introduce new physical hypotheses. The first technique consists in keeping all the basic assumptions of CED, correcting its wrong logical consequences by subtracting all undesirable quantities. This procedure can be carried out in a very sophisticated way, especially when the more or less arbitrary cut-offs and renormalizations are done covariantly. But since such manipulations have no physical correlate, they give rise to no physical theories proper: subtraction physics is actually mathematical patchwork and symptom-healing. It can acquire as much predictive power as PTOLEMY's game of epicycles and excentrics but it gives no representation of physical reality and should therefore be regarded as a makeshift.

The second approach consists in making all the necessary assumptions at the start and putting up with their logical consequences. It is a harder way and there is no reason to suppose that it can succeed. So far, the most successful attempt in this direction is DIRAC's classical relativistic electron theory (DIRAC, 1938; SCHÖNBERG, 1946; ROHRLICH, 1965). This theory of pointlike (structureless) particles avoids infinities by taking suitable linear combinations of the retarded and advanced fields (see 1.3.2 and 1.3.3); and it avoids self-accelerating (runaway) solutions by adding plausible asymptotic conditions on the solutions. Unlike most subtraction theories, this one is logically consistent and mathematically clean; and, in SCHÖNBERG's version, it is also semantically fairly complete, for '$\frac{1}{2}(F^- + F^+)$' is interpreted as the field attached to the particle and '$\frac{1}{2}(F^- - F^+)$' as the field emitted by it. But in none of its formulations it is physically satisfactory because, as a result of the occurrence of advanced potentials with their usual interpretation, the particle suffers a preacceleration, i.e. it is accelerated before the external field arrives at it. In other words, according to the theory the present dynamical state of the particle is determined in part by the future states of its environment and even of itself. But this violates the antecedence principle, often miscalled causality — a principle that is respected in the rest of physics, including the quantum theories of "elementary" "particles". Moreover these violations are experimentally undetectable, which saves the theory from experimental criticism and by the same token makes it violate the principle of testability. The impossibility of measuring the hypothetical and magical preacceleration explains why the theory works in other respects but is no excuse for remaining satisfied with it: scandals remain scandals even if no one notices them.

Different theories of CED are called for, both for charged particles that can to a first approximation be regarded as electromagnetically structureless (e.g., the electron and the μ-meson) and for extended systems with a charge distribution and spin (e.g., the proton). If fairly satisfactory classical theories were available they could guide the construction of the corresponding quantum theories. But no such theories are in sight — partly because it is not usually acknowledged that electrodynamics, both classical and quantal, are in a sad state.

1.5. Phenomenological Macroelectromagnetism

There are two possible approaches to macroelectromagnetism: a direct or macrophysical and a microphysical one. Both deal with field-body pairs: their referent is $\Phi \times \Sigma$. But whereas the former concerns compound systems $\varphi \dot\times \sigma$ (interpenetration), representational macro-electromagnetism builds every $\varphi \dot\times \sigma$ as a statistical average over a large number of $\varphi + \sigma$'s (juxtaposition, as characterized in Ch. 2, 5). The global electromagnetic theory is a nearly black box theory, in the sense that it does not explain how certain properties, such as polariz-ability, come about: it takes them for granted. It can be formulated right away in a general covariant form in terms of the following specific primitives in addition to M^3, T, Φ, Σ, $\mathscr{B}$ and X:

$F = (E, B)$: The e.m. field tensor

$\mathfrak{G} = (D, H)$: The e.m. induction pseudotensor

χ: The constitutive tensor

j: The current density

$\mathfrak{T} = \mathfrak{T}_{em} + \mathfrak{T}_{me}$: The total energy-momentum-stress tensor (e.m. + mechanical). It is assumed that $\langle F, \mathfrak{G} \rangle \triangleq \varphi$ and that F and $\mathfrak{G}$ satisfy the field equations

$$\partial_{[\lambda} F_{\mu\nu]} = 0, \qquad \partial_\nu \mathfrak{G}^{\mu\nu} = \frac{4\pi}{c} j^\mu, \tag{4.20}$$

the constitutive equations

$$\mathfrak{G}^{\mu\nu} = \tfrac{1}{2} \chi^{\mu\nu\sigma\tau} F_{\sigma\tau}, \tag{4.21}$$

and the equations of motion

$$\nabla_\nu T^\nu_\mu = j^\nu F_{\nu\mu}. \tag{4.22}$$

It is easily seen that macroelectromagnetism subsumes microelectro-magnetism. And when the constitutive tensor degenerates into (ε, μ^{-1}), the field equations become

$$\nabla \times E = -\frac{1}{c} \frac{\partial B}{\partial t}, \qquad \nabla \cdot B = 0, \tag{4.23}$$

$$\nabla \times H = \frac{1}{c} \frac{\partial D}{\partial t} + \frac{4\pi}{c} j \qquad \nabla \cdot D = 4\pi\varrho. \tag{4.24}$$

These relations between general macroelectromagnetism and its two subtheories shows clearly wherein the novelty of the former resides: it lies in $\mathfrak{G}$ and in the equations (4.21) that express the way matter affects the field. These are generalizations of the equations $D = \varepsilon E$ and $H = (1/\mu) B$ holding for homogeneous and isotropic materials. The constitutive tensor χ is of course the great unknown. In CEM one can only hypothesize its components and see what happens, i.e. deduce some logical consequences and find out what kind of materials they fit. In other words, one can invent kinds of materials (e.g., the simple dielectric, for which $\varepsilon =$ const and $\mu = 1$) and then look around for their possible real referents. Such a procedure is obviously hopelessly slow. The determination of the χ that characterizes the electromagnetic properties of the various kind of materials is best made by trying nonphenomenological hypotheses concerning the structure of materials. This task is in fact being performed by solid state physics on the basis of the quantum theory. In this way not only known classical properties are being explained but also the behavior of newly studied materials such as semiconductors and superconductors. The success of this programme shows that it pays to take the deepest possible approach.

Notice that the induction $\mathfrak{G}$ was assumed to be a tensor density not a tensor. Correspondingly j is a vector density not a vector, and ϱ a scalar density. If $\mathfrak{G}$ were assumed to be a tensor one would get undesirable consequences, such as the generation of electric charges merely by rotating the frame of reference. Which shows, incidentally, that the tensor character of physical magnitudes is not altogether arbitrary but must be assumed taking the eventual empirical consequences into account. Notice also that macroelectromagnetism assumes that bodies cannot only be charged and be the seat of electric currents but can also be in a state of electromagnetic stress. Unfortunately the precise form of the stress tensor is known only in simple cases, as it depends not only upon field variables but also upon the structure of matter and is therefore beyond the reach of macroelectromagnetism. $\mathfrak{T}$ and $\mathfrak{G}$ are the two lids through which the results of the application of atomic theories to macrosystems can be poured. But they must be there if only in outline for atomic theories to be applicable to classical physics and by the same token experimentally tested. For, after all, experiments are conducted with macrosystems.

1.6. Representational Macroelectromagnetism

The macrophysical Maxwell equations in material media can be deduced from microelectromagnetism by adding the hypotheses that (a) a body is — as far as CEM is concerned — a stable collection of charged particles and (b) the bulk field is the average of the microfields associated with those particles. The first hypothesis concerns the constitution of bodies; it was proposed long before the first atomic theory was conceived. The second hypothesis is a nonspecific statistical assumption amounting to the statement that the detailed structure or configuration of the collection of particles making up a body is immaterial, as the details are averaged out: it is not a hypothesis of randomness but one of irrelevance of details. The first hypothesis can be implemented by adopting the formulas (4.17) for the four-current; the second, by suitably averaging the resulting Maxwell equations over the particle positions and velocities (DE GROOT and VLIEGER, 1965).

In that way, which is essentially LORENTZ', macroelectromagnetism is shown to emerge from the microlevel. This illustrates the process of deriving a macrophysical theory from a microphysical one in conjunction with additional assumptions concerning the constitution and/or structure of the entities concerned. But it should be realized that the derivation employs an assumption that is plainly false in CEM, namely that charged particles can be packed into stable systems, i.e. systems that will neither collapse nor explode nor radiate. (It is possible but not necessary to assume that there are stable atoms: only the system as a whole need be stable in order to build macroelectromagnetism on the basis of microelectromagnetism.) Since this stability is just a pretence, strictly speaking microelectromagnetism not only does not entail macroelectromagnetism but is incompatible with it. Only quantum mechanics and quantum electrodynamics can be hoped to entail macroelectromagnetism — once they have got rid of their own difficulties.

1.7. Nonfield Theories of E.M.

MAXWELL's theory did not have an easy success, both because it collided with the corpuscular cosmology so successfully implemented by NEWTON and his followers, and because it was so obviously transempirical (nonoperational). The first obstacle was mainly psychological and was eventually surmounted: after all, one had theories of continua in mechanics and in gravitation. The second obstacle was philosophical and therefore harder to overcome, so much so that no sooner were MAXWELL's and EINSTEIN's field theories accepted, that attempts were made to dispense with the field concept altogether, or to assign it a purely auxiliary (computational) role. In these neo-Amperian theories of e.m., so close to the heart of the electric circuit engineer, charges and currents alone are the real thing because they are wrongly supposed to be directly measurable. Therefore the question "What happens at x, t?" is here equivalent to "What happens to charged bodies at x, t?" Nothing is assumed to happen in the space in between — although, strangely enough, space is assumed to exist somehow; if something were to happen in midspace, it would be observable and therefore it would not be a region of free space.

The most complete direct interparticle action theory of electromagnetism is WHEELER and FEYNMAN's (1949). Its central postulate is FOKKER's action principle for a collection of point charges:

$$I = - \sum_i m_{0i} c \int ds_i + \sum_{j \neq i} (e_i e_j/c) \int \int dX_{i,\nu} dX_j^\nu \cdot$$
$$\delta[(X_i^\mu - X_j^\mu (X_{i,\mu} - X_{j,\mu})] = \text{extremum},$$

(4.25)

from which the equations of motion for charged particles follow. As can be seen, only particle variables occur in this theory, which is an electrodynamics in AMPÈRE's sense and therefore the opposite of a unitary field theory. (The theory is mathematically equivalent to an "adjunct" field theory, but this adjunct field, far from being free, is attached to the particles.) The theory has been abandoned by its creators because it is inconsistent with the quantum theory, in particular with the photon hypothesis. Yet it continues to appeal to many not only because it smacks of mechanism and operationalism, but also because no self-forces appear in it and consequently no infinite self-forces (and the corresponding self-accelerations). But this is too small an advantage compared with its shortcomings even at the classical level. Indeed, the theory makes use of special relativity — as is apparent from its central axiom (4.25), yet it is inconsistent with it for SR borrows from CEM the assumption that e.m. signals propagate in a vacuum with a constant speed c and moreover lead an existence which, while it lasts, is independent of both emitters and absorbers. In other words, by denying the existence of free radiation fields propagating in space, the action at a distance formulation of CEM contradicts its own basis: it is logically inconsistent.

Morals. (1) Do not employ formulas — e.g., those of relativity — without ascertaining what they commit you to. (2) Do not give up lightly principles which, like the ones of contiguity and antecedence, are at the basis of the whole physics (see Ch. 2, 1). (3) Do not fall in love with a theory just because it lacks some defects other theories have: it may have worse defects. (4) Do not try to revive views which, like corpuscularism, have failed because they were too simple. (5) Do not fall in the operationalist trap.

1.8. Testability of CEM

CEM cannot be admitted by empiricism because it postulates an occult substance, the e.m. field. Thus electrostatic fields are unobservable by themselves: to test the surmise that there may be such a field in a certain region we must introduce a test body in it, watch its behavior, and interpret (explain) it with the help of the equation "$m\ddot{X} = eE$" or some refinement of it — which does not belong to electrostatics but to one of the auxiliary theories employed in the test of electrostatics (see Ch. 1, 5.1). In other words, although electrostatics is not about the motion of test bodies, it is tested with their help and with the assistance of further theories. Similarly with e.m. waves: we never observe travelling e.m. waves, much less stationary ones. What we can observe is, say, pointers indicating maxima and minima of photo-

currents at definite places. We do not even see light: we see only illumi-
nated bodies. We use light and certain ideas about it — e.g., the law
of rectilinear propagation — as body objectifiers, and we control those
ideas by watching the behavior of bodies. Whenever we wish to detect
or measure light we must distort or even destroy it completely — e.g.,
by absorbing it in our retina.

In order to give an operational description of the propagation of a
light ray (or rather light pencil) along a line (or rather strip) we would
have to cover the path with detectors of diaphragms that would disturb
the original ray or even kill it. For this reason BRIDGMAN (1927) and
before him ARISTOTLE (*De anima* 418*b*) held that it is wrong to speak
of light as something travelling. Partly for the same reason the absorber
theory of radiation — in which there is no radiation proper — was
proposed (see 1.7). *Eppur si muove:* every adequate theory of light
assumes that light travels through empty space in a continuous way —
hence according to partial differential equations in x and t — even
though we cannot possibly verify this assumption pointwise. (Moreover,
light velocities would be unmeasurable if it were meaningless to talk
about them.) Yet the testability of CEM is, precisely for this reason,
much higher than the testability of any action at a distance electrodynamics for, while the former requires exploring limited regions of
spacetime (owing to the principle of contiguity or near action), theories
of action at a distance force us to scan the whole universe (PLANCK,
1932).

This does not mean that every statement of CEM is accurately testable: for one thing, its continuity assumptions are not. For another
thing, its constitutive equations, being phenomenological hypotheses,
are weakly testable — and this holds for the constitutive equations of
continuum mechanics as well. Indeed, suppose a theorem in either of
these theories is found to be empirically inadequate: what part of the
theory shall we suspect: the general schema or the specific constitutive
equations? Clearly we shall suspect the latter since these depend on the
particular kind of material and are ultimately dependent on its particular
microstructure, the study of which is the subject of different theories
(e.g., solid state theories). There is a reason for this choice: whenever
the particular constitution of the system is irrelevant, constitutive
equations do not occur and the general schema alone is employed. Yet
one should not blame every failure of the theory on the constitutive
equations, for otherwise the acceptance of the basic field equations
would be a matter of convention. In any case microelectromagnetism
is better testable than macroelectromagnetism because (*a*) it does not
involve hypotheses concerning the response of materials to external
fields and (*b*) it concerns systems that can be insulated more perfectly.

And either of them is far better testable than action at a distance theories, which have less to say, hence less to check, and yet would have us being on the lookout for unexpected actions coming from every corner of spacetime.

The best empiricist is the one who realizes in time that empiricism is unfeasible not only antitheoretical. And the best modern physicist is the one who acknowledges that neither classical nor quantum physics are cut and dried, both being full of holes and in need of a vigorous overhauling not only to better cover their own domains but also to join smoothly so as to produce a coherent picture of the various levels of physical reality.

2. Special Relativity

The special theory of relativity (SR) can be formulated in a number of ways depending on the choice of primitives and axioms. Here we shall keep quite close to the historical origins of the theory as far as its leading axioms are concerned. That is, we shall keep in mind that EINSTEIN titled his founding paper "On the Electrodynamics of Moving Bodies". In this way we hope to avoid certain widespread misinterpretations, chiefly the beliefs that SR owes nothing to electrodynamics (kinematical interpretation), that it deals only with material systems not with every kind of physical system (mechanism), that it concerns solely readings of yardsticks and clocks (operationalism), that it is based on a redefinition of the simultaneity concept (conventionalism), and that the coordinates occurring in the Lorentz transformation are those of a geometrical continuum which, after all, can be studied in a coordinate-free way (formalism).

2.1. Background and Heuristic Cue

SR presupposes CEM, hence shares the latter's formal and material background. In particular, CEM supplies SR the concept of electromagnetic radiation field as well as with the assumption that this field, once in existence, is indifferent to both its emitter and its eventual absorber. This assumption originated with MAXWELL's theory: indeed, in free space $\left(\frac{1}{c^2}\frac{\partial^2}{\partial t^2} - \nabla^2\right) A = 0$ holds, which does not contain material coordinates and in which c is a constant. (If c were to depend on the field source, as it does in W. RITZ's 1908 theory, CEM would be Galilei covariant.) A second concept CEM supplies SR is that of inertial frame, as a frame relative to which MAXWELL's equations hold (see 1.1.3, Df. 9). One could also use a more restricted concept of inertial frame in SR, namely this: A reference frame is called *inertial* iff all light rays propagate in vacuum along straight lines relative to it. The electromagnetic concept of inertial frame enables us to state the *principle of relativity* in a mechanics-free fashion, namely thus: "The basic laws of physics ought to be the same in (relative to) all inertial reference frames" — where 'inertial' means "such that MAXWELL's equations are satisfied in it". Once CEM is asserted, it is only natural to look for further laws, not contained in it, that are consistent with it and in particular that are invariant under the same transformations which preserve MAXWELL's equations. Now classical mechanics is not invariant under the group that preserves CEM: the transformation properties of the two theories are very different. Hence if the unity of physics is to be restored, of the two theories, CM and CEM, the former must be sacrificed. This purely theoretical requirement motivated the invention of SR.

In the building of SR the principle of relativity played an important heuristic role: it did not entail SR but it did help weed out all the basic statements that did not comply with it. Yet it is often regarded as a constitutive axiom of SR, although it involves the metatheoretical concept of basic law statement. A source of this misunderstanding is that the principle is often misstated. A common misstatement is this: "The reference frame makes no difference to physical events". This is false: we can always escape an unpleasant noise by riding a supersonic frame. The relativity principle refers to patterns, not to events: it is "a restriction principle for natural laws" (EINSTEIN, 1949). Another common misstatement is this: "The results of any experiment performed in an inertial system are identical with those obtained in any other galilean frame." In fact, experiments are a subset of facts, and facts are not frame-invariant. A fortiori, it is false that "Physics is the same for all observers". What is held is that the basic laws ought to hold no matter what the speed of the reference frame may be, as long as it is inertial. But nothing is said about observers, particularly when speaking about unobservable objects such as laws of nature, hence the operationalist interpretations of SR are a forgery. The principle of relativity is, in short, (a) a heuristic principle and (b) a metalaw statement — and a normative one not a declarative metanomological statement for it does not say what is but what ought to be the case (BUNGE, 1961 b).

The basis of SR is relativistic kinematics (SRK). Let us then start by reconstructing SRK.

2.2. Basis of Relativistic Kinematics

2.2.1. Building Blocks. The primitive base of SRK is this:

Σ: The set of physical systems (material or not)

S: The set of electromagnetic signals

I: The set of inertial reference frames

T: The range of the local time function

E^3:   A manifold mapping local space

$\{X\}$: A family of functions representing the position of a system $\sigma \in \Sigma$

c: A positive real number representing the magnitude of the speed of e.m. signals in vacuum.

A member σ of Σ symbolizes an arbitrary physical system, whether a portion of matter or a portion of field. $S \subset \Sigma$ and $I \subset \Sigma$, which are here primitives, are characterized in CEM. Any system with fixed geometrical and chronometrical properties, e.g., a cubic clock, qualifies as a member of I as long as it moves in such a way that CEM holds relative to it. But the I's are not limited to laboratory devices: SRK is intended to refer to systems placed both within and beyond measurement. The only requirement on the members of Σ and I is that they be connectible by some members s of S. Indeed, the object of SRK is the set $\Sigma \times S \times I$ of triples physical system — e.m. signal — inertial frame. As to the remaining primitives, we are familiar with them from CEM and protophysics. Note that no specific mechanical concept occurs among the primitives of SRK except the one of inertial system: indeed fields do not qualify as inertial systems. The functions X may localize any localizable system whatever: the tip of an X will fall either on a particle or on wave front. Finally, the concepts of measuring rod and clock do not occur in our list. This removes an inconsistency occurring in the usual formulations of SRK and noticed by its inventor: "strictly speaking rods and clocks would have to be represented as solutions of the basic equations (objects consisting of moving atomic configurations), not, as it were, as theoretically self-sufficient entities" (EINSTEIN, 1949).

We shall now introduce two abbreviations:

Df. 1. Spacetime: $E^{3+1} \stackrel{\text{df}}{=} E^3 \times T$

Df. 2. Cotime: $X_0 \stackrel{\text{df}}{=} ct$

The symbol 'E^{3+1}' suggests that time retains its individuality in SR although it becomes indissolubly linked to space. The much exaggerated equivalence of space and time in SR will be discussed later on, but let us anticipate this: space and time are mutually independent (not interdefinable) concepts in SR just as in classical physics. The fact that durations can be reckoned in distance units does not prove that time has been spatialized — as the philosopher H. BERGSON deplored — but that the two concepts can be related.

Let us finally introduce two kinematical concepts.

Df. 3. If $\sigma \in \Sigma$, $\iota \in I$ and $t \in T$, then $V(\sigma, \iota, t) \stackrel{\text{df}}{=} dX(\sigma, \iota, t)/dt$

Df. 4. If $\sigma \in \Sigma$, $\iota \in I$ and $t \in T$, then: σ is in uniform rectilinear motion w.r.t. $\iota \stackrel{\text{df}}{=} V(\sigma, \iota, t) = \text{const}.$

2.2.2. Axioms of SRK. We shall give the axioms without much detail because protophysics and CEM take care of most of our primitives: indeed, SRK is the offspring of preexisting but previously disconnected hypotheses. EINSTEIN's merit was to bring them together and not to retreat in front of their revolutionary consequences. (He did not know as much as LORENTZ and, unlike POINCARÉ, he regarded the axioms of mechanics as hypotheses not as disguised definitions.)

$SR\ 1$ (a) $\Sigma \neq \emptyset$. (b) Every $\sigma \in \Sigma$ represents a physical system.

$SR\ 2$ (a) $S \neq \emptyset \wedge S \subset \Sigma$. (b) Every $s \in S$ represents an e.m. signal.

$SR\ 3$ (a) $I \neq \emptyset \wedge I \subset \Sigma - S$. (b) Every $\iota \in I$ is an inertial reference frame. (See CEM.) (c) For every $\iota \in I$ there is a basis $e = \langle e_0, e_1, e_2, e_3 \rangle$ in E^{3+1} such that $e \triangleq \iota$.

$SR\ 4$ (a) E^3 is a tridimensional Euclidean space with inner product. (b) E^3 represents ordinary space relative to ("as seen from") any given $\iota \in I$.

$SR\ 5$ (a) T is an interval of the real line. (b) T is the range of the time function satisfying the axioms of the local time theory (see Protophysics). (c) Every $t \in T$ represents an instant of ι-time.

$SR\ 6$ (a) $\{X\}$ is a nonempty family of functions. (b) Every $X \in \{X\}$ is a function from $\Sigma \times I \times T$ to R^3. (c) $X(\sigma, \iota, t)$ represents the position of a point system $\sigma \in \Sigma$, referred to the frame ι, at the instant t relative to ι. (d) For every point event there exists a septuple $\langle \sigma, s, \iota, X_0, X_1, X_2, X_3 \rangle \triangleq$ event.

$SR\ 7$ Every $s \in S$ propagates in vacuum, relative to any $\iota \in I$, with uniform rectilinear motion at the speed c.

$SR\ 8$ For every $\iota, \iota' \in I$, the associated cotimes are such that $\partial X_0 / \partial X_0'$ exists and is positive.

Remarks. (1) The first postulate makes it clear that SR is not just about bodies, much less restricted to measuring devices, but intends to span the whole class of physical systems. (2) The concept of e.m. signal (S) is supposed to be characterized by the chapter of CEM dealing with e.m. waves; since we place CEM in the background of SR nothing more need be said for the moment. (3) The concept of reference frame is characterized by protophysics and more precisely by CM. Here we need only recall that, to qualify as a member of I, a material frame must be closed or nearly so, for any interaction with other systems would accelerate it. In the treatments influenced by subjectivist philosophies, the concept of observer occurs in place of I and the whole theory is presented as a game of actual and fictitious observers filling up the whole universe. No such fictions can be accepted in a physical theory. (4) By $SR\ 4$, SR adopts a classical concept of space as a container which "remains always similar and immovable" (NEWTON) but restricts it to every single frame. E^{3+1} is not the space of events $\Sigma \times S \times I \times E^{3+1}$ (see SR 6d). While E^{3+1} is Euclidean, hence homogeneous, the space of events will turn out to be pseudo-Euclidean and inhomogeneous in the sense that it has an intrinsic arrow of time. (5) By SR 5, there are as many times as inertial frames: time is frame-dependent. In this sense time is relative. Moreover, in our formulation relativistic time is also relational, in the sense that it is a triadic relation between pairs of events and frames (see Ch. 2, 3). But one might keep NEWTON's absolute time, which "flows equably without relation to anything external", in relation with every inertial frame; such a ghostly (but relative) time would indeed be consistent with SR. (6) In contrast to the usual view, which identifies spacetime quadruples with events, we require some physical system to be there for an event to occur, and we do not equate an event with a certain septuple but make the semantical assumption that the septuple in question represents the event. (7) SR 7 is the leading postulate of SRK. It can be restated in infinitely many negative ways — e.g, "The propagation velocity of e.m. signals in vacuum is independent of place (homogeneity), direction (isotropy), source, and sink". In other words, e.m. signals are independent entities rather than states of a transmitting medium — e.g., LORENTZ' stationary aether. No wonder that the same mind advanced also, at the same time, the photon hypothesis: in either case light was regarded as a thing rather than as a state of a substratum. Yet as a thing very different from a particle although the photon is superficially similar to a particle. For one thing the principle of autonomy of e.m. signals rivets the matter/field dichotomy introduced by the Faraday-Maxwell theory, by treating the e.m. radiation field as something whose motion can be neither compounded with that of a piece of matter nor transformed away by the choice of a suitable frame. (But the velocity of light is not a general invariant: thus a light ray coming from a star spirals down to Earth rather than propagating in a straight line relative to it: our favorite reference frame is not inertial.) The principle calls for the surrender of most analogies between the e.m. field and matter, as well as of many pictures of the behavior of matter. It is highly counterintuitive and the ultimate source of the various puzzling (but true) formulas of SR. If theories were rejected when failing to be intuitable, plausible, or simple, SR would not have been adopted — and as a matter of fact it was initially criticized precisely on those grounds by scientists and philosophers who had forgotten that intuition grows with familiarity — until it becomes so ingrained that it turns into an obstacle to further progress. Anyhow SR 7 is, happily, one of the few postulates that can be subjected to fairly direct experimental test. For example, the velocities of gamma rays produced by electron pairs in swift motion and nearly at rest are the same within experimental error (SADEH, 1963). (8) SR 8 is equivalent to this other statement: the L_{00} element of the matrix of the Lorentz transformation exists and is positive. In ordinary language: should the direction of time be reversed in one frame, it would "flow backwards" in every other frame as well. This restricts the Lorentz group to the orthochronous subgroup.

2.3. Some Logical Consequences

2.3.1. Relativity of Simultaneity. We shall see that simultaneity, though relative, is not conventional as usually taught.

Thm. 1. Let σ_1 and σ_2 be two point sources of spherical e.m. signals located respectively at X_1 and X_2 in $\iota \in I$, and let σ be a receiver at the midpoint of the segment joining the two sources. Then if the two sources emit each a signal at the time t_1 relative to ι, the signals arrive simultaneously at σ, at the instant $t_2 = t_1 + (1/2c)|X_2 - X_1|$. *Proof.* Use SR 7 and adjoin the special assumptions (idealizations) that the emission and absorption take no time.

Thm. 1 is the basis of

Criterion 1. It can be inferred that two distant point events in vacuum are *simultaneous* in $\iota \in I$ iff they consist of or are accompanied by e.m. signals arriving at the same time at the midpoint of the straight line segment passing through them.

Remarks. (1) No measurements proper need be made in order to apply Criterion 1 as long as the two points X_1 and X_2 are accessible. If they are not, a computation is in order. In such a case Thm. 1 enables us to compute the local time t_1 at which the events occurred, provided their mutual distance and c are known. The same holds for a single distant event: if it is recorded at t_2 then it is inferred to have occurred at $t_1 = t_2 - c^{-1}|X_2 - X_1|$. This again, far from being a convention, is an elementary consequence of SR 7. (2) Criterion 1 is usually presented as a definition of simultaneity. But there is no need for such a definition as long as it is postulated (SR 5) that local time is mapped on the real line (see Ch. 2, 3). Criterion 1 does not serve the purpose of elucidating the concept of simultaneity but the one of deciding, in an experimental situation, whether two distant events are simultaneous relative to a given frame. (3) The experimental problem of finding out whether two distant events are in fact simultaneous in a given frame cannot be solved unless the events are accompanied by the emission of e.m. signals. But this does not prevent two "silent" (e.g., purely mechanical) events from being simultaneous events in a given ι: only, the observer attached to ι should abstain from asserting that they are simultaneous — or that they are not.

Thm. 2. Let σ_1 at X_1 in $\iota \in I$ emit at t_1 an e.m. signal that bounces off at σ_2 at X_2 and t_2 — both events referred to the same frame ι — and returns to σ_1 at t_3. Then $t_2 = \frac{1}{2}(t_1 + t_3)$. *Proof.* Use *SR* 7 and make the same simplifying assumptions as for Thm. 1. In this way one gets: $t_2 - t_1 = t_3 - t_2$.

Remarks. (1) If the point X_2 at which the reflector σ_2 is placed is inaccessible, then t_2 cannot be measured in ι; but it can be computed from clock readings by means of Thm. 2. In other words, Thm. 2 is not restricted to clock readings although it must be tested through clock readings. (2) Thm. 2, too, usually passes for a definition. If it were, it could be exchanged for any other one; thus it has been suggested that the factor $\frac{1}{2}$ could well be replaced by an arbitrary constant $0 < \varepsilon < 1$ (REICHENBACH, 1924). But Thm. 2 follows from *SR* 7; moreover, it can in principle be tested. What is true is that Thm. 2 constitutes the ground of the following

Df. 5. A periodic process in $\iota \in I$ is said to be *synchronous* with another periodic process in the same frame just in case their time phases satisfy Thm. 2.

Remarks. (1) This Df. applies in particular, but not exclusively, to clocks. And, far from being arbitrary, it is grounded on *SR* 7. (2) The concept of synchronicity is as relative as the one of simultaneity; no wonder that different time keeping devices, even if exact, can keep different times.

2.3.2. Lorentz Transformations. Consider Maxwell's equations (4.1) and (4.2). By Remark 4 on *CEM* 2 (see 1.1.3), when referred to a frame moving with speed $u \neq$ const relative to an arbitrary inertial frame, the field equations acquire additional terms involving u. In other words, CEM entails

Lemma 1. Every $\iota \in I$ is in uniform rectilinear motion w.r.t. any other $\iota' \in I$: $V(\iota, \iota') =$ const.

Remark. The e.m. and the mechanical concepts of inertial frames are now seen to overlap in part. But Lemma 1 is their only thing in common: EINSTEIN's, not NEWTON's mechanical equations hold in an electromagnetically inertial system, i.e. one satisfying Df. 9 of Sec. 1.

Lemma 2. If ι, $\iota' \in I$, then the physical coordinates X_0', X' of a physical system $\sigma \in \Sigma$ relative to ι' are related to its coordinates X_0, X referred to ι, by a linear relation: $X_\mu' = L_{\mu\nu} X_\nu + D_\mu$, with μ, $\nu = 0, 1, 2, 3$. *Proof.* Apply the transformation $X_\mu \to X_\mu'$ to the frames ι and ι' themselves, which by Lemma 1 move in straight lines relative to one another, and recall that the affine group is the only one that preserves straight lines.

Remarks. (1) The transformation involved in the previous Lemma does not represent a physical process — e.g. the passage of a thing from a state of rest to a moving frame, much less the jump of an observer to a moving frame: it is just the relation between two different representations of events. By the principle of relativity all such representations are equivalent as regards the formulation of basic laws. (2) These are not yet the Lorentz transformations, since the elements of L and the components of D are so far indeterminate. The Lorentz transformations are those special homogeneous transformations that satisfy the following

Lemma 3. The law of propagation of e.m. signals is preserved under the general linear transformation of Lemma 2. *Proof.* Apply the transformation of Lemma 2 to D'ALEMBERT'S wave equations under the assumption $SR\,7$ of the frame-independence of c.

Corollary. A spherical e.m. signal propagating in vacuum with the speed c relative to $\iota \in I$, retains its form and speed in every other $\iota' \in I$. I.e.

If $X_1^2 + X_2^2 + X_3^2 = X_0^2$ in ι, then $X_1'^2 + X_2'^2 + X_3'^2 = X_0'^2$ in $\iota' \neq \iota$

Thm. 3. Let $\iota, \iota' \in I$ such that, at $t = t' = 0$, ι and ι' coincide and, for $t > 0$, ι' moves along the x-axis of ι with uniform motion $V(\iota', \iota) = u = \text{const}$. Then the coordinates of one and the same physical system $\sigma \in \Sigma$ in the two frames for $t > 0$ are related by

$$X_0' = \frac{X_0 - \dfrac{u}{c} X_1}{(1 - u^2/c^2)^{\frac{1}{2}}}, \qquad X_1' = \frac{X_1 - \dfrac{u}{c} X_0}{(1 - u^2/c^2)^{\frac{1}{2}}}, \qquad X_2' = X_2, \qquad X_3' = X_3. \quad (4.26)$$

Proof. Use Lemmas 2 and 3 and axiom $SR\,8$. Notice (a) that the inhomogeneous terms in the general linear transformation vanish because the frames are assumed to coincide at $t = 0$; (b) that X_2 and X_3 must remain unchanged for otherwise ι' would rotate or strech, contrary to hypothesis; (c) that L_{00} must be positive because it represents the rate of a periodic process at ι' as reckoned in ι and such a rate is positive by assumption $SR\,8$; (d) that $L_{11} > 0$ because by hypothesis ι' moves along the positive direction of x; (e) by the corollary of Lemma 3, the coefficients of L satisfy the relations

$$L_{10} - k L_{00} = 0, \qquad L_{11}^2 = (1 - k^2)^{-1}, \qquad L_{00}^2 - L_{10}^2 = 1, \qquad k \overset{\text{df}}{=} L_{01}/L_{11};$$

whence $L_{00} = L_{11}$ with the help of $SR\,8$ and $L_{10} = L_{01} = k L_{00}$; ($f$) if the resulting relations are carried to the general linear transformation we get

$$X_0' = (1 - k^2)^{-\frac{1}{2}}(X_0 + k X_1), \qquad X_1' = (1 - k^2)^{-\frac{1}{2}}(X_1 + k X_0);$$

(g) the value of k is determined by applying these relations to ι and $\iota' = \sigma$: in this case $X_1 + kX_0 = 0$ whence, by differentiation, $k = -\,\mathrm{d}X_1/\mathrm{d}X_0 = -\,u/c$, which leads to (4.26).

Remarks. (1) What gets transformed is not a point in E^{3+1} but the physical coordinates of some physical system. If a single reference frame is at stake and σ happens to move in it by itself, then $X(\sigma, t) = X(\sigma, 0) + v(\sigma) \cdot t$. In other words, the Lorentz transformations do not apply within a given frame but *relate any two inertial frames*. (2) The Lorentz transformations break down for $u \geqq c$: for $u = c$ the physical coordinates in ι' become indeterminate, and for $u > c$ they become imaginary, in violation to *SR 6b*. Consequently if SRK is true, there are no reference frames that attain, let alone surpass, the velocity of e.m. signals in vacuum. This does not preclude the possibility that there be physical entities, not qualifying as reference frames, which do overtake e.m. signals. (3) Since u is the relative velocity of two reference frames, it is sometimes said that SR does not apply to accelerated motions. This is mistaken: SR applies to arbitrary motions as long as they are referred to inertial frames: only, it cannot be expected to hold for accelerated frames. (4) The Lorentz transformations are not directly testable. What can be tested are some of their logical consequences. Also, we can check mathematically whether a given statement is Lorentz invariant; if it is and in addition it has been corroborated in any given frame, then we count this as indirect evidence in support of the Lorentz transformation formulas. (5) The Lorentz group can be obtained in alternative ways. In particular, it is entailed by the retarded action ("causality") principle when formulated in the Minkowski spacetime (ZEEMAN, 1964).

2.3.3. Invariants and Covariants. The idea of a Lorentz transformation enables us to introduce

Df. 6. A magnitude is called *Lorentz invariant* or *absolute* iff its values are unchanged under a Lorentz transformation.

Df. 7. A statement is called *Lorentz covariant* or *absolute* iff its Lorentz transform holds even though every one of the magnitudes occurring in it is not Lorentz invariant.

Remarks. (1) The solutions of equations of motion and field equations cannot be Lorentz covariant. Absolutes are found among basic laws not among all of their logical consequences. Reason: fundamental equations, unlike their solutions, contain no secondary information involving special frames. (2) Lorentz covariance is required not only in macrophysics but also in microphysics. It is not known, though, whether all microsystems qualify as inertial frames. It is furthermore possible that the basic microlaws are not locally Lorentz covariant but that they are Lorentz covariant in the large.

Df. 8. A theory is called *special relativistic* iff all its fundamental laws are Lorentz covariant.

Remark. A theory may be covertly rather than manifestly covariant. The actual variance of a theory can best be ascertained upon reformulating it in a four-dimensional way. This is how it was found that CEM had been covariant *avant la lettre*. The covariant reformulation of a given theory may be far from trivial and it may exhibit previously hidden traits.

Thm. 4. The element of distance in the spacetime of events, i.e. $ds^2 = dX_0^2 - dX^2$, is Lorentz invariant. *Proof.* Apply the Lorentz transformations.

Corollary. If $\langle \sigma, s, \iota, X_0, X_1 \rangle$ and $\langle \sigma', s, \iota, X_0', X' \rangle$ are two events in ι, such that $\pm c(X_0' - X_0) > |X' - X|$, then the second event is *later* (earlier) than the first one in every other inertial frame as well — i.e. the order of events lying within a light cone is invariant. *Proof.* The inequality is equivalent to $c^2(X_0' - X_0)^2 - (X' - X)^2 > 0$, which by Thm. 4 is invariant.

Df. 9. A positive (negative) spacetime interval is called *timelike (spacelike)*.

Remarks. (1) According to the corollary of Thm. 4, events separated by a timelike interval are absolutely (universally) later or earlier to one another. Conversely, events separated by a spacelike interval are not ordered in an absolute way: in one frame an event may be later than another while in a different inertial frame the converse may hold. Some authors (e.g., REICHENBACH, 1927) have inferred from this that the region of the spacetime of events outside a light cone is "indeterminate as to the order of time". The right inference is that the order of events in that zone is frame-dependent (not absolute). Likewise the length of a body is not indeterminate for being relative rather than absolute. (2) If two events are causally related then their time order is absolute. Conversely, if they are not related by the inequalities of our corollary, i.e. if they do not lie within a light cone, then they are causally unrelated. This follows from the assumption that c is an ultimate signal velocity, for then events separated by a spacelike interval cannot influence, much less cause one another. This does not render SR indeterministic: it just excludes certain events from the set of events that can be causally related. SR does not eliminate causality but restricts its extension (BUNGE, 1959a). (3) Even if two events do not satisfy the inequalities in our corollary, the concepts of earlier and later remain meaningful and objective — only, they are now relative to a frame.

Caution: do not confuse "relative" with "subjective" (pertaining to a subject or observer). (4) The invariance of the interval in the spacetime of events gave POINCARÉ and MINKOWSKI the idea of casting SR in the language of a four-dimensional geometry. This geometrical method has the following virtues: (a) it renders computations easier (in particular, by using $X_4 = ict$ we can write, say, the square of the four-vector ∂_μ as the sum $\partial_\mu \partial^\mu$); (b) it simplifies the covariant reformulation of a theory; (c) it proves the logical consistency of SRK, for it is a model of it (see Ch. 1, 4.1.6). But it has been misleading in some respects. For example, it has fostered the belief that the world is a 4-dimensional continuum of which time is but one of its dimensions. This rests in turn on the wrong identification of an event with a quadruple $\langle X_0, X_1, X_2, X_3 \rangle$. The central concept of physical system, without which there is no physics, is thereby bypassed: we are left with an empty mathematical object, namely a pseudo-Euclidean 4-space. The peculiarities of time are then wiped out: time ceases to be anchored to becoming, the absence of universal time reversal (SR 8) is not even mentioned, and it is forgotten that the time variable occurs as an argument in several functions, notably in the physical coordinate X (systematically confused with the geometrical coordinate x). In short, this formalistic interpretation (or rather deinterpretation) of SR exhibits it as a net of formal relations among points in spacetime rather than as a network of possible physical connections, through e.m. signals, among physical systems. A correct understanding of SR rests on the recognition that its object is $\Sigma \times S \times I$ and that the space of events is not E^{3+1} but $\Sigma \times S \times I \times E^{3+1}$ which is a partially ordered manifold (see 2.5.4).

2.3.4. Relativity of Distance and Duration. Let us recall quickly the transformations to simultaneity and to spatial coincidence. *Thm.* 5: If two events are separated by a spacelike interval in $\iota \in I$, then there exists at least one other $\iota' \in I$ such that those events are simultaneous in ι'. The loss of universal simultaneity is compensated for by the ever-present possibility of finding a frame in which two causally unrelated events occurring in succession relative to some frame, coincide in time in the new frame. *Thm.* 6: If two events are separated by a timelike interval in $\iota \in I$, then there exists at least one other $\iota' \in I$ such that the events take place at the same point in ι'. This possibility is absent from nonrelativistic physics because the spatial distance is invariant in it.

Thm. 7. If σ_1 and σ_2 are two systems at rest in ι', and if their separation is L_0 in ι' and L in ι, then $L = L_0 (1 - u^2/c^2)^{\frac{1}{2}}$. *Proof.* Set $t_2' = t_1'$ and apply the Lorentz transformation to $|X_2' - X_1'| \equiv L$. Caution: in the unprimed frame $t_2 \neq t_1$.

Remarks. (**1**) The Lorentz contraction is reciprocal: in the primed frame the spatial separation between σ_1 and σ_2 will be shortened. This is occasionally expressed in the following way: "Each observer finds that the length measurements made in his own reference frame are longer than those made by other observers". Not so fast: by hypothesis the values X_1' and X_2' of the positions of σ_1 and σ_2 were taken at the same local time ($t_1' = t_2'$). But in the unprimed frame the two events are no longer simultaneous: the length contraction is accompanied by the time dilatation $t_2 - t_1 = u L_0/c^2$. This time shift makes it impossible to measure the distance $|X_2 - X_1|$ in ι. If we do measure the distance in ι' then we cannot measure the separation between the same things in ι at the same time — and conversely. Consequently we cannot compare measurement results: there is one result at a time. If we measure the distance in ι we can compute the distance in ι' but then we shall not be comparing two measured values. The Lorentz-Fitzgerald contraction theorem ruins operationalism. (**2**) The Lorentz-Fitzgerald formula is usually said to apply to a rigid rod or yardstick. True but narrow: Thm. 7 applies to the distance between any two objects at rest in ι'; it may as well apply to the distance between two parallel light rays. In fact no assumption concerning the nature of σ_1 and σ_2 was made to derive Thm. 7. (**3**) The Lorentz-Fitzgerald formula does not apply to distances between geometrical points in free space: indeed, the physical coordinates occur in it. (**4**) The length value L in the unprimed frame is often called the *apparent* length, which suggests that the proper length L_0 is the real one. This implies that the contraction is not real. The word 'apparent' is however out of place in theoretical physics because it points to some subject, and no such concept occurs among the primitives of SR. Moreover, as we saw in Remark 1, a distance L cannot appear to an observer attached to ι if it is measured in ι' and conversely. The two distance values, L_0 as assumed or measured in ι', and L as computed in ι, are real in the sense that they refer to a pair σ_1, σ_2 of real physical entities. Only, distances are not absolute: length is a triadic relation in SR. (**5**) SR supplies no contraction mechanism: SR is independent of special assumptions regarding the structure of matter. Moreover, there is no contraction process to be explained, just as there is no source of the Coriolis force aside from a change of frame. The Lorentz transformations and their consequences, being kinematical relations, defy every causal interpretation. (**6**) The Michelson experiment did not exude SR: Einstein did not seem to know about it when he advanced SR, and MICHELSON always opposed SR (SHANKLAND, 1963, 1964). On the other hand Thm. 7 is sufficient to explain the null result of the Michelson experiment. SR and GR were not laboratory children (HOLTON, 1966). (**7**) Owing to Thm. 7, shapes are not Lorentz invariant: the geometry of any complex physical system is Euclidean in every single $\iota \in I$ but it is not universally Euclidean — unless it happens to be an e.m. wave front — another reminder of the privileged place of the e.m. radiation field in SR.

Thm. 8. If e_1 and e_2 are two successive point events occurring at the origin of ι', and their time separation in ι' is T_0, then their time separation in ι is $T = T_0 (1 - u^2/c^2)^{-\frac{1}{2}}$. *Proof.* Set the position coordinates of the events at ι' equal to zero and apply the inverse Lorentz transformation to $t_2' - t_1' = T_0$. Since you are at it prove that the time dilatation is reciprocal.

Remarks. (1) Despite the mathematical symmetry of the physical coordinates and time in the Lorentz formulas, the space and time spans behave inversely — this being why the interval in the spacetime of events is invariant (Thm. 4). One more proof that space and time are not on an equal footing in every respect. (2) Just as Thm. 7 is not restricted to yardsticks, so Thm. 8 is not limited to clocks: by hypothesis it applies to the time interval between any two events in a frame, whether or not they are timed by a clock; in particular, it applies to the period of a wave. (3) Unlike the length contraction formula, time dilatation can be checked in a fairly direct way on unstable particles. (4) T_0 is called the proper time and T the relative time. The designation 'apparent time' for T is incorrect both because time dilatation is reciprocal and because appearances require the presence of observers, which are not involved in Thm. 8. Both times are as real as time can be. (5) The relativity of duration is counterintuitive and gives rise to a number of paradoxes which bear the collective name of *clock paradox*. The simplest member of this class is the following. An observer attached to ι judges that a clock in ι' is slower than his own; but so does an observer attached to ι' with regard to the readings on a clock in ι. If every observer thinks in terms of universal time, there is contradiction. This dissolves as soon as one recalls that the various readings refer to different frames and therefore need not be identical. None of the observers is entitled to conclude "Your clock is slow"; each must say "In my reference frame (or measured from my observation platform) your clock is slower than mine" — and these two statements are mutually compatible. In SR the democracy of durations is not more paradoxical than the democracy of extensions, yet no one speaks of the yardstick paradox.

Thm. 9. If a system $\sigma \in \Sigma$ moves in $\iota \in I$ with speed $v = \langle v_1, 0, 0 \rangle$, then it moves in $\iota' \in G$ with velocity $v_1' = (v_1 - u)/(1 - v_1 u/c^2)$, where $u = V(\iota, \iota')$ *Proof.* Differentiate the Lorentz transforms of the position and time coordinates and compute dX'/dX_0'.

Remarks. (1) If σ happens to be a wave, the theorem holds only for its group velocity. (2) The Galilei and the Einstein laws of composition of velocities do not satisfy the law of transformation of a vector under a change of coordinate system: the latter is homogeneous, whereas the former are inhomogeneous. A change of frame can create motion or transform it away; a change in coordinate system cannot. One more reason for distinguishing physical frames from coordinate systems. (3) Should ι be ridden by an observer then he could measure v and u and compute v' by means of Thm. 9; but the formula is supposed to hold even for uninhabited frames. (4) The preceding formula breaks down for $uv = c^2$, which is satisfied in particular by $u = v = c$. Worse: if $uv > c^2$ and $v > u$, then $v' < 0$. In this case a system σ moving forward in ι would be moving backward in ι', i.e. to the left $(dX' < 0)$ if σ moves to the right in ι, or towards the past $(dX_0' < 0)$ if σ moves to the right in ι. The last motion (communication with the past) is excluded by axiom SR 8. The first motion would eventually end up in a collision of the system with itself, which contradicts the assumption that σ is a single thing. Therefore uv cannot surpass c^2. In other words Thm. 9 and the Lorentz transformations, which entail it, are restricted by the conditions $u < c$ and $v \leqq c$. That is, if SR is true then no inertial frame ever catches up with an e.m. signal in vacuum, and no physical system, whether material or field-like, overtakes an e.m. signal in vacuum. If superlight velocities were found, SR would not hold for them; but SR does not prohibit the existence of such nonrelativistic entities.

Let us now have a glimpse of how SRK can be applied to the rest of physics.

2.4. Relativistic Physics

SR physics is the set of Lorentz covariant physical theories (see 2.3.3, Df. 8). The purpose of relativizing a theory is to enhance its objectivity and universality by freeing its basis from the peculiarities involved in the prediction and observation of facts, which require the adoption of a particular frame and the introduction of singular statements concerning specific physical systems in specific states. The process of universalization through the discarding of particulars (e.g., observers) is started in nonrelativistic physics with the search for basic laws and is crowned in general relativity with its requirement of general covariance (not restricted to inertial frames).

Given an arbitrary physical theory containing at least one physical coordinate and time, the theory can be relativized by applying the following method: (a) pick the fundamental laws of the given nonrelativistic theory and assume they hold in the rest (primed) frame; (b) reformulate those law statements in MINKOWSKI's 4-dimensional language; (c) check whether the new statements are in fact covariant; (d) append to them the transformation formulas for the basic magnitudes of the theory; (e) check whether the formulas containing the relative velocity approach the corresponding nonrelativistic formulas for $c \to \infty$ — and keep calm if some of them do not. In short, Relativistic physics = Nonrelativistic physics + Four-dimensional language + SR kinematics. This is not a method for constructing theories from scratch but a (fallible) recipe for relativizing given nonrelativistic theories by the adjunction of relativistic kinematics: it is a recipe for joining any given classical theory with SR kinematics. It works only with nonquantum theories: quantum relativistic theories are not obtained in that way.

There are no restrictions on the theory that is to be relativized except that it must contain at least one physical coordinate and time. If this requirement is not fulfilled, but the theory is essentially nonlocal (e. g., elementary thermodynamics), the relativization is either impossible or ambiguous. This is why relativistic thermodynamics has been in a deplorable shape for nearly 60 years, to the point that nearly all of its transformation formulas were wrong (OTT, 1963; ARZELIÈS, 1965; SUTCLIFFE, 1965). These mistakes could not have been avoided by making measurements. Which does not show that relativistic thermodynamics is meaningless (operationalist objection) but only that not every physical formula can be checked directly. Relativistic thermodynamics can be checked indirectly, chiefly through CM and statistical mechanics.

Let us outline a few applications of SRK. The order of the applications is up to a point arbitrary, but it seems best to adopt the partial ordering suggested by the relation of presupposition. From this point of view, a natural arrangement is this: CEM — Optics — CM — Statistical mechanics — Thermodynamics. We shall do no more than sketch the relativization procedure in a few cases, leaving the axiomatization *ab ovo* to the interested reader.

2.4.1. Macroelectromagnetism. Postulate the field equations for a medium at rest in an arbitrary inertial frame; introduce the 4-vector density j^μ, the 2nd rank covariant antisymmetric tensor F and the contravariant pseudotensor $\mathfrak{G}$:

$$\langle F_{01}, F_{02}, F_{03} \rangle = E, \qquad \langle F_{23}, F_{31}, F_{12} \rangle = B$$

$$\langle \mathfrak{G}^{01}, \mathfrak{G}^{02}, \mathfrak{G}^{03} \rangle = D, \qquad \langle \mathfrak{G}^{23}, \mathfrak{G}^{31}, \mathfrak{G}^{12} \rangle = H$$

Now rewrite the field equations in a Lorentz covariant way in terms of these new field variables:

$$\partial_\sigma F_{\mu\nu} + \partial_\mu F_{\nu\sigma} + \partial_\nu F_{\sigma\mu} = 0, \quad \text{briefly Curl } F = 0, \qquad (4.27)$$

$$\partial_\nu \mathfrak{G}^{\mu\nu} = \frac{4\pi}{c} j^\mu, \quad \text{briefly Div } \mathfrak{G} = \frac{4\pi}{c} j. \qquad (4.28)$$

Finally compute the Lorentz transforms of the field components:

$$E_1' = E_1, \qquad E_{2,3}' = (1 - u^2/c^2)^{-\frac{1}{2}} \left[E + \frac{u}{c} \times B \right]_{2,3}, \qquad (4.29\text{a})$$

$$B_1' = B_1, \qquad B_{2,3}' = (1 - u^2/c^2)^{-\frac{1}{2}} \left[B - \frac{u}{c} \times E \right]_{2,3}, \qquad (4.29\text{b})$$

$$D_1' = D_1, \qquad D_{2,3}' = (1 - u^2/c^2)^{-\frac{1}{2}} \left[D + \frac{u}{c} \times H \right]_{2,3}, \qquad (4.29\text{c})$$

$$H_1' = H_1, \qquad H_{2,3}' = (1 - u^2/c^2)^{-\frac{1}{2}} \left[H - \frac{u}{c} \times D \right]_{2,3}. \qquad (4.29\text{d})$$

(Notice that the coupling of the electric and the magnetic components persists to order u/c: it is not a relativistic feature.)

This is not a new theory but a reformulation of CEM, which was relativistic without knowing it. Yet the explicit (manifest) covariant reformulation of CEM brings several important traits to the fore: (a) the natural pairings (E, B) and (D, H), (b) the frame-dependence of the splitting of the field into electric and magnetic components, and (c) the redundancy of the aether. Moreover, the classical aether hypothesis was more than redundant and untestable: it was also self-contradictory (TRUESDELL and TOUPIN, 1960). But, contrary to a popular belief, SR did not kill every sort of aether: it just deprived the aether (a) of the peculiar mechanical properties the older theories had assigned it — such as, e.g., being constituted by particles moving along definite trajectories, and (b) of its former role of support of the e.m. field. A non-mechanical aether, though redundant in CEM and SR, is compatible with either and might, if necessary, be postulated by other theories. As a matter of fact GR and quantum electrodynamics, by assigning space physical properties, have reintroduced the aether concept by the back door (EINSTEIN, 1920). Only, in either case one speaks of space rather than of a special substance spread over space — or, if preferred, one speaks of gravitational fields in vacuum and of the zero e.m. field in the case of QED.

Since the magnetic field of a charged body in uniform rectilinear motion can be transformed away by a Lorentz transformation, it is sometimes thought that we could revert to AMPERÈ's idea that the magnetic field is a subproduct of the electric field and even a dispensable one. But (a) the magnetic field cannot be eliminated in a covariant way: the only way in which $B = 0$ holds in every frame is by setting $E = 0$ in every frame; similarly for the inductions H and D; (b) the Lorentz transforming away does not work for an accelerated charged body; (c) the neutron has a magnetic moment even in its rest frame.

In short, CEM was cleansed and clarified by SR: there was no question of a basic change in the formulas because SR was the crowning of CEM.

2.4.2. Optics. FRESNEL's drag formula, the Doppler effect, and stellar aberration seem to have been the strongest motivations for the invention of SR (SHANKLAND, 1963, 1964), because they were inconsistent with classical kinematics, in particular with the classical formula for the composition of velocities. This inconsistency aroused no criticism of CM: authority can sustain a conspiracy of silence over half a century, even in science. The corresponding formulas are now elementary consequences of SRK.

Thm. 10. If an e.m. signal propagates in a material medium, relative to $\iota \in I$, with speed $v = \langle c/n, 0, 0 \rangle$, then it moves in $\iota' \in I$ with velocity $v' = (c/n - u)/(1 - u/nc) \cong c/n - (1 - n^{-2})u$. *Proof.* Make the indicated substitutions in Thm. 9 and expand in series retaining only first order terms.

Remark. The above consequence of EINSTEIN's composition of velocities formula was originally obtained (to first order in u) in the elastic theory of light; its confirmation by FIZEAU in 1854 was regarded as a triumph of that theory but not as a defeat of classical kinematics.

Thm. 11. If a monochromatic plane e.m. wave of frequency v falls on ι in the direction $\langle \cos \vartheta, \sin \vartheta, 0 \rangle$, then it propagates in the moving frame ι' with the following characteristics:

$$v' = \frac{\left(1 - \dfrac{u}{c} \cos \vartheta\right)}{(1 - u^2/c^2)^{\frac{1}{2}}} \, v, \qquad \tan \vartheta' = \frac{(1 - u^2/c^2)^{\frac{1}{2}} \sin \vartheta}{\cos \vartheta - (u/c)} .$$

Proof. Prove first that the wave phase $\varphi = \dfrac{2\pi v}{c} X_0 - kX$ is Lorentz invariant and apply the Lorentz transformation to the space and time physical coordinates occurring in φ.

Remarks. (1) The first formula accounts for the Doppler effect, the second for aberration, which in classical optics are totally independent effects. (2) The second formula shows that, in SR, kinematics and the geometry of physical systems are interdependent; moreover, that physical systems obey a motion-dependent geometry. It can be shown that the space of events is Lobachevskyan (FOCK, 1964).

2.4.3. Particle Mechanics. To build relativistic particle mechanics start by postulating the Newton-Euler equation for a particle with mass m_0 in a frame in which the particle is "momentarily at rest". Since in the rest frame $dt' = dT_0 \equiv d\tau$, the equation of motion can be written: $d(m_0 dX'/d\tau)/d\tau = F'$. Introduce now the 4-velocity $u^\mu \overset{\mathrm{df}}{=} dX^\mu/d\tau$ and the Minkowski force K^μ and generalize the law of motion to all four coordinates: $d(m_0 u'^\mu)/d\tau = K'^\mu$. Finally refer these equations to the unprimed inertial frame by performing a Lorentz transformation. Since the two sides of the equation are 4-vectors, the form of the equation is preserved:

$$\frac{d}{d\tau}(m_0 u^\mu) = K^\mu. \qquad (4.30)$$

In order to relate K^μ to known magnitudes and so interpret it, consider first a spatial part of (4.30):

$$\frac{d}{d\tau}(m_0 u^1) = \frac{dt}{d\tau}\frac{d}{dt}\left(m_0 \frac{dt}{d\tau}\frac{dX^1}{dt}\right) = (1 - v^2/c^2)^{-\frac{1}{2}}\frac{d}{dt}\frac{m_0 v}{(1 - v^2/c^2)^{\frac{1}{2}}} = K^1.$$

This equation can be rewritten in the Newtonian form "$\dot{p}=F$" on condition that the following identifications be made:

$Df.$ 10. $$F \overset{df}{=} (1 - v^2/c^2)^{\frac{1}{2}} K \qquad (4.31)$$

$$m = (1 - v^2/c^2)^{-\frac{1}{2}} m_0. \qquad (4.32)$$

The first is a definition of the Newtonian force in terms of the space part of the Minkowski 4-force and the relativistic factor. It is not a transformation formula but the relation between K and F, hence theoretically dispensable: it is needed only to relate SR mechanics to CM and so test the former. The second formula is not a definition but a transformation law: it is a hypothesis and moreover one that has been well corroborated in the laboratory. Nevertheless it could be false in some domain; and even the hypothesis "$m_0 = $ const" might not hold for certain particles.

In order to determine K_0 in terms of old acquaintances, multiply (4.30) by u_μ and use the lemma: $u_\mu u^\mu = -c^2$ and its corollary: $u_\mu(du^\mu/d\tau) = 0$. This yields

(a) $K^0 = (1 - v^2/c^2)Fv/c$,   whence $d(mc^2)/dt = Fv$ (b)   (4.33)

A comparison with PM suggests that K^0 be interpreted as the power $dE/dX_0 = F dX/dX_0 = Fv/c$. A similar comparison enables us to rewrite (4.33 b) as $dE = d(mc^2)$, whence

$$E = mc^2 + \text{const} = (1 - v^2/c^2)^{-\frac{1}{2}}m_0 c^2 + \text{const}. \qquad (4.34)$$

Nothing has so far been assumed concerning the nature of the forces and no such assumption was necessary because we have dealt with the law of motion not with a law of force. Yet in SR not every function will qualify as a force; in particular, hypotheses concerning actions propagating with superlight velocities will have to be discarded within SR, and instantaneous actions are of course of this kind. In classical physics there are essentially two cases to which SR dynamics applies: direct collisions and interactions mediated by an e.m. field. This does not prove that the Lorentz covariant equations of motion (4.30) have a smaller extension than NEWTON's corresponding equations, but rather that the latter was wrongly assumed to hold for every possible force hypothesis, while in fact it holds only when the retardation effects are negligible.

It will be noticed that the preceding construction of the foundations of SR dynamics has been heuristic rather than axiomatic, and that it differs from the standard one, which is based on the assumption of momentum conservation in elastic collisions. The latter is too restricted: examples play a heuristic role but they cannot constitute general theories. As to the axiomatic reconstruction of SR dynamics, it is a comparatively simple enterprise. In its elementary (non-hamiltonian) formulation, both m and the 4-force K will occur as primitives whereas E will be defined.

The novelties of SR dynamics with respect to CM are the following: (a) it reinterprets T, L and m as relative and local magnitudes, all of them dependent on the state of motion of the body concerned; (b) it abandons the tacit assumption of the invariance of forces and it restricts them to those which do not involve a universal time. From SR kinematics we are familiar with some of these changes: the genuine novelties concern m and F. The reinterpretation of the primitive concept of mass as a relation between a mechanical system and a reference frame [see (4.32)] rather than as an intrinsic property of the former constitutes a revolution and it limits NEWTON's interpretation of mass as the quantity of matter: indeed from now on only the proper mass m_0 — an invariant — may be regarded as a measure of the amount of substance. As to the noninvariance of forces, it entails (a) that if two forces balance in a given reference frame they need not balance in every other frame — i.e. the action and reaction law is not Lorentz covariant and is therefore alien to SR; (b) that the splitting of forces into central and noncentral (in particular velocity-dependent) is noncovariant as well, hence secondary.

2.5. Disputed Questions

SR teems with unsettled questions, partly because it offends common sense, partly because it has never been adequately axiomatized, and partly because so many philosophical schools have tried to force it into their prefabricated molds. Let us examine a few of them.

2.5.1. $E = m\,c^2$. Of all the misinterpretations of SR the most famous victimizes theorem (4.34) of SR dynamics — the so-called *equivalence* between mass and energy. It is usually regarded as holding in every branch of physics, i.e. for every kind of system, whence it is often called the law of the *inertia of energy* — wrongly so because it is a theorem in SR mechanics, hence referring only to material systems. To realize this recall how it was obtained: one of the premises entailing it was the generalization (4.30) of the law of motion of bodies, of NEWTON and EULER. The neglect of this logical point is responsible for the following queer ideas: (a) that inertia has an e.m. origin (hypothesis of the e.m. mass); (b) that light is deflected by gravitational fields because it carries mass and is therefore subject to NEWTON's law of gravity (Newtonian explanation of general relativity), and (c) that mass and energy are logically equivalent, so that one of them, and particularly the mass concept, can always be eliminated.

The numerical equivalence of m and E holds only for systems that are assigned a mass to begin with; and strictly speaking it is as little an equivalence as the linear relationship between force and displacement in HOOKE's law. Consider these two elementary particle processes: (a) the conversion of a proton-antiproton pair into a pair of pions and (b) the destruction of the former pair with emission of a γ-photon. In both processes the total energy is conserved; but the total mass is conserved only in the first while it disappears from the picture in the second process: unlike pions, photons are massless — they are not mechanical systems. The second process is not a case of "conversion of matter into energy", just as the mechanical theorem "$E = mc^2$" is not to be interpreted as the "interconvertibility [let alone the identity] of matter and energy". Matter, i.e. the class of material systems, must not be confused with energy, a property of physical systems, whether material or not. The theorem concerned says nothing about the conversion of particles into radiation or conversely: it just states a relation between two properties, E and m, of a material system. Only if taken in conjunction with the hypothesis that the energy of a closed system is conserved in every inertial system, can it be used to compute the energy of the "annihilation" photon. Conversely, given the energy of a "creation" photon, the formula can be used to compute the total mass of the particle pair.

The identifications of matter with mass, of energy with e.m. radiation, and of mass with energy, lead to absurd consequences such as attributing photons and neutrinos a mass, thereby forgetting (a) that these entities are not supposed to "obey" dynamics but quite different theories, and (b) that there can be no relative m unless there is a proper mass m_0 to begin with — but by the principle of the autonomy of e.m. signals ($SR\ 7$), photons have no rest mass because there is no frame in which a photon can be at rest. Once that initial blunder is made several amusing consequences follow, such as that the "proper" wavelength of the photon is indeterminate ($\lambda/0$) and that the photon energy is infinite. To crown the performance, the same author who attributes photons a mass is apt to regard this property (following MACH) as an acceleration ratio —

which of course is pointless in the case of e.m. signals in the absence of gravitational fields.

What then is the meaning of the "equivalence mass" of radiation energy and other kinds of energy? And why cannot we interpret every expression of the form 'E/c^2' as a mass, particularly since it has the right dimension? The answer is found upon looking at the referent, that Cinderella of operationalist physics: m is always the mass of a body, consequently 'E/c^2' can be interpreted as the mass of a body provided E happens to be the energy of that body. When a body is heated, magnetized, or electrically polarized, its increase ΔE in energy is accompanied by an increase in mass $\Delta m = \Delta E/c^2$. But if the radiation in a hollow cavity increases in energy by ΔE, the field in it gains no mass although the walls do lose mass by the amount $\Delta m = \Delta E/c^2$. Similarly, if a body emits radiation with a flux S, then it recoils with a momentum which is numerically the same as the field momentum S/c^2. But whereas we can put $p = m_b v_b$ for the emitting body we cannot put $S/c^2 = m_r v_r = m_r c$ for the radiation, simply because the mass concept is as out of place in the theories of the radiation field as the concept of grandfather. In general: numerical equalities do not entail the identity of the predicates involved.

2.5.2. Relative and Subjective. Another popular misinterpretation of SR is that it relativizes everything — as suggested by its name. It is true that SR shows that certain physical magnitudes, such as distance, duration, and mass, are relational rather than intrinsic, and that events (whether or not they appear to someone) are relative to the reference frame in a much more conspicuous way than assumed by prerelativistic physics. But, like every other physical theory, SR has a number of absolutes or frame-free magnitudes (e.g., electric charge and entropy) as well as hypotheses (all the Lorentz-covariant statements). SR is the theory of invariants under the Lorentz group (KLEIN, 1910).

To stress the relative traits of SR at the expense of the absolute ones is as misleading as the converse emphasis (PLANCK, 1933). Thus although it is true that in SR motion is relative, so it is in CM — but in neither case is this relativity of motion unrestricted in the sense that any motion can be transformed away with the choice of a suitable frame. For example, an oscillatory motion is not transformed away by a Lorentz transformation. Consequently an oscillating charged particle that in one reference frame emits an e.m. wave emits the same wave, though with changed v and λ, in a different inertial frame. Similarly, the electric and the magnetic polarization of rotating bodies are qualitative changes that cannot be cancelled by suitable frame choices. In general, qualitative changes such as the emission and absorption of radiation, the spontaneous decay of particles, and chemical and nuclear reactions, are not eliminated by changes of frame: only their tempo varies. SR, then, does not assert that every change is relative, let alone that every change can be transformed away, but only that any rectilinear uniform motion is equivalent to, hence indistinguishable from, any other such motion as long as no qualitative changes occur along it. Finally, it is clear that SR exempts e.m. radiation from the relativity of motion as regards the absolute value of its velocity.

A related and equally popular mistake is equating "relative" with "subjective". *Relative* is that which does not exist *per se* but in relation to something else; and *absolute* is whatever is not relative — in particular, to reference frames. On the other hand *subjective* is anything that depends in some way upon a cognitive subject and disappears with him. These are just matters of definition. Everything subjective is relative (to some subject) but not conversely. The relative objects handled by physics, e.g. forces, are not more subjective than the absolute ones, e.g. charges: otherwise they would pertain to psychology not to physics. Moreover certain absolutes, such as the spacetime interval, are made up of relatives. And, whether relative or absolute, physical ideas should be *objective*, i.e. subject-independent as regards their meaning and truth value. Relativity facilitates the attainment of objectivity by pointing out to absolutes (invariants and covariants) as candidates for ideas representing objectively existing things, properties and patterns, but it constitutes no guarantee of objective truth: an assessment of truth value requires also suitable empirical tests. (It is perfectly possible to fabricate a Lorentz covariant theory of ghosts.) In sum, invariance is a symptom of objectivity but not its warrant (BUNGE, 1959b).

The confusion between "relative" and "subjective" is unavoidable in operationalism, which refuses to admit unobserved physical events and patterns, and interprets every reference frame as an observer and every time value as a clock reading. So much the worse for operationalism: having competent and well equipped observers, indispensable as it is to test SR, is neither necessary nor sufficient to outline its physical meaning, if only because (*a*) not every inertial frame need or can be inhabited, and (*b*) no human observer has so far traveled in a perfectly inertial frame.

2.5.3. SR and Observers. We must insist on the observer-independence of SR because this theory is frequently presented as the child of observationalism. (EINSTEIN himself fell in this trap when young but got out of it in time to build GR: EINSTEIN, 1949, 1950.) SR makes no reference to observers and measurement operations: look at its basic concepts (see 2.2.1). This does not prevent the application of SR to

experimental arrangements and measurement acts *qua* physical events —
as long as the whole of relativistic physics is called in to cooperate, for
every experimental event is at the same time mechanical, electromagnetic, optical, and thermodynamical.

SR is not based on measurements but the other way around: space
and time measurements must be designed with the help of SR — and
GR — or at least no measurement result contradicting prerelativistic
physics should be rejected provided it agrees with SR. Far from summarizing experiments, SR has suggested a number of them, such as the
determination of the mass-velocity dependence and the time dilatation.
But neither SR nor any other single theory can cover any single experiment, because experiments involve macrofacts having a number of
sides with no universal significance. Invariants and covariants express
essential and universal properties and patterns, in contrast to the
accidental and local traits that come up in every observation act. For
these and other reasons a consistent subjectivist (e.g., operationalist)
interpretation of SR is impossible. Thus the identification of reference
frames with observers would render SR inapplicable even in terrestrial
laboratories, which are the only inhabited frames — but are also noninertial ones. And it would lead to science-fiction pieces like "m_0 is the
mass of a meson flying at a relativistic speed as seen by an observer
moving with it". SR cannot be interpreted in terms of observers and
their operations — much less of fictitious observers — because the very
gist of SR is that it attempts to eliminate the observer and his peculiarities
from the physical picture as far as the foundations are concerned. The
observer will come back anyway through empirical tests — but he will
come armed with SR rather than ignorant of it. For similar reasons
it is mistaken to speak of *apparent* masses, distances, times etc., instead
of the respective *relative* magnitudes. One and the same event may
present itself differently in different frames — and not at all in some;
a fortiori it will appear differently to different observers. The relativity
of appearances is just a particular case of the relativity of events. The
concern of SR is not with appearances but with universal patterns
(laws); if anything, relativity has demoted appearances and therefore
phenomenalism in favor of realism (BUNGE, 1967b).

2.5.4. What is SRK About? If SRK is not about appearances, what
class of physical objects does it refer to? Some take SRK to be a theory
of spacetime, others for a theory about the behavior of yardsticks and
clocks, still others for a new-fangled mechanics. If we look at the set
of primitive concepts employed in building SRK (see 2.2.1) we must
reject either opinion. The preceding axiomatization of SRK shows quite
clearly that it is about possible objective connections, via e.m. signals,

among physical systems of any kind, in particular physical reference frames. In short, the referent of SRK is the class of triples: system — e.m. signal — reference frame, i.e., $\Sigma \times S \times I$. The kind of system, i.e. Σ, is specified in every branch of physics by adding specific assumptions — e.g., mechanical ones — whereby the extension of the theory is decreased. In a region where no e.m. signals are possible, the inter-system connections envisaged by SR are absent, whence the theory becomes pointless (not false) in regard to that region. Such a switching off of e.m. fields would not stop every interaction among physical systems nor, in particular, the transmission of information: both direct body collisions and weaker signals, such as neutrino beams and gravitational waves, could still occur, but their respective theories need not be Lorentz covariant. In brief: no e.m. radiation field, no SR. It is not just that the test of SR involves light signals, but that the concepts S of e.m. signal and c of its velocity are involved in the very foundations of SR.

The above characterization of SR, as a logical (not only historical) outgrowth of CEM, should prevent one from teaching SR before CEM, as well as from trying to derive CEM by applying the Lorentz transformation to COULOMB's law — which does give interesting results precisely because by postulating the Lorentz transformations (instead of proving them) one smuggles in a fragment of CEM though without being able to understand what does 'c' mean in those formulas. By starting from this end one gets back many formulas of CEM — which appear as representing second order effects — thereby falling a victim to the illusion that CEM, and particularly the field concept, are dispensable. This is paid for by being unable to interpret many of the symbols gotten in this way. The awareness that SRK is a child of CEM precludes such circularities and in addition it dissolves two puzzles: (a) how is it possible that SRK is a theory proper while classical kinematics is a set of loosely related formulas? and (b) why should the velocity of light occur in purely mechanical formulas? — unless, of course, there are no purely mechanical facts but everything occurs against an e.m. background.

The standard mistakes in interpreting SR can be avoided by respecting a few metascientific rules:

(1) *Unearth the presuppositions* — in this case CEM and all that comes with it.

(2) *Watch the referent* (in this case $\Sigma \times S \times I$): do not apply your formulas blindly but find out what they are about — e.g., do not apply SR mechanics to photons.

(3) *Spot the primitives* and do not mistake relatedness for interdefinability — e.g., from the existence of relations among m, e and L do not conclude that at bottom all magnitudes are lengths.

2.5.5. Inequivalence of Space and Time. It is often claimed that SR has wiped out the difference between space and time and even between what has been and what may be: that it has spatialized time and that it pictures the world as a block given once and for all, so that nothing ever happens: everything would exist already in some region of the Minkowski space, which would be thoroughly homogeneous and isotropic. This is preposterous. SR cannot even be stated without the notion of e.m. signal, and an e.m. signal is a process (a sequence of events), not a static being. The welding of space and time in E^{3+1} does not rob time of its privilege, for (a) E^3 and T are mutually independent (irreducible) concepts and (b) the changes of state of physical systems are described by equations in which t (or τ) occur as independent variables.

It is true that any point in E^{3+1} is as good as any other one: E^{3+1} is in this sense homogeneous and isotropic. But E^{3+1} is only an abstraction of the space of events $\Sigma \times S \times I \times E^{3+1}$, which is inhomogeneous and anisotropic, for every point in it has its own light cone. And every light cone effects (a) an absolute (frame-independent) separation between timelike and spacelike vectors, and (b) a relative (frame-dependent) separation between past and future. In other words, the space of events, in which the future-directed e.m. signals exist, is not given for all eternity but is born together with happenings, and it has the arrow of time built into it. This is confirmed by the recent discovery that the Lorentz group is a consequence of the retarded action ("causality") principle, according to which the points in the space of events (not in E^{3+1}) are partially ordered (ZEEMAN, 1964).

2.5.6. Relativization and the Nonrelativistic Limit. As we saw in 2.4, there exists a recipe for the relativization of nearly every classical theory. But the recipe is fallible and moreover it requires that a theory be on hand before it can be relativized: it is a theory reconstruction procedure not a method for building a relativistic theory *ab ovo*. In the most interesting cases the theoretician will have to supply the egg himself — as in the cases of CEM and DIRAC's electron theory.

What is the relation between a nonrelativistic theory T and its relativistic extension RT? One usually says that T is contained in RT in the sense that RT goes over into T for $v \ll c$. (There is no proper limit for $c \to \infty$ since, by hypothesis, $c = \mathrm{const.}$) Thus the spatial components of the 4-velocity approach the ordinary velocity in the nonrelativistic limit. But the time component $u^0 = \mathrm{d}X^0/\mathrm{d}\tau$, which does not occur at all in CM, approaches c instead of vanishing in the nonrelativistic limit. Similarly the total energy of a body does not vanish but approaches the proper or rest energy $m_0 c^2$. And the spacetime interval approaches $\mathrm{d}s = (c - v^2/2c)\,\mathrm{d}t$, which does not equal the Euclidean space interval.

This is why it was said in 2.4 that one should not be alarmed if, in the "nonrelativistic limit", there is a relativistic remainder. SR is not just a question of large velocities or high energy: several of its basic concepts are structurally different from the classical homologues. Consequently, neither T⊂RT nor the converse relation hold. Rather, T and RT have a nonempty intersection that does not contain all of T. The relations between any given T and the corresponding RT are far from simple and are not well known. One more field that could use some more FR.

3. General Relativity

EINSTEIN's 1916 theory of gravitation (GR) is a good example of a fundamental theory on whose foundations no two authorities agree. If one could succeed in showing that what invites controversy over GR are chiefly the principles believed to have guided its construction, rather than its constitutive postulates (see Ch. 2, 1); that those heuristic principles, though helpful, were insufficient to create the theory; and that the theory stands now without the original scaffolding, then one might divert the controversy to the more profitable area of the proper postulates of GR. Our first aim will therefore be to spot and analyze the regulative principles of GR; once this has been done it will be easier to recognize, analyze and order the actual assumptions of GR.

3.1. Heuristic Components

GR is often said to be based upon five assumptions: (1) the equality of inertial and gravitational mass, (2) the principle of equivalence, (3) the principle of general covariance, (4) EINSTEIN's field equations, and (5) the geodesic postulate. As a matter of logic the first of these hypotheses cannot even be stated in the language of GR, while the second and the fifth are now theorems of GR although the first and the second did take part, along with the third, in the construction of the theory. There were several other heuristic cues: that the field equations should be covariant generalizations of the Poisson equation, that for weak gravitational fields the line element should go over into the SR expression, that the energy-momentum-stress tensor of matter should be conserved, that SR should be recovered for freely falling frames, and so on. Since these are usually acknowledged to be heuristic (theory formation) principles, we need not insist on them. Let us concentrate on those which usually pass for postulates of GR.

3.1.1. Inertial and Gravitational Mass. Discussions on the foundations of GR often begin by distinguishing two mass concepts: the inertial and the gravitational mass — and sometimes even two kinds of gravitational mass, the active and the passive ones. This distinction seems

to have been unknown to the founders of CM and the classical gravitation theory. In fact, in the basic equations of the Newtonian theory of gravitation, as modernized by POISSON,

$$m\ddot{X} = -m\nabla\varphi, \qquad \nabla^2\varphi = 4\pi\gamma\varrho, \qquad \varrho \overset{dt}{=} dm/d^3x, \qquad (4.35)$$

a single mass occurs. And if the distinction does not appear in the foundations of the theory it should occur nowhere else in it unless different mass concepts could be explicitly defined in terms of the single mass concept figuring in (4.35). Strictly speaking, neither the inequality nor the equality of inertial and gravitational mass can even be stated within classical gravitation theory.

Yet mass does seem to play two different roles; or, better, the mass concept connotes two different properties: inertia and gravitation. Hence it might well have happened — anthropomorphically speaking — that Nature had wished to do things differently, endowing every piece of matter with two different masses relatively to a single reference frame. In such a case the classical theory of gravitation would have to be modified: two different mass concepts, designated by m_i and m_g, would have to be introduced as mutually independent primitives entering the basic equations in the following way:

$$m_i\ddot{X} = -m_g\nabla\varphi, \qquad \nabla^2\varphi = 4\pi\gamma\varrho_g, \qquad \varrho_g \overset{dt}{=} dm_g/d^3x. \qquad (4.36)$$

Moreover, if the principle of the equality of the action and the reaction were to fail even for static gravitational interactions, we would be justified in tentatively splitting the hair into three, by replacing 'm_g' by 'm_p' in the equations of motion and by 'm_a' in the field equations, where 'm_p' and 'm_a' would stand for the passive and the active gravitational masses respectively (BONDI, 1957).

But it so happens that experiment has never detected any difference between these hypothetically different masses; in other words, the hypothesis that there are two (or three) different masses is experimentally unwarranted. This is usually expressed by saying that the inertial and the gravitational masses are equal, a statement sometimes called the *equivalence principle*. But this statement is misleading insofar as it suggests that there *are* two different masses, only they happen to be numerically equal rather than identical. Such an interpretation is bolstered by the operationalist demand to "define" every physical concept by means of a set of laboratory operations. According to this philosophy, since there are different mass measurement techniques there must be different mass concepts. Thus a mass measurement based on the law of momentum conservation will "define" inertial mass while a measurement using a gravitation law will "define" gravitational mass. (Actually different kinds of operation are associated with different law statements rather than being theory-free, but this is overlooked or else it is declared that the laws of physics are just rules of operation.) By proceeding in this fashion it is overlooked (a) that measurements do not define anything but help determine (assign) some numerical values of magnitudes, whether primitive or defined in a given theory, and (b) that the most important concepts in any theory are not the defined ones but those which help define — namely the primitive or undefined concepts. In any case the proper setting for conceptual analysis is the conceptual framework itself, i.e. theory, rather than the laboratory; in particular, one does not analyze the mass concept in the laboratory.

Since experiment has so far refuted the non-Newtonian conjecture "$m_i \neq m_g$" (ROLL et al., 1964), we infer that "$m_i = m_g$" holds — but we cannot make this statement in the context of the Newton-Poisson theory, whose language does not contain the terms 'm_i' and 'm_g'. In brief, the statement is an outcast and therefore does not count as a law statement (see Ch. 1, 3.1.2). Moreover, to say that the equality of the inertial and the gravitational masses is a law of nature is very much like announcing that the day the Morning Star was found to be identical with the Evening Star an important law of nature was discovered. Science is certainly built by advancing hypotheses and trying them out, but if one were to regard the negate of every unsuccessful conjecture as a law of nature one would unnecessarily overpopulate physics. Just think of distinguishing as many kinds of lengths, durations, or pressures as measurement techniques (see Ch. 1, 2.1.3).

In short, the conjecture that there might be two different masses has been shelved (perhaps only temporarily) and consequently the original unity of the (theoretical and classical) mass concept has been restored after much waste of time caused by a dusty philosophy. The role played by GR in this has been to help with this restoration by pointing out that, if there were two kinds of mass, then GALILEI's law would not hold and consequently the equivalence principle would fail as well. (Incidentally, the equivalence principle is not equivalent to the so-called law of the equality of the inertial and the gravitational masses: the latter is at best necessary but not sufficient to escape the local effects of gravitation by riding a freely falling cage. In fact if the law of motion were much different from NEWTON's second law, the local equivalence between a constant homogeneous gravitational field and a uniformly accelerated frame might not hold.)

In short it seems doubtful that the equality of the inertial and gravitational mass need or even can be stated in a meaningful way, even as a heuristic guide to the construction of a gravitation theory. All one does assume about mass when building such a theory are the following mutually independent hypotheses: (1) "Where there are bodies there is gravitation" — the converse being true in the classical theory but false in GR. Sometimes the formula "Mass is the source of the gravitational field" is advanced, but this is obviously a misfit: whole things, not their separate properties, can originate other things. (2) "The motion of a structureless particle in a purely gravitational field does not depend on the particle mass or on any other parameter specifying the nature of the particle". I.e., the gravitational field, unlike every other known field, acts regardless of the nature of the particle as long as the latter is simple (to the point of not existing). These two hypotheses are found in both the classical and the relativistic theories of gravitation: the first statement is an ordinary language (hence ambiguous) version of the Poisson and the Einstein field equations, while the second statement is a corollary of the Newton and the Einstein equations of motion. They can function as vague guides in building a gravitation theory, whether relativistic or not. In particular, statement (2) seems to have suggested Einstein the lift gedankenexperiment. But, being a negative statement, (2) is compatible with infinitely many statements and provides therefore an exceedingly vague guidance: it cannot function as a postulate for it furnishes no positive indication concerning the variables that do occur in the equations of motion. By no stretch of the imagination does it suggest, say, that these equations should contain the affine connection. In short, (1) and (2) are heuristic not constitutive principles.

In summary (a) the lack of difference between the inertial and the gravitational mass does not constitute a law of nature: the statement is just the negate of the tentative conjecture that there might be such a difference and it does not belong to any accepted physical theory; (b) if there were two different mass kinds, the motion of a body in a gravitational field would depend on their ratio and the equivalence principle proper would fail — but this is another story.

3.1.2. Equivalence of Gravitational Effects and Acceleration. GALILEI's

law, $\ddot{X} = g$, covers (in the slow motion approximation) motions determined by a genuine static and homogeneous gravitational field, as well as free motions related to a uniformly accelerated frame. This suggests but does not entail that, at least locally and from a purely kinematical viewpoint, the effects of a static homogeneous gravitational field are the same as those of a suitably accelerated frame. And if they are the same then they are indistinguishable. (Physical identity implies empirical indistinguishability not vice versa.) In other words, in the neighborhood of any point in a static homogeneous gravitational field, the effects of this field on matter and on e.m. radiation are identical with those

3.1. Heuristic Components

connected with an accelerated reference frame in a field-free region. This is one of the various versions of the equivalence "principle" — actually a theorem of GR (see 3.2.4).

An immediate consequence of this hypothesis is that all bodies gravitate alike regardless of their mass — which is not surprising as this was the heuristic source of the "principle". A second consequence is that an observer riding an "equivalent" frame, such as EINSTEIN'S freely falling lift, will detect none of the effects of the field as long as it is strictly static and homogeneous. But, of course, the "principle" is supposed to hold even when nobody is looking: after all, it is a physical assumption not a statement about observers. Moreover such comoving observers are not always employed and they are not needed to test the "principle". Thus the bending of starlight in our star's gravitational field — which follows qualitatively from the "principle" — is measured by observers temporarily at rest relative to the center of the field (the sun).

Contrary to popular belief, there is no equivalence between gravitation and acceleration but only a local equivalence (i.e. one restricted to "small" spacetime regions) between (a) the ponderomotive effects of a static homogeneous gravitational field on either matter or e.m. radiation, and (b) the motion of either referred to an accelerated frame (e.g., BERGMANN, 1962). Thus one can imitate a ballistic trajectory by referring a free motion to a freely falling elevator. On the other hand the collision of the projectile with a second body cannot be mimicked in that way. Further, there can be a gravitational field in a region unaccompanied by an acceleration because there are neither bodies nor photons in it: in this case the "principle" is vacuously true (pointless). In short, the "principle" does not assert that gravitation and acceleration are the same: it suggests neither a kinematical theory of gravitation nor a gravitational kinematics. Nor does it state that it is impossible to distinguish, by observation alone, between a constant and homogeneous gravitational field and an Einstein cage: this is a corollary.

The "principle" works in the very special circumstances obtaining in terrestrial measurements: every such operation is local and performed in an approximately static field. But this is another way of saying that by terrestrial measurements alone a static uniform gravitational field cannot be distinguished from the kinematical effects associated with a noninertial reference frame. To make the distinction we need something else. A *theoretical criterion* for telling genuine ("true") from fictitious gravitational fields is this: "A gravitational field is genuine iff it cannot be eliminated globally (all over) by a transformation of coordinates — a transformation which is always possible locally". The application of this theoretical criterion need not involve any measurement: what is called for is an investigation of the field equations

14*

and their solutions or, equivalently, an examination of the metric tensor. In particular, a gravitational field vanishes over a given region just in case the full curvature tensor $R^\tau_{\mu\nu\sigma}$ vanishes at every point in the region. Since the latter constitutes a tensor equation, it holds relative to every frame: it is absolute, and consequently no tricks with lifts and observers will create a field out of nothing. Moving frames will only mimic the kinematical effects of a uniform static gravitational field — one which, strictly speaking, exists nowhere. On the other hand if the curvature tensor is different from zero over a certain region in any one frame, then the corresponding gravitational field is independent of the frame: the field will be out there in an absolute way (see, e.g., SYNGE, 1960). If experiment is insufficient to distinguish the local kinematical effects of the field from a change of frame, so much the worse for experiment: this only shows that experiment, though necessary, is insufficient to gather reliable and relevant information.

The value of the equivalence "principle" is chiefly heuristic: it helps in a vague way to generalize the equations of SR to curved space. Indeed, since in the immediate neighborhood of any given point the kinematical effects of a static homogeneous gravitational field can be duplicated by a suitable change of frame, and since in the absence of a gravitational field we assume that SR holds, we can hypothesize SR for that frame and then perform a coordinate transformation. Moreover the "principle" strongly suggests the general covariance principle — another tool in the transition from SR to GR. In fact if, in the small (yet not microscopic), physics is the same whether referred to a frame at rest in a static field or referred to a freely falling frame, then why not make one more step and require that the laws of physics be altogether frame-independent?

But, like every other heuristic guide, the equivalence "principle" does not determine the generally covariant equations in a unique way (see, e.g., TOLMAN, 1934 and TRAUTMAN, 1965). Moreover it is possible to reformulate Newtonian mechanics and the Newton-Poisson equation so as to comply with the equivalence "principle" (HAVAS, 1964), whence the latter is not peculiar to GR. What is more important, the equivalence "principle" is irrelevant to the nucleus of GR, which is a system of partial differential equations referring to ("describing") the gravitational field. Indeed, how could the "principle" entail or even suggest any field equations if it holds that the field can be neglected in a small region as far as its kinematical effects are concerned and as long as it is static and homogeneous? In summary, the "principle" did fulfil a limited heuristic function in EINSTEIN's early gropings (1911) but its present status is that of a theorem — not to speak of the confusion its various misstatements have created.

3.1.3. General Covariance. Just as the irrelevance of mass to motion in a gravitational field suggested the equivalence "principle", so the latter, in conjunction with other considerations, suggested the widest of all the regulative principles of GR: general covariance. But before we can state this principle we must distinguish between basic and derived law statements (see Ch. 2, 1). A *basic law statement* is one containing no individual constants and entailing other general statements intended to represent objective patterns. Field equations are such basic laws while their solutions — which require the choice of a reference frame — are derived laws. The covariance principle applies only to the former.

The *general covariance principle* (= general relativity principle) is stated in a number of ways, not all of them equivalent. One of its versions reads: "The basic law statements of physics should hold (= be true) in any frame of reference and in any coordinate system — or, equivalently, they should be preserved under arbitrary changes in coordinate systems, whether or not the latter mirror different physical reference frames". For this it is sufficient but not necessary that the fundamental law statements in question be tensor or pseudotensor equations. Notice the following points. (1) The principle refers to law statements not to physical events: it is a metanomological statement not to be placed in the same class with physical laws proper (BUNGE, 1961b). (2) It does not refer to every physical law but only to basic laws; hence it need not be obeyed by the logical consequences of those laws, notwithstanding a widespread opinion (see e.g., HILBERT, 1915, 1917). In particular, it is not obeyed by the equations of motion, whose form depends on the coordinates used (BERGMANN and BRUNINGS, 1949). And it is not obeyed by the boundary conditions either. Equivalently: physical properties and events, in particular observable ones, need not be covariant. (3) The principle does not say that every basic law so far found is in fact generally covariant but rather that every statement of that kind ought to be formulated in a generally covariant way: it is a normative metanomological principle even if its normative status is sometimes disguised by stipulating that general covariance is to be satisfied by "properly formulated" law statements (BUNGE, 1961b). (4) The principle is not restricted to a specific theory, such as the theory of gravitation, but is supposed to be implemented everywhere in fundamental physics. (5) The principle invites the observer to withdraw from the basic physical picture of nature: it entails, in fact, that his choice of a reference system is external to the objective physical pattern and that it should make no difference — so far as the form and content of the basic laws are concerned — whether or not a reference frame is ridden by an operator as long as his peculiarities and deeds do not show up in those statements: he must be at most a sleeping partner. On the other hand most experimentally accessible predicates and statements are frame-dependent: experimental data are seldom invariant.

The principle is therefore blatantly inconsistent with operationalism: far from placing the observer and his idiosyncrasies in the center of things, general covariance enjoins him not to count himself among the referents of physical theory, however important he may become in the empirical test of it. In a sense the principle of general covariance is a technical implementation of the age-old desideratum of objectivity and universality, as well as a device for pushing aside what is only apparent, i.e. dependent on some cognitive subject. (Actually it constitutes a necessary though insufficient condition for objectivity — just as the weaker requirement of Lorentz covariance.) In all of the five above mentioned respects the general covariance principle is the heir to the Poincaré-Einstein principle of restricted relativity, whence the name GR is partly justified (see however FOCK, 1964). Only partly, because 'relativity' refers to the transformation properties of the theory not to its physical referent — the gravitational field. A better name for EINSTEIN's 1916 theory is therefore that of *general relativistic theory of gravitation*. This is what 'GR' will here be taken to stand for.

General covariance is not the same as *invariance of form* under coordinate changes. Form invariance, though desirable, is neither indispensable nor always attainable. What is the point of preserving, say, the form of a wave equation when "going" from a rectangular to a spherical coordinate system? What we want is that the equation as a whole be preserved even though every one of its terms changes: these individual changes should compensate one another. Thus if $\partial_\mu \partial^\mu \psi = 0$ holds in cartesian coordinates, we require that it holds also in spherical coordinates even though its form will alter; for example, the ϑ-term will be $(1/r^2)\frac{\partial^2 \psi}{\partial \vartheta^2}$ not $\frac{\partial^2 \psi}{\partial \vartheta^2}$. (This partial change in form is necessary for dimensional homogeneity.) In short, we want covariance rather than invariance of the basic laws, and indeed with respect to both coordinate systems and reference frames. If interpreted as form invariance, then general covariance ceases to constitute a partial implementation of the desiderata of objectivity and universality — i.e. absoluteness — of the basic laws: it appears as a purely mathematical requirement (KRETCHS-MANN, 1917; TOLMAN, 1934; FOCK, 1964; HAVAS, 1964). Moreover the requirement of form invariance restricts the class of coordinate systems (and therefore also of coordinate transformations) to the subset of holonomic coordinates (POST, 1967) — a troublesome restriction with no physical ground. In short, form covariance cannot be general. Equivalently: if 'covariance' is taken to mean "form invariance", then the expression 'general covariance' is self-contradictory. If only for this reason the more comprehensive sense of 'covariance', i.e. preservation of meaning and truth value upon coordinate transformations, will be adopted here. In our interpretation the principle of general covariance is neither a formal principle nor a physical law but a metalaw and as such a regulative rather than a constitutive assumption: it entails nothing by itself but serves as a control of other statements — in particular to eliminate the peculiarities of individual frames and observers.

That general covariance should not be understood in a purely formal way has been suggested at various times (HILBERT, 1917; SCHRÖDINGER, 1954; IKEDA, 1959). In fact one should not accept arbitrary coordinates and coordinate transformations but should exclude all those quaint coordinate systems that do not allow the distinction between spacelike and timelike vectors, and all those coordinate transformations that do not preserve this dichotomy, or that obliterate physical differences. This restriction to *physically admissible* coordinate systems and coordinate transformations shows both that GR does not blot out the distinction between space and time, and that general covariance is not properly construed as form invariance. Our axiom system for GR (see 3.2) will involve such restrictions as well as boundary conditions at spatial infinity, which further restrict the invariance group(oid) of GR and recall that it is simply not true that relativity, whether special or general, boils down to recognizing that there is no difference between the time coordinate and the spatial coordinates.

The restricted democracy of coordinate systems that holds at the level of principles is further limited at the level of theorems. In fact it is not true that all frames are equivalent and therefore the choice of frame is conventional: while for stating a basic law we may choose any physically admissible frame (in HILBERT's sense), the solutions of those fundamental equations are best interpreted if referred to the frame associated with the more massive system — the sun in the case of the Schwarzschild solution — which is the one determining the salient features of the metric. Thus if we are to describe planetary motions we had better choose a heliostatic frame, whereas if we are to describe the motion of a cosmic ray beam a geostatic frame recommends itself. In either case the choice of frame is determined by the referent or object of study, and if possible we choose the more nearly inertial frame because it happens to be associated with the main field and it does not introduce irrelevant complications: thus the sun can be regarded as freely falling in the galactic gravitational field, and neutral cosmic rays are likewise Einstein cages in the terrestrial field. As long as we keep the distinction between basic laws and their logical consequences there is no incompatibility between general covariance (or general relativity) and the preference for certain privileged frames such as the geodesic one. And even at the level of axioms we may single out a class of coordinate systems as long as it is infinite — e.g., the class for which Det $\|-g_{\mu\nu}\| = $ const.

This applies, in particular, to the heliocentric-geocentric dispute. Mathematically the terrestrial (Ptolemaic) axes are as good as the solar (Copernican) ones. But the former are definitely inadequate from a physical point of view: kinematically, if only because relative to terrestrial axes the fixed stars acquire superlight tangential velocities, and the path of the light they emit is helical; dynamically, because it is actually the sun which, by modifying the structure of the adjacent space, guides the motion of the planets and not conversely — as shown by the amazing success of the Schwarzschild formula applied to the sun-planet system schematized as a central field in which a passive test particle moves. Therefore the attempts to justify the enemies of GALILEI in the name of GR are ludicrous (GIORGI and CABRAS, 1929; SYNGE, 1960; BUNGE, 1961a; FOCK, 1964). The blunder might have been avoided by a more careful statement of the general covariance principle, one emphasizing its reference to basic laws. It is only when consequences are deduced from the basic laws, e.g. in order to describe motions, that a particular reference frame and associated items, such as boundary conditions, must be chosen, whence the description of events becomes relative to such a choice. Another way of putting it is this: coordinate systems, being conceptual objects, are all equivalent; but reference frames, being physical things, are not all alike and therefore whenever a coordinate system happens to represent a physical frame, the choice ceases to be arbitrary or conventional — and this is the case with the low-level laws. Moreover, it is conceivable that certain physical systems should have their own coordinate grids engraved in them. Such intrinsic coordinates could be possessed not only by bodies but also by fields. In the latter case we could claim to have disclosed a set of intrinsic coordinates provided they were uniquely determined by local characteristics of the field itself (BERGMANN, 1962). It would be only natural to use such natural or intrinsic coordinates preferably to others, as they would introduce no accidental (e.g., operational) features into the physical picture. This would be just as effective a means for attaining objectivity as the requirement of general covariance. And if intrinsic coordinates could be determined in every case, they might render general covariance pointless, for this requirement presupposes that any given coordinate is as arbitrary (nonphysical) as any other one. But this is just a possibility (BERGMANN, 1961): so far physics proceeds as if nature did not wear coordinates.

Let us finally investigate three questions: is general covariance peculiar to GR, is it testable, and what is its estimated life span? The principle does not characterize GR uniquely but can be implemented in every branch of classical physics: in particular, in CM (CARTAN, 1923, 1924), in CEM (POST, 1962) and in the Newton-Poisson theory of gravitation (HAVAS, 1964). This alone shows that general covariance has no deductive power and is therefore unfit to count as an axiom proper: its function is not to generate theorems but to prune physical statements from operator-dependent features. The answer to the second question is likewise in the negative: the general principle of relativity is not empirically testable, for it directs us to write the basic laws of physics precisely in such a way that no observation could possibly detect any difference between the law as referred to one frame and the same law as referred to another frame. In other words, by stating the basic laws in a general covariant way we ensure that, if they hold in any one frame, then they hold in every other frame — i.e. irrespective of changes in point of view and mode of observation. Moreover, the principle is irrefutable: no experiment can refute it because it does not refer to facts but to statements. And no failure to implement it in a given field will suffice to persuade us to give it up. In the face of such a failure we will first blame our own wits, then the set of formulas we want to reformulate in a generally covariant way. In any case with general covariance, as with other metanomological principles having a deep philosophical root, it seems always too soon to abandon them (BUNGE, 1961 b). At most we shall conclude that they have deserted us. All of which confirms the metatheoretical character of the principle. Finally: can general covariance be expected to survive the theories that satisfy it? The complaint is sometimes heard that general covariance has rendered no remarkable service in microphysics, where Lorentz covariance is usually felt to be sufficient (and workable). And even this more restricted covariance is sometimes looked upon with suspicion either because it seems to be designed for macrophysics or because it often brings with it redundant solutions, i.e. solutions with no physical meaning, as in the case of the Klein-Gordon and the Kemmer-Pétiau equations. Whether this shortcoming should be blamed on any or all of the covariance principles rather than — as seems more plausible — on some of the specific traits of the hypotheses concerned, has not been established. If anything, it would seem that Lorentz and general covariance will stay unless natural or objectively privileged spacetime grids are found, and that not so much because they have helped create SR and GR — after all we know they are heuristic principles — but because they help attain two inescapable goals of science: objectivity and universality.

In summary: (*a*) general covariance is a valuable heuristic principle (EINSTEIN, 1916 and 1918); (*b*) although general covariance was born with GR it is not exclusive of it, whence even though GR is general relativistic, it should not be called by this name alone; (*c*) the principle is supposed to apply to basic laws in any chapter of physics; (*d*) it expresses the bold hope that the basic laws are the same throughout the universe: that they can be "displaced" anywhere in spacetime even though their constituents cannot; (*e*) it renders particular frames and observers irrelevant to the basic laws. General covariance is, in short, a valuable principle, one which GR abides by, but not one which can be placed at the basis of either GR or any other theory, because it is regulative, not constitutive (see Ch. 2, 1). The constitutive principles of GR shall be dug out in the following.

3.2. Basis of General Relativity

3.2.1. Background. The *formal background* of GR consists essentially of PC=, semantics, algebra, topology, analysis, and manifold geometry — in particular the theory of Riemann spaces. This formal background supplies the units out of which the formalism of GR is built. For example, it provides the relation between the affinity Γ and the metric tensor g in a Riemann space:

$$\Gamma^{\lambda}_{\mu\nu} = \frac{1}{2} g^{\lambda\sigma} \left(\frac{\partial g_{\sigma\mu}}{\partial x^{\nu}} + \frac{\partial g_{\sigma\nu}}{\partial x^{\mu}} - \frac{\partial g_{\mu\nu}}{\partial x^{\sigma}} \right). \tag{4.36}$$

The *material or physical presuppositions* of GR are more difficult to trace because of the traditional intertwining of constitutive and heuristic elements. In any case they seem to be the following: (*a*) the theory of local time and the general systems theory, as well as some elements of physical geometry — the nonspecific ones (see Ch. 2, Secs. 3, 4 and 5); (*b*) SR, whose function is to supply the asymptotic values of the coefficients of the metric and the matter tensors. SR fulfils, in addition, a regulative function: it constitutes the starting point of some generalizations, it provides a check of the GR formulas, and it suggests the meaning of some of the symbols of GR, notably of the coordinates. But so does the Newton-Poisson theory of gravitation.

3.2.2. Basic Concepts. The axioms of GR are built with concepts borrowed from its background and the following specific primitives which GR either newly introduces or endows with a new meaning:

M^4: A differentiable manifold representing spacetime.

$\{g\}$: A family of tensor fields representing the gravitational field potentials.

Σ: The set of gravitational fields.

$\bar{\Sigma}$:   The set of physical entities other than gravitational fields.

$\{\mathfrak{T}\}$: A family of tensor fields representing the momentum and stress of the members $\bar{\sigma}$ of $\bar{\Sigma}$.

K: The class of reference frames.

$\{X\}$: A family of functions representing physical coordinates of $\bar{\sigma}$'s.

$\varkappa$:   A positive real number representing the coupling of σ with $\bar{\sigma}$.

As usual in physics, every primitive concept has both a definite mathematical structure and a fairly definite meaning: GR is in no way more mathematical than other physical theories; in particular, it does not constitute a geometrization of physics.

3.2.3. Defined Concepts. The preceding primitives suffice, in conjunction with generic concepts taken from the background of GR, to derive the following additional concepts.

Df. 1. A nonempty, simply connected and finite region of M^4 is called V.

Df. 2. If x^μ and x'^μ are local coordinates in the neighborhoods of some points x, $x' \in M^4$ and these neighborhoods have a nonempty intersection, and if f^μ are $1:1$ functions of the class C^r, then $x'^\mu = f^\mu(x^\nu)$ is called a *coordinate transformation* of class C^r.

Df. 3. Any primitive or function of primitives that remains unchanged under an arbitrary coordinate transformation with nonvanishing Jacobian is called *generally invariant*.

Df. 4. Any relation or statement involving any of the primitives or functions of them, that is satisfied when subjected to an arbitrary coordinate transformation with nonvanishing Jacobian, is said to be *generally covariant*.

Df. 5. *Curvature tensor*

$$R^\tau_{\mu\nu\sigma} \overset{\mathrm{df}}{=} \frac{\partial \Gamma^\tau_{\mu\sigma}}{\partial x^\nu} - \frac{\partial \Gamma^\tau_{\mu\nu}}{\partial x^\sigma} + \Gamma^\tau_{\alpha\nu} \Gamma^\alpha_{\mu\sigma} - \Gamma^\tau_{\alpha\sigma} \Gamma^\alpha_{\mu\nu}. \tag{4.37}$$

Df. 6. *Contracted curvature tensor*

$$R_{\mu\nu} \overset{\mathrm{df}}{=} R^\sigma_{\mu\sigma\nu}. \tag{4.38}$$

Df. 7. *Scalar curvature*

$$R \overset{\mathrm{df}}{=} g_{\mu\nu} R^{\nu\mu}, \quad \text{with} \quad R^{\mu\nu} = g^{\mu\alpha} g^{\nu\beta} R_{\alpha\beta}. \tag{4.39}$$

Df. 8. *Einstein tensor*

$$G^{\mu\nu} \overset{\mathrm{df}}{=} R^{\mu\nu} - \tfrac{1}{2} R g^{\mu\nu}. \tag{4.40}$$

Df. 9. A body whose gravitational field does not distort appreciably a given gravitational field acting on it is called a *test body*.

The first eight definitions are purely mathematical: only Df. 9 introduces a physical concept. Yet the concept of invariant is sometimes identified with that of observable (BERGMANN, 1962). This designation is objectionable: (a) observability is not an intrinsic property of physical objects but depends also on the means of observation; (b) most observables proper, like position and brightness, are not invariant; (c) most invariants, like the spacetime interval, are not observable; (d) one does not attain objectivity and universality by sticking to what can be observed but by raising above it: invariant quantities and covariant statements may be attributed to the physical situations themselves precisely because they are observer-independent. As to Df. 9, it is deliberately vague: a clear cut definition would restrict a priori the applicability of GR.

Several additional concepts will occur in the following which are neither among the primitives nor are defined in terms of them but are introduced into GR through the matter tensor $\mathfrak{T}$ — the door through which the whole of physics can enter GR, which renders it the most sociable of physical theories. Among those concepts are mass, electric charge, and e.m. field component. As far as GR is concerned there might not exist bodies, electrically charged particles, electromagnetic fields, or de Broglie waves: GR does not deny their existence but it does not deal with them except in so far as they are represented in bulk by $\mathfrak{T}$. In other words, GR leaves the detailed study of matter and nongravitational fields to other theories: these can profit from GR but must supply their own premises. GR is interested in $\mathfrak{T}$ as an over-all representative of systems of $\bar{\sigma}$'s, and only insofar as their interaction with gravitational fields is concerned. Consequently GR should not be expected to say anything new about mass, charge, and other predicates that characterize nongravitational entities. This goes to show that GR is a pluralistic theory (see Ch. 1, 4.2.2). Any theory which, like geometrodynamics (WHEELER, 1962), attempts to construct $\mathfrak{T}$ out of g in a unitary fashion, goes far beyond GR. Let us now bind the previously introduced concepts into initial assumptions.

3.2.4. Axioms. The leading axiom of GR and therefore the only one that is usually mentioned, is the set of field equations — the last to occur in our presentation. The initial assumptions that precede it will, as usual, set the stage for the appearance of the main character, one whose scope and beauty are staggering. We shall first list the axioms and then comment on them.

Axiom group 1: referents

GR 1.1 (a) Σ is a nonempty set. (b) Every $\sigma \in \Sigma$ represents a gravitational field.

GR 1.2 (a) $\bar{\Sigma}$ is a nonempty set disjoint from Σ. (b) Every $\bar{\sigma} \in \bar{\Sigma}$ represents a physical system other than a gravitational field.

Axiom group 2: spacetime

GR 2.1 M^4 is a 4-dimensional manifold of class C^2. [Roughly: (a) for every point $x \in M^4$ there exists a neighborhood of x over which a local coordinate system can be given ("defined"); (b) the functions f^μ that map neighboring coordinate sets onto one another (see Df. 2) are continuous and have continuous derivatives up to 2nd order in M^4.]

GR 2.2 (a) All coordinate transformations in M^4 are bijective maps. [I.e., they are 1:1, onto and continuous. Rationale: coordinate transformations should preserve the dimensions of M^4.] (b) The Jacobian of every coordinate transformation vanishes nowhere. [Rationale: areas and volumes should not vanish upon coordinate transformations, for otherwise things themselves would appear to vanish by sheer paper and pencil operations.]

GR 2.3 For any point $x \in M^4$ there exists a neighborhood of x over which a local coordinate system can be given such that x^0 is a (local) time coordinate and x^1, x^2 and x^3 are (local) spatial coordinates. [Roughly: the tangent space of M^4 at any $x \in M^4$ is spacetime.]

Axiom group 3: metric

GR 3.1 (a) $\{g\}$ is a nonempty family of symmetric tensor fields, of valence $(0,2)$, on M^4. (b) The components of g and their first and second derivatives w.r.t. the x^μ are real valued and bounded in every $V \subset M^4$ with the possible exception of a proper subset of V. [Rationale of the exception: to make room for singularities such as those of the Schwarzschild solution.] (c) M^4 is endowed with a metric $ds^2 = g_{\mu\nu} dx^\mu dx^\nu$ with $\mu, \nu = 0, 1, 2, 3$.

GR 3.2 (a) $g_{00} > 0$ and $g_{ij} < 0$ $(i, j = 1, 2, 3)$ everywhere. (b) $g_{ij} - (g_{0i} g_{0j}/g_{00}) < 0$ everywhere. [Rationale: to ensure that x^0, or rather $(g_{00})^{\frac{1}{2}} x^0/c$, be a time coordinate and the others be space coordinates as required by GR 2.3: see HILBERT, 1915, 1916 for similar conditions.]

GR 3.3 (a) The components of g approach constant values as the $x^i (i = 1, 2, 3)$ tend to infinity. [Rationale: the metric tensor, which solves the field equations, should represent an isolated system.] (b) Outside every $V \subset M^4$, the asymptotic forms of the components of g are those of outgoing spherical waves. [Same rationale as before: see FOCK, 1964.]

GR 3.4 For every $\sigma \in \Sigma$ there exists a metric tensor g whose associated metric affinity Γ [see (4.36) in 3.2.1] represents the strength of σ. [Equivalently: the $g_{\mu\nu}$ are the gravitational potentials.]

Axiom group 4: frame and physical coordinates (pointless for $\mathfrak{X}=0$)

GR 4.1 (a) $K \neq \emptyset \wedge K \subset \overline{\Sigma}$. (b) Every $k \in K$ is a physical reference frame. (See Ch. 2, 4.) (c) For every $k \in K$ there exists an $x \in M^4$ such that over a neighborhood of it there exists a tetrad $e = \langle e_0, e_1, e_2, e_3 \rangle$ such that $e \triangleq k$.

GR 4.2 (a) For every $i = 1, 2, 3$, $\{X_i\}$ is a nonempty family of functions. (b) Every $X_i \in \{X_i\}$ is a function from $\overline{\Sigma} \times K \times T$ to R, continuous w.r.t.t. (c) $X(\overline{\sigma}, k, t)$ represents the position of a point system $\overline{\sigma} \in \overline{\Sigma}$, referred to k, at the instant t relative to k.

Axiom group 5: matter tensor

GR 5.1 (a) $\mathfrak{X}$ is a contravariant symmetric tensor field on $V \subset M^4$. (b) $\mathfrak{X}$ represents the energy, momentum and stress of all the systems $\overline{\sigma} \in \overline{\Sigma}$ in V.

GR 5.2 (a) The components of $\mathfrak{X}$ are real valued functions of $\overline{\sigma}$, k, X, and t. (b) The components of $\mathfrak{X}$ vanish outside V.

Axiom group 6: field equations

GR 6.1 $\varkappa$ is a positive real number with dimension $L^{-1}M^{-1}T^2$.

GR 6.2 For every $x \in V$, every $\sigma \in \Sigma$, every $\overline{\sigma} \in \overline{\Sigma}$ and every $\mathfrak{X} \in \{\mathfrak{X}\}$, there is a $g \in \{g\}$ such that

$$G = \varkappa \mathfrak{X}. \tag{4.41}$$

GR 6.3 Two solutions of (4.41) which can be transformed into one another by an admissible coordinate transformation refer to the same field.

The preceding axiom system is both p-complete (it characterizes all the primitives) and d-complete (it yields all the desired theorems). But it is hardly comprehensible without some comments. Here they go.

3.3. Comments

Remarks on the 1st axiom group. (1) Before GR the class of physical systems was partitioned into those endowed with mass, and e.m. fields: the gravitational field became an independent entity only with GR, which as regards gravitation puts all other physical entities in the same bag. (2) The statement that there are nongravitational entities (GR 1.2) should not be understood as holding that every region $V \subset M^4$ of interest is populated by some members of $\overline{\Sigma}$: like every other theory save cosmology, GR applies to those physical systems we care to study.

Remarks on the 2nd axiom group. (1) By definition of "differentiable manifold", GR 2.1 (a) maps every open set of M^4 on R^4 and (b) it ensures that M^4 can be flattened in the neighborhood of every point, thereby warranting the existence of a coordinate transformation that brings the

metric locally to the Minkowski form. (2) The right interpretation of
GR 2.3 requires both the local time theory (Ch. 2, 3) and Euclidean
physical geometry (Ch. 2, 4): indeed, they tell us what the local coordinates mean. (3) If a given x^0 is not interpretable right away as a time
coordinate in the sense of LT (Ch. 2, 3), GR 2.3 ensures that there exists
an x'^0, function of all four original x^μ, such that x'^0 can be interpreted
as a local time — and similarly for the other coordinates — for in the
new coordinates x'^μ the metric takes (locally) the Minkowski form. In
brief, the physical interpretation of the general (curvilinear) coordinates
is noncovariant. From the lack of covariance of the physical meaning
of the x^μ it has been concluded (HILBERT, 1915, 1916; WIGNER, 1957)
that in GR coordinates are physically meaningless — a statement which
is itself noncovariant. (4) The physical meanings assigned by GR 2.3
are nonoperational. It is impossible to have an observer field — a continuous distribution of clocks and graduated rods over spacetime. Even
if it were possible it would be of little help because of the lack of covariance of the physical meaning of the x^μ. And introducing ideal clocks,
yardsticks and observers does not make the theory more concrete but
spooky.

Remarks on the 3rd axiom group. (1) The affine connection is not
postulated: it follows from the existence of a metric and it may be
regarded as a mere abbreviation (formula 4.36) of a lengthy expression — i.e. as a set of Christoffel symbols. But it is convenient to have
it (*a*) because the components of the affinity represent the gravitational
field strength (which is not a vector field) and (*b*) because the Newton-
Poisson theory of gravitation can be formulated in a covariant way in
an affine nonmetric space (HAVAS, 1964). (2) The line element can be
interpreted physically only locally and by comparison with the corresponding Minskowski form, as recalled in the previous set of remarks.
(For SR as a source of meaning for GR, see SCHRÖDINGER, 1954.) In
any case it will not do to say that before a local flattening by means
of a suitable coordinate transformation ds/c is the time recorded by a
clock carried by a test particle, let alone the time measured by an
observer moving with a particle: (*a*) because in GR ds^2 need not be a
bilinear form of particle coordinate differentials and (*b*) because clocks
may not be able to ride particles — not to speak of their influence
upon the very metric they are supposed to probe into. As recognized
by BRIDGMAN (1949) himself, operationalism does not tally with GR.
(3) GR 3.2 restricts the coordinate system and consequently the coordinate transformations allowed in GR. The need for such a restriction was
mentioned in 3.1.3 and is apparent upon considering that two causally
related events (lying then on a timelike line) should not become simultaneous upon a coordinate transformation (HILBERT, 1915, 1916). If

these restrictions are accepted, GR is not covariant under totally arbitrary coordinate transformations but only under those which transform timelike lines into timelike lines and therefore keep a sharp distinction between space and time. (4) The choice of adequate boundary conditions is still an open question (BERGMANN in Infeld, ed., 1964). (5) Since the coefficients $g_{\mu\nu}$ of the metric are functions of space and time, they seem to be interpretable as the components of a physical field; the red shift formula, i.e. $\dfrac{\Delta\nu}{\nu} = -(1/2)\Delta g_{00}$, would seem to reinforce this interpretation. But if g were taken as the gravitational field representative, it could not be locally transformed away as required by the equivalence "principle"; on the other hand the vanishing of the affinity in one frame does not entail its vanishing in every other frame, because Γ is not a tensor. (6) The above axioms are insufficient to determine in detail the geometry of spacetime: they just sketch the framework M^4. It is only the totality of the axioms of GR in conjunction with special assumptions concerning the matter tensor that determine the metric tensor. In this respect GR departs from the rest of physics, where the geometry is hypothesized a priori.

Remark on the 4th axiom group. A more detailed characterization of the concepts K and X is left to protophysics (Ch. 2, 4). *GR* 4 becomes pointless for any hollow region.

Remarks on the 5th axiom group. (1) *GR* 5 restricts our formulation of GR to a finite region and to an isolated system. Applications to an infinite universe must be done in a piecemeal way or they require modified postulates. (2) $\mathfrak{T}$ must be hypothesized or computed: it cannot be measured. Now, GR makes no provision for computing $\mathfrak{T}$: it borrows it from other theories, thereby getting the concepts of mass, charge, e.m. fields and so on into the bargain. This shows that GR is neither unitary nor self-sufficient: it is parasitic on every other physical theory. For this reason, GR is bound (a) to clash with other theories to the extent to which they are not relativistic and (b) to solve no problem with complete accuracy. More on this in the next batch of remarks.

Remarks on the 6th axiom group. (1) The system (4.41) of 10 nonlinear and inhomogeneous differential equations makes the effective computation of the fundamental tensor g in principle possible and constitutes the leading hypothesis of GR: all other assumptions, though logically independent from it, are subsidiary to it. (2) The field equations can be obtained from a single variational principle with the known advantages of this method (see Ch. 1, 3.3). But since the physical meaning of such equations is still far from clear, we abstain from taking that course. (3) It is sometimes claimed that (4.41) is a definition of matter in geometrical terms; at other times, that it constitutes a definition of the spacetime structure in terms of matter. But these are logical mistakes: the (4.41) are equations not identities; moreover they are supposed to express an objective pattern not a convention. And they are so handled: in fact $\mathfrak{T}$ is not determined by g but is hypothesized, at least in first approximation, on the basis of theories other than GR. If $\mathfrak{T}$ were totally unknown the (4.41) would be indeterminate: 10 equations for 20 variables. That G and $\mathfrak{T}$ are mutually independent in the logical sense (see Ch. 1, 4.1.6), can be proved by setting $\mathfrak{T} = 0$, which leaves us with $G = 0$, a set of 10 equations with infinitely many solutions g. On top of being logically independent, G and $\mathfrak{T}$ are semantically independent: indeed, they have different referents. (4) By GR 6, GR welds gravitation with physical geometry. This does not amount to a geometrization of gravitation — nor to a gravification of geometry. The formal tools of a physical theory do not determine its physical meaning and do not characterize it uniquely. Thus axiom GR 6.2 might have been replaced by an action principle, which would not prove that gravitation is a matter of variational principles; and the classical theory of gravitation can be formulated with the help of a curved spacetime (CARTAN, 1923, 1924 and HAVAS, 1964). (5) GR does not build the concepts of mass, electric charge, e.m. field, and others in terms of the metric tensor but takes them over from alternative theories: what it does is to introduce new relations between those concepts and the geometry of spacetime. But the latter happens to be a physical geometry rather than a purely mathematical geometry. Consequently GR, far from constituting a geometrization of physics, utilizes several concepts — chiefly $\mathfrak{T}$ — that cannot be defined in geometrical terms. And even a genuinely unitary field theory which were to build every physical property in terms of a chronogeometry (WHEELER, 1962) would constitute no reduction of physics to mathematics but rather a representation of physical objects with the help of geometrical concepts having a field theoretical meaning. As with every other physical theory, some mathematical formulas are here attached an objective referent, i.e. a physical meaning. The converse is impossible: the semantical relation of interpretation goes only one way (see Ch. 1, 1.3.2). (6) The preceding remarks do not preclude the possibility of giving a purely geometrical interpretation of the formalism of GR — i e. of GR deprived of its physical content. This could be done by assigning the matter tensor an *ad hoc* geometrival meaning. Such an interpretation would serve the metatheoretical purpose of proving the consistency of GR — just as H. MINKOWSKI unawarely proved the consistency of SRK by showing that it has a pseudo-Euclidean model. (7) It is sometimes said that GR equates space with gravitation. Yet one computes the field strengths at a point in M^4, and by GR 3.3 the vanishing of the field in a region $V \subset M^4$ does not entail the disappearance of spacetime in it: everything that happens to the fields happens some-where in M^4. Spacetime is affected by things but does not coincide with them. (8) It is also said that the field equations show that matter deter-mines the structure of space, which is one of the versions of the highly ambiguous "Mach principle" (EINSTEIN, 1918). This is not quite so: matter only *co*determines the structure of space, as shown by setting $\mathfrak{T} = 0$. (9) RIEMANN, CLIFFORD and their modern followers have conjec-tured that matter is just a warping of space (or spacetime). This may well be so but it is not what GR holds: this theory states only that matter and gravitation are associated. This association is as loose as the one between charged bodies and e.m. fields: in fact although whenever there is matter there is a field (because the metric deviates then from the flat form), the converse is as false in GR as in CEM (see 1.3.3). Indeed, on setting $\mathfrak{T} = 0$ we still get an infinite set of gravitational fields, because $G = 0$ does not entail the vanishing of the affinity (only the converse being true). Not even the vanishing of the full curvature tensor corre-sponds to a zero field but refers to a constant homogeneous field. In short $\mathfrak{T}$ codetermines the metric; the boundary and initial conditions on the metric, which are independent of $\mathfrak{T}$, are as important as $\mathfrak{T}$. Finally GR provides no mechanism for the matter-field association. (10) EINSTEIN's field equations refer by hypothesis to a pair $\langle \sigma, \bar{\sigma} \rangle$; they do not refer immediately to a piece of the real world but rather to a model of such a chunk of reality. In particular, the equations do not refer necessarily to the whole universe. Unless this is granted it may be required that $\mathfrak{T}$ refers to (or as is sometimes said "contains") every actual physical system — which is like requiring that the F occurring in CM represents all the forces actually acting on a real body. GR is not an all-inclusive theory: like every other known physical theory it deals with problems in a piecemeal way, i.e. it restricts its referents (see however HAVAS, 1965). If it be rejoined that GR must refer to the whole universe because the gravitational field cannot be screened, it may be recalled that the e.m. fields, too, can extend indefi-nitely, Faraday cages being very unusual. (11) The field equations are epistemologically disturbing: given $\mathfrak{T}$, they allows us to determine, by integration, the coefficients of g. But if $\mathfrak{T}$ is prescribed outside GR, as it usually is, then it cannot be exact since it will neglect the influence of the metric on the state variables occurring in $\mathfrak{T}$. Even in the simplest case of the incoherent fluid, the metric shows through $\mathfrak{T}$. An exact knowledge of $\mathfrak{T}$ would require an exact knowledge of G — and so we are caught in a circle. In other words, if posed exactly the problems of GR involving matter are indeterminate. As usual with problems of this kind, circularity is avoided by employing a method of successive approxi-mations: write the matter tensor for flat space and solve the field equations; use the resulting metric to rewrite the matter tensor and introduce the latter into the field equations — and so on. Upshot: except for $\mathfrak{T} = 0$, GR cannot give exact solutions. Mind, this has nothing to do with mathematical difficulties, say the ones associated with the nonlinearity of the field equations or the convergence of the successive approximations procedure: it is a fundamental difficulty. Its source is that GR is not a self-sufficient theory encompassing every possible aspect of the physical systems enclosed within the region V under consideration. This difficulty is far tougher than the ones besetting CED: in this case, too, exact solutions can be found only for inexactly posed problems — those in which either the field or the motions are assigned; but in this case the main difficulty is purely computational, since one has all the necessary equations — quite apart from the question of their truth. On the other hand in GR one does not know one of the two sides of the question, namely $\mathfrak{T}$, even if told by some other theory — wherefore Einstein called $\mathfrak{T}$ an *asylum ignorantiae*. One can therefore easily understand EINSTEIN's tireless efforts to build a unified theory, as well as WHEELER's and LICHNEROWICZ's: it is not just the dream of the all-encompassing formula but an attempt to fill in a genuine gap, namely the right-hand side of the field equations of GR. Whether and if so how this problem will ultimately be solved remains to be seen, but it is unlikely that it will be tackled by as many people as needed unless it is realized that, for all its richness and beauty, GR is an essentially open (incomplete) theory — that, except for the case $\mathfrak{T} = 0$, it depends on every other physical theory and requires in turn their general covariant reformulation. (12) It is possible to generalize GR 6.2 in several directions. But it can be shown (CARTAN, 1922) that only the slight extension $G' = G - \Lambda g$, with $\Lambda = \text{const.}$, satisfies the conditions (*a*) that G' depends only on g and its first and second order derivatives, and (*b*) that G' be conservative. But there is of course no reason why these theory construction desiderata should be satisfied. (13) Axiom GR 6.3 (proposed by BERGMANN, 1949 in a slightly different form) enables one to establish equivalence classes of solutions of the field equations: any two members of any such class are semantically identical no matter what their mathematical differences are. This ensures the recognition of those elements which, far from being inherent in the given field, characterize the coordinate system that happens to be singled out. This axiom makes therefore an important contribution to the objectivity of GR. (14) Nothing has been assumed about the coupling constant $\varkappa$ except that it is a positive real number. Its precise value is then determined either by comparing the geodesic equation with the Newtonian equation of motion for a test particle, or by some more direct procedure. The former course yields $\varkappa = 8\pi\gamma/c^4$, where '$\gamma$' stands for NEWTON's universal gravitational constant. Since the meaning of the constants c and γ is specified outside GR, we see once more that GR is semantically dependent on the rest of physics. Therefore the fashionable dictum that GR is outside the main current of physics is plainly false.

Now some remarks about our axiom system as a whole. Clearly, it is not the sole possible axiomatization of GR: alternative organizations of the same material should be tried, subject to the condition that they retain the field equations, their characteristic consequences, and the assumptions that render them mathematically definite and physically meaningful. Yet the proposed axiom system seems to satisfy the conditions of consistency, independence, primitive independence, p-completeness and d-completeness (see Ch. 1, 4.1.6 and 4.2.3). It will also be noticed that none of the heuristic crutches employed in building the theory occurs in our postulate system. In particular, general covariance is automatically satisfied by it owing to the tensor form of the central hypothesis GR 6.2. Notice finally that every primitive save the coupling constant (which is nonreferential: Ch. 1, 2.2.1) has been assigned a physical meaning by a semantical assumption such as the one according to which ' x^0 ' represents time. A different set of interpretive postulates will characterize a different theory — eventually a thoroughly false one; and if this set is empty, as it is usually the case, then one has a mathematical skeleton not a physical theory. The semantic components of our axiom system make no mention of measurements or operators — let alone of fictitious "observers at infinity" or "observer fields", and this for the following reasons. Firstly, GR happens to be a fundamental theory not an application to measurement procedures; moreover, its referents Σ and $\bar{\Sigma}$ are supposed to exist objectively, whether or not they are subject to observation. Secondly, GR is a field theory, and fields are not observable: only some of their effects on molar systems are observable; in particular, solutions for $\mathfrak{T} = 0$ are meaningful though untestable. In short GR, like every other basic theory, is impregnable to operational interpretation. On the other hand, like every basic theory, GR is indispensable to design and interpret operations aimed at testing it (see 3.4).

3.4. Some Representative Theorems

We shall give some of the theorems entailed by the axiom system in 3.2.4, in order to discuss some further controversial points.

Thm. 1: Reformulation of GR 6.2. Under the conditions of GR 6.2,

$$R^{\mu\nu} = -\varkappa \left(\mathfrak{T}^{\mu\nu} - \tfrac{1}{2} g^{\mu\nu} T \right), \quad \text{where} \quad T \overset{\text{df}}{=} g_{\mu\nu} \mathfrak{T}^{\mu\nu}. \tag{4.42}$$

Proof. Contract (4.41) and use the orthogonality of the covariant to the contravariant components of g, and Dfs. 7 and 8.

Corollary. For every $x \in V$ such that $\mathfrak{T}(x) = 0$, and every $\sigma \in \Sigma$,

$$R^{\mu\nu} = 0. \tag{4.43}$$

Remark. This theorem is usually interpreted as meaning that in the absence of matter spacetime is flat. Actually for M^4 to be flat over an extended region all 20 components of the uncontracted curvature tensor or all 32 components of the affinity (4.36) must vanish.

Thm. 2: Continuity equation. Under the conditions of *GR* 6.2,

$$V_\nu \mathfrak{T}^{\mu\nu} = 0. \tag{4.44}$$

Proof. The covariant derivative of G vanishes identically (the Bianchi identities).

Remarks. (1) The equation of continuity must be sharply distinguished from the infinitely many conservation-like equations that can be formed in GR, without assuming the field equations, by simply applying NOETHER's theorems to the general coordinate transformations (BERG-MANN, 1958 and TRAUTMAN, 1962). As we saw in Ch. 1, 3.4.1, although such conservation-like equations are suitable mathematical forms to pour conservation laws into, they are not entitled to a physical interpretation unless they are entailed by field equations (or equations of motion as the case may be). (2) The physical interpretation of Thm. 2 requires its integration over a finite space region. But the covariant divergence equals the ordinary divergence plus terms whose volume integrals do not transform into surface integrals and consequently no globally conserved quantities are obtained. In other words, there are no global conservation laws in GR — something that can of course be traced to the lack of symmetries characterizing Riemannian spaces. What one can do is to restrict the system of interest (the referent) to a "small" region over whose boundary the affinity vanishes and consequently the covariant derivatives degenerate into the ordinary ones. In other words, by adopting a geodesic coordinate system that mirrors a local inertial frame, a conserved quantity is obtained for that neighborhood over which that geodesic coordinate system can be given ("defined"). In brief, $\mathfrak{T}$ is conserved locally not globally.

Thm. 3: Asymptotic expression. In the linear approximation and in harmonic coordinates ($g^{\mu\nu} \Gamma^\alpha_{\mu\nu} = 0$), the field equations (4.41) become

$$\frac{1}{2}\left(\frac{\partial^2}{\partial x_0^2} - V^2\right) g^{\mu\nu} = -\varkappa\left(\mathfrak{T}^{\mu\nu} - \frac{1}{2}g^{\mu\nu}T\right). \tag{4.45}$$

Proof. Use *GR* 3.3b (condition of outward radiation) and the condition of harmonic coordinates, neglect all quadratic terms, and take the Minkowski values for the coefficients of the metric but not for their derivatives.

Corollary: Leading equation. For an incoherent nonrelativistic fluid,

$$\left(\frac{\partial^2}{\partial x_0^2} - \nabla^2\right)g^{00} = -\varkappa\varrho. \qquad (4.46)$$

Proof. Introduce in Thm. 3 the values $\mathfrak{T}^{00} = T = \varrho$ given by SR for a fluid with no pressure and small velocity.

Remarks. (1) By analogy with CM and CEM, (4.46) can be interpreted as a wave field travelling with light velocity in a nearly flat background space. This is not a theorem but a possible interpretation based on an analogy. (For further analogies see FORWARD, 1961 and DEHNEN, 1964.) And it is by no means a firmly established one: (*a*) certain gravitational "waves" can be transformed away by suitable coordinate changes (INFELD and SCHEIDEGGER, 1951); (*b*) not only the metric field but also the harmonic coordinates themselves obey wave-like equations. Whether there are in fact gravitational waves must yet be decided by experiment; the quantization of GR may be very helpful in this enterprise by predicting new effects. (2) Unless special kinds of coordinates are chosen, the contracted curvature tensor does not reduce to one-half the d'Alembertian of the metric tensor. Among the preferred coordinate systems, those for which Det $\|-g_{\mu\nu}\| = $ const. (EINSTEIN, 1916) and the harmonic ones (FOCK, 1964) stand out. This shows that, though particular coordinate systems should be irrelevant at the level of principles, they become conspicuous at the level of theorems, both for the latter's statement and for their physical interpretation. But it does not show that Nature wears one kind of coordinates preferably to others.

Thm. 4: Schwarzschild solution. Outside a spherically symmetric and static distribution of matter, the metric of M^4 is given by

$$ds^2 = \left(1 - \frac{r_0}{r}\right)dx_0^2 - \left(1 - \frac{r_0}{r}\right)^{-1}dr^2 - r^2d\vartheta^2 - r^2\sin^2\vartheta\,d\varphi^2 \qquad (4.47)$$

with $r_0 \in R^+$. *Proof.* Solve the field equations (4.42) with the conditions $\mathfrak{T} = 0$ and that the g be static and spherically symmetric. Upon comparison with the Newtonian solution it turns out that $r_0 = 2\gamma M/c^2$. (This is not a definition.)

Remarks. (1) The popular belief that in ordinary GR particles are field singularities is dispelled once more by looking at the singularities of the Schwarzschild metric field, which occur over a sphere with radius r_0 — usually much smaller than the corresponding body but nonvanishing even for a point particle. Incidentally, the Schwarzschild radius r_0 would seem to have no physical meaning. (2) All three classical tests of GR are tests of the single Thm. 4: in fact the perihelion precession, the light bending and the red shift are computed assuming that the solar gravitational field is determined by (4.47), all other fields involved being negligible. Since that particular line element and others similar to it occur in several other gravitation theories, these can claim the same empirical support as GR as far as the classical tests are concerned. But none of them can claim to have predicted those effects: in fact they were all contrived in the empiricist style, i.e. *ex post facto*. Conclusion: empirical confirmation, though necessary, is insufficient as a truth criterion (BUNGE, 1961a, 1963a, 1967b).

Thm. 5: Equivalence "principle" I. There exists at least one local coordinate system in which a static homogeneous gravitational field vanishes (i.e., can be transformed away). *Proof.* By *GR* 2.1 (local flattening), the nonhomogeneous transformation law of the affinity, and *GR* 3.4 (meaning of the affinity).

Df. 10. A local reference frame relative to which a static gravitational field vanishes [e.g., an Einstein lift] is called an *inertial frame*.

Remarks. (1) Thm. 5 can be reformulated thus: "It is always possible to make a coordinate transformation such that in the new frame (a geodesic frame) a static homogeneous gravitational field vanishes. Since this frame can be given only over a neighborhood of the envisaged point, the field can be transformed away only locally." (2) This "principle" of equivalence has become part and parcel of GR and at the same time it continues to play a heuristic role in it (see however SYNGE, 1961). (3) We have now three different definitions of "inertial frame": one in CM, another in CEM and SR, and finally Df. 10. Each characterizes a different concept; in particular, neither free particles nor light rays move along straight lines in an Einstein cage. (4) With the help of Df. 10, Thm. 5 can be rephrased: "There exists at least one local inertial frame".

Thm. 6: Law of motion. For every $\sigma \in \Sigma$ and every $\bar{\sigma} \in \bar{\Sigma}$ such that $\bar{\sigma}$ represents a structureless test particle (Df. 9) with physical coordinates X^{μ},

$$\frac{d^2 X^{\lambda}}{d s^2} + \Gamma^{\lambda}_{\mu\nu} \frac{d X^{\mu}}{d s} \frac{d X^{\nu}}{d s} = 0. \tag{4.48}$$

Proof. Use Thm. 2 and specify $\mathfrak{X}$ to represent a monopole that does not contribute to the gravitational field.

Remarks. (1) The 4 equations (4.48) constitute a generalization of GALILEI's 3 equations $\ddot{X} = g$ but, like every other relativistic law of motion, its zeroth component has no classical analog and does not vanish in the nonrelativistic limit. (2) Since the force concept does not occur explicitly in (4.48), it is usually said that it does not occur at all in GR. Yet the 2nd term of (4.48) can be interpreted as a force typical of GR since the affinity measures the deviation from rectilinear uniform motion. Moreover, (4.48) can be rewritten in a form similar to the Lorentz equation (DEHNEN, 1964). (3) Thm. 6 is a physical interpretation of the geodesic equation. The interpretation was obtained by endowing the coordinates with a physical meaning with the help of GR 4. (4) It is often believed that, by virtue of Thm. 6, GR includes dynamics and is therefore a self-contained or unitary theory of matter and fields. Yet strictly speaking (4.48) holds only for the simplest test particles, with neither spin nor multipole moments. And even when some of these restrictions are removed for a system of material points (HAVAS and GOLDBERG, 1962), the resulting equations of motion contain the un-determined metric tensor — and with such an unknown it cannot be claimed that GR includes dynamics. All one can conclude is that in GR the laws of motion are not logically independent of the field equations, whence no arbitrary motions can be assigned. (5) Thm. 6 was originally postulated and called the geodesic postulate (EINSTEIN, 1916). It was based on the following heuristic reasoning: in a neighborhood where spacetime is flat, i. e. in a geodesic frame, a free particle moves along a straight line (a Euclidean geodesic). Since geodesics are independent of the coordinate system, the same equations will hold for an arbitrary frame. Finally the assumption is added that the same equations hold also in a curved space. This strategy of first conjecturing a tensor equation in a geodesic frame and then betting that it holds generally, is commonly used in GR: it is an extension of the relativization technique employed in SR, where one starts by postulating a semiclassical equation for some 4-vector in the rest frame and then rewrites it in a Lorentz covariant form (see 2.4). In neither case is the inference deductive or inductive.

Thm. 7. Equivalence "principle" II. There exists at least one local reference frame in which the 4-acceleration of a test particle vanishes. *Proof.* From Thm. 5 (equivalence "principle" I) and Thm. 6 — òr merely from the latter and noticing that, since Γ is not a tensor, it can always be transformed away.

Remark. We have two equivalence "principles" in GR and none of them is a principle: they are theorems. (See also ROHRLICH, 1963.) While the first talks about gravitational fields, the second talks about particles; and while the first states that certain gravitational fields can be locally transformed away, the second states that certain trajec-tories can be straightened out. (By choosing a parameter other than s, Thm. 7 can be restated for a light ray.)

Now that we know some theorems we can ask how does GR fare experimentally.

3.5. Empirical Tests

According to the physicist's religion, measurement is the basis and source of theory and therefore the exposition of every theory must be preceded by a study of the measurements on which the theory is "based".

Yet, it is only once we have got a theory that we know what is to be tested for. Besides, the theory must be worked out before it can be tested, as only some of its logical consequences can be confronted with experimental data. And in this processing of the initial postulate basis a number of subsidiary assumptions, as well as simplifications, are introduced which may go as far as disfiguring the face of the theory. Finally, every measurement aiming at testing the theory requires that very theory in conjunction with a number of other theories external to it and concerning the measurement technique: thus it is one thing to measure red shifts with astrophysical techniques and quite a different thing to measure them with the Mössbauer technique (see Ch. 1, 5.1). For all these reasons analyses of measurements are made in the light of theories not conversely; in particular, an examination of the procedures for measuring the metric coefficients requires both GR and a number of other theories.

Before mentioning any actual tests of GR we should consider its testability. In the absence of an exact theory of testability we shall rest content with a few imprecise remarks. We begin by recalling that no theory is fully testable (a) because it contains infinitely many statements and (b) because the most interesting formulas may have to be tested deviously and with the help of other theories. This holds particularly for GR because it is a field theory and due to its complicated formalism, which has been largely responsible for the slow growth of the body of theorems. (For a number of closed-form solutions see EHLERS, 1962 and PETROW, 1964.)

A satisfactory — yet not decisive — test of GR would require the checking of fairly rigorous solutions to all the 10 field equations for every one of a reasonable number of widely different examples — not just the Schwarzschild solution as was the case with the three classical tests. Otherwise (a) any favorable data would also favor a dozen rival theories — e.g., those of E. WIECHERT, A. N. WHITEHEAD, G. BIRKHOFF, G. GARCÍA, and F. J. BELINFANTE, and (b) any discrepancies with the data might be blamed on the poor approximations made in getting the solutions to the equations. Therefore the opinion that it is not necessary to find new solutions to the field equations is an invitation to stop both the growth of GR and its tests. The more thoroughly the theory is worked out — in particular, the more solutions of the field equations and the equations of motion become available — the more its testability will be enhanced. Even so, it will remain limited because we can scan only limited regions of spacetime. Why then do most physicists believe that GR is true or at least far truer than any of the existing *ad hoc* systematizations of the known facts? Because, despite their public philosophy, they do not act as if they really believed theories are data packages: they have estimated the degree of truth of GR not only on the strength of its coverage but also on its power to accurately predict unorthodox effects (see Ch. 1, 5.2.3), as well as on its depth and unity (BUNGE, 1961a).

The conceivable tests of GR can be grouped into two sets: the direct and the indirect ones. The former would consist in measuring space-like and time-like intervals leading to the determination of the metric. The indirect tests consist in checking some of the infinitely many logical consequences of the postulates of GR — e.g., the action of the spacetime curvature on the strain in an elastic body (WEBER, 1961), the focusing of light rays by gravitational lenses (LIEBES, 1964), or the slowing down of light rays in strong gravitational fields (SHAPIRO, 1964). Presently the performance of direct measurements seems impossible not only for technical reasons but also as a matter of principle: every measuring device distorts the very field it is supposed to probe into. Even if these distortions could be accurately corrected for, the value of such measurements would be slight because every such measurement of the metric tensor coefficients presupposes the field equations and some of their consequences — e.g., that the world lines of test particles are indeed geodesics (LEVI-CIVITA, 1929). In other words, GR itself is used as the basic conceptual tool for exploring spacetime and, due to the apparent impossibility of screening gravitation, that circularity seems inescapable. For these reasons indirect checkings are likely to remain the best tests of GR. They have the additional advantage that they do not require a particular GR theory of measurement to account in detail for the behavior of measuring instruments in a gravitational field but can employ nonrèlativistic theories to this end — an approximation direct measurements could not afford to make. Finally, indirect measurements supply collateral evidence for the particular metric hypothesized in each case. But the facts that in most cases the very particular Schwarzschild metric has been used, and that this metric can be imitated by any number of *ad hoc* metrics — i.e. tensors that do not solve EINSTEIN'S field equations — show that the available tests, though striking, are insufficient. New kinds of tests, i.e. tests of further predictions of GR, are badly needed (INFELD, 1964 and CHIU and HOFFMANN, 1964).

In conclusion, the empirical tests of GR seem to be extremely roundabout — which is only natural given the depth of the theory. But they are also, though impressive, far too few — and this is serious. Nevertheless GR is temporarily accepted and rightly so, not just because of the mass of data it covers — most of which concern tiny effects and could be gotten much cheaper with phenomenological theories — but because of its depth, its boldness (no adjustable parameters), its logical unity, its serendipity — and its beauty, matched only by MAXWELL'S theory and quantum mechanics. But beware: in science a thing of beauty is not a joy for ever.

This closes our review of the foundations of GR, one of the most controversial physical theories. It is hoped that the defects of the axiomatization proposed in 3.2 will elicit better logical organizations of the available material. This is needed not only in order to clarify many obscure points but also to facilitate the quantization of GR, which is indispensable since the matter and the nongravitational fields represented by the inhomogeneities in the field equations are best described by the quantum theory — to which we now turn.

Chapter 5
Quantum Mechanics
Introduction

Bodies and fields are now behind and for good: the things with which the quantum theories are concerned cannot be pictured as classical entities. They are *sui generis* entities deserving a special name — say *quantons* — this being why the quantum theories themselves are *sui generis*. True, the words 'particle' and 'field' are retained even in the names of some quantum theories, but they designate concepts differing from the classical ones. In particular, a "particle" is in the new context anything that, whether localizable or not, is endowed with mass: it is the referent of quantum mechanics, the basic theory of matter — to be abbreviated QM. Massless entities like photons and neutrinos (which of course are not particles since they have neither a mass nor a narrow localization) are studied by quantum field theories of QFT for short. But these latter theories also study the "particles" at a deeper level (second quantization). Therefore QFT covers the whole realm of quantons and QM is just the elementary part of the theory and the one applying to microsystems endowed with mass. Yet QM gives rise to most of the interpretation puzzles that have ridden the field for nearly half a century. This, and the lack of space, justify our concentration on it.

One of the goals of this study will be to justify our claim that quantons are neither particles nor waves. We shall also find out that dualism is not the right answer either, for at the quantum level there seem to be no corpuscular and undulatory properties but rather particle-like and wave-like ones — in short the "particles" and "waves" of QM are metaphorical not literal, and we had better stop trying to picture the new in old terms. Yet the central goal of this chapter is not to debunk dualism or complementarity (for this see BORN, 1949; MARGENAU, 1950; BUNGE, 1955b; POPPER, 1959; LANDÉ, 1965), but to propose a consistent and thoroughly physical formulation of the foundations of ordinary QM. For the usual doctrine, proposed by BOHR and HEISEN-BERG and often called the *Copenhagen interpretation* (e.g. GEORGE, ed., 1953; BOHR, 1958; HEISENBERG, 1958; PAULI, 1961), is inconsistent and psychophysical rather than physical, on top of which it is usually held and defended in an uncritical way. Let us briefly substantiate these charges, for they motivate the proposed logical reorganization and physical reinterpretation of QM to be expounded shortly.

Inconsistency of the orthodox interpretation. (a) It reads every quantum formula concerning a quanton in terms of variables absent from the formula itself and added *ad hoc*, namely those characterizing a usually nondescript experimental arrangement and eventually the observer's mind as well. If such extra variables are not injected in the initial assumptions — e.g., in the form of the hamiltonian of the quanton — then they are smuggled into the theorems derived therefrom. This hyperinterpretation renders the usual QM semantically inconsistent (BUNGE, 1965 and Ch. 1, 4.1.6). (b) It claims that QM and QFT refer solely to actual or possible laboratory operations — which are macroscopic and therefore classically describable. Yet at the same time it holds that q.m. properties (the "observables") are not classically describable, much less picturable. It can therefore be summed up in the self-contradictory statement: "Quantum theories are about classically (= non-quantum-mechanically) describable laboratory operations". (c) The theory is said to be "deduced" from actual empirical data as well as from a critical analysis of measurement processes in a few typical examples. But the theory contains a number of unobservables (e.g., the Hilbert space); it talks about the superposition of states for a single quanton although this is supposed to be found experimentally in a single state at a time; and the same theory is finally applied to the analysis of measurements, which would be pointless if it did deal with them to begin with. (d) Microsystems are not allowed to lead an autonomous (subject-free) existence but are allegedly shaped or even conjured up by observation operations. Yet those very entities are also assumed to constitute those measuring instruments which, they, must be out there for all to see and touch if they are to be used at all. (e) Despite the alleged observer-dependence, all the q.m. systems studied by quantum statics (i.e., QM with time-independent hamiltonians) are assumed to be closed, i.e. cut off from every other system, in particular the ominous apparatus, ever ready to disturb (or even generate) the quanton. And when open systems are studied (quantum dynamics), the incoming disturbance is always another physical entity — never an observer — save in popularization articles.

Semiphysical character of the usual interpretation. In this doctrine it is never clear what is physical and what mental, what is objective and what subjective. In any case the observer, loaded with his intentions, expectations and uncertainties, is inextricably mixed up with the observed object and even with the totally observer-free Helium atom in the next galaxy. This defect stems from a philosophy that was fashionable in the 1920's and 1930's but has ever since been superseded by the developments in semantics and methodology. Indeed, the effort to interpret theoretical symbols exclusively in terms of human operations rather than in terms of physical objects can be traced back to a confusion between the physical meaning of a symbol and the way a formula is tested by means of empirical operations — a confusion christened "the verifiability theory of meaning", now abandoned by most philosophers. We now think that only a previously interpreted formalism can be put to the test, i.e. that testing presupposes meaning — not the other way around — and that meanings need not be empirical to be physical (see Ch. 1, 1.3).

Uncritical character of the orthodox attitude. (a) It is naively assumed that infinitely precise measurements are possible and moreover that they are needed to formulate QM; it is moreover alleged that such measurements can always be direct, i.e. in no need of intermediary physical processes and intermediary theories that bridge the microphysical and the macrophysical, the unobservable and the observable (see Ch. 1, 5.1 for the cause-symptom relations). Otherwise the doctrine could not claim, e.g., that the eigenvalues of a q.m. operator are, even when non-fractionary, exactly those any measurement of the corresponding "observable" would yield. (b) The function of experiment is said to supply the actual frequencies of the theoretically calculated possible values of an "observable". These possibilities — i.e. the eigenvalue spectra — are regarded as fixed by the theory once and for all. In this regard the theory is incorrigible, hence metaphysical. (c) The system-apparatus interaction is declared to be unanalyzable (BOHR, 1932, 1935; WIGNER, 1963). Even if the analysis is admitted as possible, it is claimed to be pointless or even misleading "since it would only obscure the essential function of the experimental arrangement in establishing the connection between the quantal system and the classical concepts indispensable for its description" (ROSENFELD in INFELD, 1964). Reason, hold back! (d) Although the theory is (wrongly) said to relate experimental data, calculations of actual quanton-apparatus interactions are seldom made and when they are made (e.g., for the single slit) they are approximate and they do not tally with key formulas of QM such as HEISENBERG's relations (BECK and NUSSENZWEIG, 1958) — which does not shatter the believer's faith. (e) QM and QFT with this interpretation are held to be perfect as far as the atomic and electrodynamic domains are concerned, criticism is either ignored or branded as heresy, and attempts to reconstruct the foundations of QM are dismissed as futile (ROSENFELD, 1961).

Now an inconsistent, half-physical and even spooky theory cannot be rendered consistent and thoroughly physical by subjecting it to small reforms: a fundamental revision is required, one that frees it from an outdated anthropomorphic philosophy — as felt by many physicists dissatisfied with some of the above-mentioned blemishes of an otherwise admirable theory (BUNGE, 1955b, 1955c, 1956b, 1962b, and 1967d). Since the basic quantum formalism has had such a marvellous success — no matter how badly new theories are needed to cope with high energy events — the good old QM deserves being depurated of its *ad hoc* operationalist interpretation dictated by a dusty philosophy rather than by any inner physical need. To this end, all we have to do is to lay bare the mathematical skeleton of the usual theory, display it orderly, and reinterpret it in a purely physical way. This will be done in Secs. 2 to 6. Needless to say, the classics of QM (chiefly VON NEUMANN, 1931 and DIRAC, 1958) will be employed although the ghosts that haunt them will be chased and a logical order will be introduced.

Our work shall be restricted to the nonrelativistic spinless one-"particle" theory in the Schrödinger formulation ("picture") and the "position" representation. (The reasons for the quotation marks will become apparent in due course.) The generalization of our version of QM to an aggregate of microsystems is straightforward but will not be carried out here although it poses interesting problems of its own, such as the one of the individuality of identical ("indistinguishable") "particles". Nor will alternative formulations and representations of QM be mentioned except incidentally. Moreover, we shall work throughout on a moderate level of mathematical sophistication, for the urgent task is to free QM from unphysical ballast. On the other hand we shall employ the axiomatic method because this alone, by enumerating the basic concepts and stating honestly the initial assumptions, avoids the uncontrolled introduction of foreign (particularly subjective) elements when convenient for philosophical speculation. (The adoption of the axiomatic method does not involve condemning the attempts to deduce QM from more basic principles, as BOHM, 1957 and LANDÉ, 1965 have attempted. Even if any such attempt should succeed, it would pay to have QM well organized on its own level.)

But then, it may be asked, why not adopt one of the available axiomatizations of QM? Answer: because there is none that meets the contemporary standards of physical axiomatics (see Ch. 1, 4.2). VON NEUMANN's formulation, which passes for an axiomatic treatment of QM, is nothing of the sort — actually it is quite untidy. Even the best available axiomatizations (SEGAL, 1947; LUDWIG, 1954; MACKEY, 1963) fail to list all the presuppositions and the basic concepts as well as the totality of the basic assumptions actually used in quantum dynamics. And they all take for granted that "observables" are literally observable — but, then, through ideal experiments — and that every eigenvalue is a measurable value; as a consequence they incur in some of the inconsistencies listed above and due to the adoption of a subjectivist epistemology. We must therefore make a fresh start. And in the first place we shall disentangle the heuristic from the constitutive principles (see Ch. 2, 1).

1. Quantum Heuristics

Like every other theory QM has been built with the help of a scaffolding. Part of this external framework is recognized as purely heuristic and is retained only to throw students into confusion; this is the case of the optico-mechanical formal analogy. But another part of it is still presented together with the theory proper, its only function being to obstruct the view of the building. We shall focus our attention on two outstanding heuristic principles of QM that usually pass for postulates: the principle of "observables" and the correspondence principle.

The *principle of "observables"* may be stated thus: "Every physical property ("observable") Q shall be represented by a linear hermitian operator $\mathcal{Q}$ [i.e. by a linear and essentially real transformation of the Hilbert space into itself]". The converse rule is sometimes also stated, namely that any linear hermitian operator our fancy may eject represents some physical property. But the latter is an unnecessarily bold assumption; moreover some hermitian operators are known to represent no "observables" (WIGNER, 1952). Whether a given theoretical construct has a real referent or not must in the last analysis be decided by experiment (real not ideal); nature need not follow our imagination: to suppose that it does is sheer Pythagoreanism.

At any rate both rules are methodological prescriptions; and the first of them is a theory construction direction, one enjoining us to look for an operator representative or conceptual image of any given physical property without however telling us what the operator looks like. Moreover it is not typical of QM since every theory, in particular classical mechanics, can in principle be handled with the operator method (KOOPMAN, 1931; VON NEUMANN, 1932). In any case it is not an axiom of QM for the simple reasons that (a) it is not specific or constructive (it does not pinpoint the property-operator correspondence) and (b) it entails nothing — i.e. no theorems follows from it. It is then a heuristic principle.

The same holds for its various partners: "If $\mathcal{Q}$ represents Q, then $F(\mathcal{Q})$ represents $F(Q)$"; "If $\mathcal{Q}$ and $\mathcal{P}$ represent Q and P respectively, then $\mathcal{Q} \pm \mathcal{P}$ represents $Q \pm P$"; (c) "If $\mathcal{Q}$ and $\mathcal{P}$ represent Q and P respectively, then $\frac{1}{2}(\mathcal{Q}\mathcal{P} + \mathcal{P}\mathcal{Q})$ represents QP". These, too, are theory construction recipes and moreover sterile by themselves: they cannot be implemented unless one specifies, say, that the position is to be represented by x (or by $i\hbar\, \partial/\partial p$) and the momentum by $(\hbar/i)\, \partial/\partial x$ (or by p). Furthermore they must be supplemented with the following rule, which in addition to "observables" involves the fundamental unobservable ψ: "Every classical subsidiary condition or constraint of the form '$F(x, p) = 0$' shall be replaced by the formula '$F(x, p)\psi(x) = 0$'". Even thus completed the "principle of observables" can break down — which could not be tolerated if it were an axiom, The rules linking physical properties and operators are fertile but fallible heuristic recipes: there are no secure quantization methods (DIRAC, 1950). Moreover these rules involve a faulty name: the properties of atomic systems should not be called 'observables' because they are imperceptible and measurable, if at all, in a roundabout fashion. One does not render a theory consistent with operationalism with the simple trick of calling 'observable' what is unobservable.

Our second example of a heuristic quantal rule is the *correspondence principle*, which has been stated in a number of inequivalent ways, whence it actually is a set of statements. The chief members of this set are: (a) "Quantum results must go over into the corresponding classical ones in the linit of large quantum numbers". This rule leaves us in the lurch in the case of nonperiodic problems, e.g. elastic scattering. (b) "In the limit as the Planck constant vanishes both the operator relations — e.g., the commutation relations — and the q.m. averages must go over into the corresponding classical formulas (e.g., the Poisson brackets)". This rule cannot be applied to formulas concerning spinning particles: some of them should go over into classical formulas concerning spinning particles but they don't. (c) "The basic equations of QM must bear a formal resemblance to those of classical physics". This, again, is often useful but not quite correct. Thus the Schrödinger equation has no classical analog although its ingredients H and ψ do have each a classical analog in the n.r. approximation. In any case these and possibly other correspondence principles as well have been instrumental in building QM yet not so much for inventing the new constitutive assumptions as for discarding unsuitable candidates — perhaps hastily sometimes. Correspondence principles — in SM, SR, GR, QM and QFT — are conceptual tests for the compatibility of those theories with less refined theories (see Ch. 1, 4.1.6 for external consistency). The correspondence principle for QM, not the complementarity dialectics, is N. BOHR's unperishable contribution to metascience; unfortunately so far it has been given no precise statement.

There are several other quantal heuristic principles, such as the superposition principle (actually a theorem), the condition that the total energy of two independent systems be the sum of the partial energies, and the invariance of probabilities under unitary transformations — in particular translations and rotations. (On the other hand the complementarity "principle" is not heuristic or path-finding but just the opposite, its role being to petrify and excuse difficulties in the usual presentations of QM.) The heuristic principles suggest some of the features of QM, they control its building, but they do not erect it. (This is general: there are no theory construction techniques: see Ch. 1, 4.3.) The theory is built so as to conform to these principles in some way or other. Thus by taking the "limit" $\hbar = 0$ one can always check whether a given formula containing $\hbar$ does go over into a classical formula. Consequently when reformulated as statements not as directions or norms, those principles can be retrieved as theorems or as meta-theorems. (The same holds for covariance principles.) But, for all their importance, they are not constitutive assumptions of QM. The latter shall be spotted in what follows.

2. Background and Building Blocks

Like every other physical theory, QM has two sets of presuppositions: formal and material. The *formal background of* QM consists of ordinary logic (PC=), semantics, analysis, and all the set-theoretical, algebraic and topological presuppositions of analysis, as well as probability theory and its application, mathematical statistics. It is occasionally claimed that QM uses a logic of its own, different from ordinary or classical logic: some say it is a multiple-valued logic, others that it is intuitionistic logic, and so on (Sec. 8). This is mistaken: the mathematical formalism of QM is made up of a number of fragments of classical mathematics, which has ordinary (two-valued) logic built into it: from a purely formal point of view every formula of QM is a formula of PC= and every valid argument in QM is an inference pattern of PC=. (Whether the arguments employed to defend the orthodox interpretation of QM are countenanced by logic, is another question — one that must be separated from a discussion of the proper foundation of QM.) The mistake regarding the so-called quantum logic has been perpetuated only because it was introduced and endorsed by a few eminent people: unfortunately authority still plays a large role in QM. A second popular mistake regarding the formal background of QM is the claim that QM violates probability theory, mainly because probability amplitudes do not obey the probability axioms. But since the squares of their moduli do obey these postulates, the claim is wrong. Moreover we shall prove that QM obeys automatically probability theory. The only off-beat formal objects of QM, the various delta "functions", have long since become mathematically respectable by being absorbed in distribution theory. Indeed, to a modern mathematician QM must look disgustingly classical.

The *material background of* QM is constituted by the whole of proto-physics (physical probability theory, chronology, physical Euclidean geometry, systems theory, analytical "dynamics" and dimensional analysis) as well as by CEM. The latter is not or rather should not be presupposed, save heuristically, by quantum electrodynamics. And from a heuristic point of view QM presupposes classical PM in its hamiltonian form: indeed, it borrows whole formulas from particle mechanics — to begin with all those relating the so-called "*c-numbers*" occurring in QM (which are of course functions not numbers). But in an axiomatic reconstruction of QM such formulas must be newly introduced, the more so since, in the new context, they acquire new meanings.

The *specific primitive concepts* that will be interrelated by the postulates (which will also delineate their structure and meaning) are:

Σ: Set (kind of physical system under consideration) of individuals σ.

$\bar{\Sigma}$: Set (kind of physical systems, other than Σ, whose members $\bar{\sigma}$ can act on members of Σ and are classically describable).

E^3: Euclidean space (configuration space) with points x.

T: Interval of the real line (range of the time function) with points t.

$\mathfrak{H}$: Hilbert space (state space of every pair $\langle \sigma, \bar{\sigma} \rangle$) with members ψ.

$\{Q\}$: Family of functions on Σ (physical properties of σ).

$\{\mathscr{Q}\}$: Family of operators in $\mathfrak{H}$ (dynamical variables concerning σ).

$\{\varphi_k\}$: Family of functions on E^3 (eigenfunctions of $\mathscr{Q}$).

$\{c_k\}$: Family of complex valued functions on T (probability amplitudes of $\mathscr{Q}$ relative to ψ).

H: Operator-valued function of $\sigma, \bar{\sigma}, x$ and V.

$\hbar$: Positive real number (PLANCK's constant divided by 2π).

M: Function from Σ to R^+ with values μ (mass of σ).

V: Operator-valued function of $\sigma, \bar{\sigma}, x$ and V.

$\langle A_0, A \rangle$: Real-valued four-vector on $\Phi \times E^3 \times T$ (e.m. potentials).

c: Positive real number (light velocity in vacuum).

$\mathscr{E}$: Function from Σ to R^+ (electric charge).

Defined concepts shall be introduced as needed. Some of them will be borrowed directly from the background of QM, others will be built with the specific primitives of QM. To begin with we shall need the following conventions.

Df. 1. *eiv* $\mathcal{Q} \stackrel{\text{df}}{=}$ eigenvalues of $\mathcal{Q}$.

Df. 2. *eif* $\mathcal{Q} \stackrel{\text{df}}{=}$ eigenfunctions of $\mathcal{Q}$.

Df. 3. *Inner product* of f and $\mathcal{Q}g$ over $D \subseteq E^3$: $(f, \mathcal{Q}g) \stackrel{\text{df}}{=} \int_D \mathrm{d}x f^*(x)\, \mathcal{Q}g(x)$.

Df. 4. *Weighted average* of the set $\{q_k\}$ of real numbers w.r.t. the probability distribution $\{\omega_k\}$:

$$\langle\{q_k\}\rangle \equiv \langle q \rangle \stackrel{\text{df}}{=} \sum_k \omega_k q_k.$$

Df. 5. *Mean standard deviation* of the set $\{q_k\}$ of real numbers w.r.t. the probability distribution $\{\omega_k\}$:

$$\Delta\{q_k\} \equiv \Delta q \stackrel{\text{df}}{=} \left[\sum_k \omega_k (q_k - \langle q \rangle)^2\right]^{\frac{1}{2}}.$$

Remarks. (1) The summations (integrations for nondenumerable sets $\{q_k\}$) extend over the whole given sets, which are usually infinite. This alone shows that they cannot be sets of empirical data. By means of data one can get only *estimates* of such stochastic parameters. Since the q_k will turn out to be *eiv* $\mathcal{Q}$, we anticipate that the full statistics of a q.m. property ("observable") cannot be determined (let alone defined) operationally. (2) All the other concepts of QM (angular momentum, etc.) must be defined in terms of the foregoing primitives and with the help of Dfs. 1—5. In particular, if one wished to introduce the concepts of apparatus, measurement, observer, or even God, one should try to construct them in terms of the above primitives: introducing them otherwise would be plain cheating. (3) This does not preclude the possibility of taking some q.m. symbols to represent *schematically* certain experimental features. Thus in scattering theory the three chief components of a scattering experiment are represented in a symbolic way: the source, the target, and the detector. But the former is taken for granted and grossly simplified, and so is the output (the outgoing wave packet). Similarly the target is represented symbolically as a scattering field. Finally the absorber is represented merely by the solid angle into which the flux falls. Nothing is said about the actual construction and operation of the gun, the scatterer, and the detector: the experimentalist is supposed to figure these things out and to make them. Much less is said about the observer's mind, a ghost of the usual version of QM. And in no case does one, in the actual practice of QM (as opposed to the "philosophical" vagaries), confuse a dynamical variable or pseudo-observable with an experimental device. Only mathematicians take such idealizations and metaphors literally — and then physicists believe them upon the word of mathematicians. (4) The reference class of QM is the set of all pairs quanton-chunk of environment, i.e. $\Sigma \times \bar{\Sigma}$. In particular, the second coordinate of an ordered pair $\langle \sigma, \bar{\sigma} \rangle$ may be void, i.e. the theory may refer to an isolated quanton. In principle σ may represent any physical system, but in practice there is usually no point in applying QM to systems other than microphysical entities or aggregates of such. (5) This must be emphasized both because textbook "observations" on microphysical systems do not render them observable *stricto sensu* and because it is often claimed that QM is a statistical (not just stochastic) theory in the sense that it refers to actual or potential aggregates of quantons. But the concept of aggregate does not occur in our list of primitives because it won't figure in our postulates. Aggregates can occur in basic QM, if at all, as environmental units, i.e. as individuals belonging to $\overline{\Sigma}$. Actual q.m. aggregates require an extension of our basic theory. This is a matter of initial assumption: we are postulating that one-system QM is about a single individual of the kind Σ — a sort to be specified by the axioms — eventually in interaction with a non-quantal system $\overline{\sigma} \in \overline{\Sigma}$. In other words, contrary to what is often stated (e.g., EINSTEIN, 1935; KEMBLE, 1935; MOYAL, 1949; BLOCHINZEW, 1953; LUDWIG, 1961), the basic QM is a *stochastic* but not a *statistical* theory: it does not refer to statistical ensembles, be they aggregates of coexistents or Gibbs ensembles of conceptual copies of a single individual. The proof of this is that quantum statistics, be it of bosons or of fermions, is a different theory built on the basis of elementary QM. The statistical concepts occurring in the latter — chiefly the averages and variances — refer to a single individual. This cannot be understood if only the frequency interpretation of probability is accepted; but it is understandable on alternative interpretations, notably the propensity interpretation of probability (POPPER, 1959a and Ch. 2, 2). (6) Statistical aggregates are absent from the foundations of QM but they occur in some of its applications, provided certain additional assumptions are adjoined. In particular, they occur in the applications of QM to measurements (Sec. 7) — e.g., when a given property of an individual system is measured a number of times in succession or when the same property is measured simultaneously on a number of identically prepared individuals. But QM is not primarily concerned with measurement — a particular case of σ-$\overline{\sigma}$ interactions and one whose study has only been sketched. (7) The concepts of particle and wave are absent from our list of basic concepts; nor will they be defined. Both are here regarded as classical concepts that may occur in the description of macroevents — e.g., measurements — but which, if taken literally in QM, give rise to paradox (HEISENBERG, 1930). The important heuristic role they have played is now nearly exhausted in QM. This does not mean that electrons and other quantons are neither particles nor waves: it just means that QM does not picture them as such and that in fact they are neither particles nor waves to the extent that QM is true.

3. Comprehensive Postulates

QM is based on a set of assumptions some of which — e.g., the Schrödinger equation with unspecified hamiltonian — are generic while others — e.g., the one about the hamiltonian of a quanton in an external e.m. field — are specific. In this section we shall give the comprehensive axioms; the specific ones will be presented in Sec. 5. As will be seen in Sec. 4, a number of important consequences can be derived from the comprehensive postulates before advancing more specific hypotheses concerning the nature of quantons.

We shall let protophysics (Ch. 2) take care of the primitives E^3 and T, particularly since elementary QM is nonrelativistic. Moreover space and time are non-quantal in QM: neither is a dynamical variable in the Schrödinger picture, and although x is a random variable t is not (it does not spread). In other words, we assume that E^3 is the 3-dimensional Euclidean space with inner product and that it represents physical space; we also assume that T is the range of the time function characterized by the axioms of the universal time theory. Here go our comprehensive postulates.

QM 1: *existence*. (a) Σ and $\bar{\Sigma}$ are nonempty denumerable sets. (b) Every $\bar{\sigma} \in \bar{\Sigma}$ is the environment of some $\sigma \in \Sigma$.

QM 2: *state space*. (a) For every ordered pair $\langle \sigma, \bar{\sigma} \rangle \in \Sigma \times \bar{\Sigma}$ there exists a Hilbert space $\mathfrak{H}$ associated with $\langle \sigma, \bar{\sigma} \rangle$.

(b) The points ψ of $\mathfrak{H}$ are complex valued functions on $\Sigma \times \bar{\Sigma} \times E^3 \times T$.

(c) For any fixed pair $\langle \sigma, \bar{\sigma} \rangle \in \Sigma \times \bar{\Sigma}$, $\dfrac{\partial \psi}{\partial t}$ and $V^2 \psi$ are everywhere finite and piece-wise continuous.

(d) For all $\sigma \in \Sigma$ and all $\bar{\sigma} \in \bar{\Sigma}$, if the spatial region accessible to the compound system $\sigma \dotplus \bar{\sigma}$ at any time $t \in T$ is $V \subseteq E^3$, then $\psi(\sigma, \bar{\sigma}) = 0$ on the border ∂V of V.

(e) For every $\psi \in \mathfrak{H}$ and at any $t \in T$, ψ and $V\psi$ are square integrable over any region $V \subseteq E^3$ accessible to $\sigma \dotplus \bar{\sigma}$.

QM 3: *dynamical variables*

QM 3.1 (a) $\{Q\}$ is a nonempty family of functions on Σ.

(b) $\{\mathscr{Q}\}$ is a nonempty family of operators in $\mathfrak{H}$.

(c) If $Q \in \{Q\}$ is a property of $\sigma \in \Sigma$, then there exists a $\mathscr{Q} \in \{\mathscr{Q}\}$ such that $\mathscr{Q} \triangleq Q$.

QM 3.2 For every $\sigma \in \Sigma$, every $x \in E^3$, every $t \in T$, every $Q \in \{Q\}$ and every $\mathscr{Q} \in \{\mathscr{Q}\}$,

(a) $\mathscr{Q}$ is a linear mapping (operator) of $\mathfrak{H}$ into itself. [I.e., $\mathscr{Q}: \mathfrak{H} \to \mathfrak{H}$.]

(b) $\{\mathscr{Q}\}$ constitutes a ring of operators. [I.e., if $\mathscr{Q}_1, \mathscr{Q}_2 \in \{\mathscr{Q}\}$ and $c_1, c_2 \in C$, then $c_1 \mathscr{Q}_1 + c_2 \mathscr{Q}_2$ and $\mathscr{Q}_1 \cdot \mathscr{Q}_2$ are in $\{\mathscr{Q}\}$.]

(c) $\mathcal{Q}$ has a complete set $\{\varphi_k\}$ of orthonormal eigenfunctions $\varphi_k \in \mathfrak{H}$ defined on $\Sigma \times E^3$:

$$\mathcal{Q}\varphi_k = q_k \varphi_k, \qquad (\varphi_k, \varphi_{k'}) = \delta_{kk'} \tag{5.1}$$

$$\psi = \sum_k c_k \varphi_k, \qquad c_k : T \to C. \tag{5.2}$$

(d) For every fixed $\sigma \in \Sigma$, every $\varphi_k \in \{\varphi_k\}$ is a bounded complex valued function, piece-wise continuous w.r.t. x.

(e) For every $\sigma \in \Sigma$, every $Q \in \{Q\}$ and every $\mathcal{Q} \in \{\mathcal{Q}\}$, if $\mathcal{Q} \triangleq Q$ then $eiv\,\mathcal{Q} \equiv q_k \in R$.

(f) If $\mathcal{Q} \triangleq Q$ then the value of $\psi^* \mathcal{Q} \psi$ at the point $\langle \sigma, \bar{\sigma}, x, t \rangle$ represents the Q-density of the compound system $\sigma \dotplus \bar{\sigma}$ at x, t.

(g) For every $\sigma \in \Sigma$ and every $Q \in \{Q\}$ and every $\mathcal{Q} \in \{\mathcal{Q}\}$, if $\mathcal{Q} \triangleq Q$, then the eigenvalues q_k of $\mathcal{Q}$ are the sole values of Q that σ takes on.

QM 4: kind and structure of system. For every pair $\langle \sigma, \bar{\sigma} \rangle \in \Sigma \times \bar{\Sigma}$ there is a distinguished element H of $\{\mathcal{Q}\}$ such that

(a) H is a function of x, t and V.

(b) H is invariant [symmetric] under spatial rotations, spatial reflections [parity transformations] and time inversions.

(c) For every pair $\langle \sigma, \bar{\sigma} \rangle \in \Sigma \times \bar{\Sigma}$, the eigenvalues E_k of $H(\sigma, \bar{\sigma})$ represent the energy values of σ when acted on by $\bar{\sigma}$.

QM 5: evolution of states. If $H \in \{\mathcal{Q}\}$ and $\psi \in \mathfrak{H}$ refer to one and the same compound system $\sigma \dotplus \bar{\sigma}$, where $\sigma \in \Sigma$ and $\bar{\sigma} \in \bar{\Sigma}$, then, for all $t \in T$ and all $x \in E^3$,

(a)
$$i\hbar \frac{\partial \psi}{\partial t} = H\psi. \tag{5.3}$$

(b) If, for a given pair $\langle \sigma, \bar{\sigma} \rangle \in \Sigma \times \bar{\Sigma}$, there exists a $\psi \in \mathfrak{H}$ that satisfies (5.3), then: up to an arbitrary phase $exp\,(i\alpha)$, with $\alpha \in R$, $\psi(\sigma, \bar{\sigma}, t)$ represents the state of σ at time t when acted on by $\bar{\sigma}$. [I.e., every state of a quanton at a given instant is represented by a ray in $\mathfrak{H}$.]

(c) If $U \in \{\mathcal{Q}\}$ is a unitary operator in $\mathfrak{H}$ [i.e., $U^+ = U$], and $\psi \in \mathfrak{H}$ and $\langle \sigma, \bar{\sigma} \rangle \in \Sigma \times \bar{\Sigma}$, then $U\psi(\sigma, \bar{\sigma})$ and $\psi(\sigma, \bar{\sigma})$ represent the same state of σ when associated with $\bar{\sigma}$.

(d) $\hbar$ is a positive real number such that $[\hbar] = L^2 M\,T^{-1}$.

On the basis of the preceding hypotheses, the following definitions make sense:

Df. 6. Any $\sigma \in \Sigma$ that satisfies the preceding axiom system QM 1—5 will be called a *quanton*.

Df. 7. Any $\bar{\sigma} \in \bar{\Sigma}$ that satisfies the preceding axiom system QM 1—5 will be called the *environment* of a quanton.

Df. 8. Any $\psi \in \mathfrak{H}$ that satisfies (5.3) is called a *state* of the corresponding quanton.

Df. 9. Any $\psi \in \mathfrak{H}$ satisfying (5.3) and such that ψ is not a linear combination of two other states is called a *pure state*. [More precisely: Any prime ray in $\mathfrak{H}$ is called a pure state.]

Df. 10. Any state that is not a pure state is called a *mixture*.

And now some extrasystematic clarifying comments.

Remark 1. (*a*) The first axiom is usually not stated either because it is taken for granted or because the existence of microsystems on their own right, without The Observer's assistance, is disclaimed as a metaphysical extravagance. In any case some assumption of the sort has to be made lest the theory be not vacuously true. An axiomatic treatment of the subjectivistic interpretation of QM would have to keep $QM\ 1\,a$ and replace $QM\ 1\,b$ and $QM\ 1\,c$ by something like "Every $\sigma + \bar{\sigma}$ is a constellation of phenomena appearing to a competent observer endowed with infinitely precise experimental devices". In any such interpretation physical symbols concern the apperceptions of an arbitrary observer, whereas in ours they are about physical systems. (*b*) The reference class of QM is, by QM, $\Sigma \times \bar{\Sigma}$, i.e. the set of all pairs quanton-environment. The environment is treated as a unit rather than analyzed, say, as an aggregate of quantons. Moreover, it may be absent, i.e. the particular member of $\bar{\sigma}$ occurring in the second coordinate of the ordered pair may be the null individual $0_{\bar{\Sigma}}$ (see Ch. 2, 5). In other words, there are formulas of QM which refer to free quantons (or aggregates of such). This platitude must be stated because the received doctrine is that every q.m. formula concerns a microsystem under disturbing observation — even when it is explicitly stated that certain formulas refer to a free quanton.

Remark 2. The Hilbert space is as important as the configuration space for, while every physical system is somewhere in space — though usually with no sharp boundary — whatever happens to $\sigma + \bar{\sigma}$ is mapped on $\mathfrak{H}$ (see $QM\ 5\,b$). Obviously $\mathfrak{H}$ lies beyond observation, whence it can be assigned no operational interpretation. Yet contrary to the usual teaching the $\mathfrak{H}$ of QM is not abstract either, for by $QM\ 2$ it is associated with the physical systems under study. It is a concept all right not a thing, but a concept representing a trait of physical things — i.e. a physical concept (see Ch. 1, 2).

Remark 3. (*a*) Thanks to $QM\ 1$ and $QM\ 2$, none of the clauses of $QM\ 3$ is purely mathematical even though most of them focus on mathematical properties of the operators $\mathscr{Q}$, their eigenfunctions or eigenvalues. In this group of axioms, $QM\ 3.1\,c$, $QM\ 3.2\,f$ and $QM\ 3.2\,g$ are the properly semantical (interpretive) hypotheses. (*b*) In the usual interpretation, the solutions q_k of the eigenvalue equation (5.1) are said to be the only possible values a *measurement* of Q on σ can yield. The reference to measurement has been discarded for one should make no dogmatic assertions concerning the only possible results of any possible experiment: tentative predictions, not cock-sure prophecies, characterize science. Besides, the measurement of physical properties on microsystems is anything but direct and infinitely precise: for one thing measurements will never yield irrational numbers, and it so happens that the spectra of many operators are continuous; for another thing measurements are not concerned with an isolated system, but with a system σ embedded in a highly complex environment $\bar{\sigma}$, the characterization of which is essential to the description of the experiment but irrelevant to most of the theoretical predictions of QM, which concern quantons in highly idealized environments. (*c*) The connection with measurements is not done at the level of principles, whether in QM or in any other fundamental theory (see Ch. 1, 5). But *QM* 3.2g, in conjunction with general principles concerning measurement, does allow us to infer that, *if* QM is true, then: if a property Q is measured on a system idealized as σ with a technique t involving an experimental error $\varepsilon_{k\,t}$, then the measurement results will be

$$(q_k)_t = [q_k] \pm \varepsilon_{k,\,t} \tag{5.4}$$

where $q_k = eiv\,\mathcal{Q}$, with $\mathcal{Q} \triangleq Q$, and where '$[q_k]$' designates a fraction near the theoretical value q_k. This statement does not concern the meaning of $\mathcal{Q}$ and it does not belong to the foundations of QM but to the theory of measurement. And the usual claim that every measurement result is a *real* number (or an *n*-tuple of reals) is necessary for an operationalist interpretation of QM, which does not allow things to have definite properties while these are not being measured. But it is also inconsistent with operationalism as well as with the most elementary considerations on measurements, since every measured value is a fractionary number — this being the only kind of number a consistent operationalist can accept. (*d*) Time does not belong to the family $\{\mathcal{Q}\}$ of operators in Hilbert space: it is not a dynamical variable but a parameter, so much so that ψ is not normalized w.r.t. t and that two ψ's differing in t are not mutually orthogonal. This situation is not changed in relativistic QM. Hence the claim that t and H are canonically conjugate variables is mistaken: if t is not an operator in $\mathfrak{H}$ then it can have no partner in it. Consequently there can be no 4th Heisenberg indeterminacy relation in the known forms of QM. (More on this in Sec. 6.)

Remark 4. The hamiltonian operator H characterizes partly the kind and structure of system — linear oscillator, simply ionized Helium atom, etc. Only partly: the boundary conditions do the rest. But notice that so far we have not specified H.

Remark 5. (a) As long as H is not specified, the Schrödinger equation $QM\ 5a$ is not an entirely definite assertion and consequently has no definite solutions. Still it can be solved symbolically: in fact (5.3) is equivalent to

$$\psi(t) = \exp\left[-\frac{i}{\hbar}\int_{t_0}^{t} dt\, H\right]\psi(t_0)\,. \qquad (5.5)$$

This equation shows that the state of a system at any time t is uniquely determined by its state at an arbitrary initial time $t_0 < t$ and by H. It also shows that the evolution of the state of a system is represented as a trajectory in the Hilbert space of the system — or rather the systems pair $\langle\sigma, \bar{\sigma}\rangle$. (b) The Schrödinger equation or its symbolic solution (5.5) is often regarded as the quantum version of the principle of causality (MARGENAU, 1950). Yet a regular unfolding of states, with no indication of the actions that bring about their emergence, is not a causal sequence (BUNGE, 1959a). And to talk of mathematical causation (BIRKHOFF and VON NEUMANN, 1936) is senseless, since causation is by definition a physical not a logical relation. The Schrödinger equation is not a comprehensive ontological principle but a comprehensive physical law — in fact the only law statement among the comprehensive postulates $QM\ 1-5$. Moreover, as we shall see in Sec. 4, it does not exemplify the principle of causality. On the other hand it does tally with the principle of antecedence (see Ch. 2, 1). (c) In classical physics a *state* of a system characterized by $n < \infty$ mutually independent properties (counting the components of tensors) is any n-tuple of values of such functions as long as it is compatible with the laws of the system. In QM the concept of state is introduced independently from the concept of property (dynamical variable): states are no longer clusters of properties. (d) Unlike initial value problems in particle mechanics, but like boundary value problems in CM, CEM, GR and other field theories, in QM one cannot enter empirically found values characterizing the initial state: $\psi(t_0)$ is not measurable but must be hypothesized. The hypothesis must of course satisfy the previous axioms and it must eventually be checked (indirectly) through its consequences — just as in classical field theories. (e) $QM\ 5b$ is often stated in one of these ways: "ψ represents in a complete way the state of the system" or "ψ yields the fullest possible information about the system". But so far we do not know, except unofficially, how to extract any information from ψ: we only assume that, *as far as* QM *goes*, $\psi(\sigma, \bar{\sigma})$ characterizes the state of a quanton σ in an environment $\bar{\sigma}$. To say that ψ constitutes a complete description of a physical system, or that it contains the sum of human knowledge about it —

in aeternum, amen — is characteristically dogmatic: it amounts to say that nonrelativistic QM is perfect — which everyone knows or suspects is not — so that neither new experimental material nor new theoretical developments can touch it. The claim that QM — or for that matter any other theory — gives a complete description of its referents is a piece of *propaganda fidei* not a physical statement. (*f*) The reservation in $QM\ 5b$ concerning the phase factor, and the equivalence of states related by a unitary transformation ($QM\ 5c$), make room for infinitely many representations of one and the same state of a σ–$\bar{\sigma}$ pair. In particular, $QM\ 5c$ is a postulate of objectivity warning us against taking all the traits of a given representation for objective properties of things. (If only for this reason one should not speak of a state ψ but of a state represented by ψ — but tradition is too strong.) (*g*) Following v. NEUMANN (1931) it is usually stated that the Schrödinger equation holds provided the system is not disturbed, e.g. measured — in which case the projection of ψ onto one of the eigenfunctions φ_k of the corresponding operator $\mathscr{Q}$ would take place. It this were so, the proper place to say it would be when specifying the properties of the hamiltonian. But no such restriction is ever placed upon H, which is allowed to contain terms representing actions of external systems $\bar{\sigma}$ on the thing σ under study. Consequently we assume that the Schrödinger equation holds for any pair $\langle \sigma, \bar{\sigma} \rangle$, whether or not $\bar{\sigma}$ happens to represent an apparatus (and even when $\bar{\sigma}$ is a nonentity), and provided H covers (in conjunction with the boundary conditions) the relevant (known or assumed) characteristics of the system. (*h*) The claim that the Schrödinger equation holds only for unobserved systems is of course inconsistent with the empiricist philosophy underlying the orthodox interpretation of QM: it takes the thing-in-itself for granted. This inconsistency in the usual formulation of QM does not occur in ours. (*i*) It is often claimed that a run of measurements of a set of commuting dynamical variables allows one to determine ψ unambiguously, so that it can be "read from experiment". For example, measuring the mutually commuting H, L^2 and L_z on a system σ in a central field $\bar{\sigma}$ allows us to pin down its precise state ψ_{nlm}. Indeed, an energy measurement gives (ideally) n, a total angular momentum measurement l (or rather $[l(l+1)]^{\frac{1}{2}}$) and a measurement of the projection of L on the z-axis yields m. But to interpret these measurements in such a way we must previously solve certain eigenvalue equations. In fact before we can make any use and even any sense of the measurements of H, L^2 and L_z, we need the following theoretical items: the formula for the energy eigenvalues (which will depend on the specific field), the formulas for the eigenvalues of L^2 and L_z, and the general form of ψ_{nlm} (a radial function times a spherical harmonic). No set of formulas yields by itself ψ_{nlm} or any other state function: one must first hypothesize H and the boundary conditions, then solve the corresponding Schrödinger equation and the relevant eigenvalue equations with the suitable boundary conditions — only then can one hope to learn from a run of experiments and enter the relevant data into the theoretical formulas. Yet the best expositions of QM give a golden rule for constructing the whole infinitely-dimensional Hilbert space of a $\sigma-\bar{\sigma}$ system out of the measured values of a set of commuting "observables". O amazing force of a dead philosophy! (j) Mathematically, ψ is a scalar field — but so are the Jacobi characteristic function for an unspecified system (see Ch. 2, 6.5), the thermodynamic temperature, and many other magnitudes. Not every mathematical field represents a physical one and not every periodic function represents a real undulation. In our version of QM there are no matter waves — nor are there information waves. There exists only a limited formal *analogy* between ψ and a classical wave field. The following points reinforce this view. Firstly, it is only for the single quanton and in the "position" representation that ψ can be visualized as a wave or as a wave packet: in alternative representations or for aggregates of quantons this analogy breaks down — which does not entail that ψ loses its physical meaning. Secondly, ψ cannot accompany, let alone guide the quanton in its motion, if only because the phase velocity of a ψ-wavelet equals half the velocity of the packet or group, which is in turn equal to the corresponding classical velocity. On the other hand it could be argued that the bilinear forms $\psi^* \mathcal{Q} \psi$ are genuine field variables. (k) None of the conditions laid so far on ψ guarantees its existence: existence theorems must be proved for every form of H. Unfortunately only too few such theorems are available. This explains the cautiousness in the statement of the semantical hypothesis QM 5b: *if* ψ exists, then its physical meaning is such and such. (l) No numerical value has been assigned to $\hbar$. This is something that cannot be assumed but must be determined by experiment — just as in the case of c and $\varkappa$. (Yet mathematicians can be found to attest that the definition "$\hbar = h/2\pi$" has been experimentally established.) In addition to this methodological reason there is a psychological motive: by forgetting, at the foundation level, what the actual numerical values of the basic constants are, one does not inadvertently introduce considerations of size and order of magnitude. Such considerations are out of place in the case of comprehensive theories like QM; they are indispensable when the theory is applied, tested, and evaluated. (m) PLANCK's constant has been interpreted variously, chiefly as the unit of angular momentum and as the bridge between the so-called particle and wave properties (e.g., in the relations $p = h/\lambda$ and $E = h\nu$). After learning about half-integer angular momenta we should be wary of the first interpretation; and as to the second one it is a classical analogy taking for granted that the q.m. momentum and energy are corpuscular properties (as if classical fields did not have momentum and energy). All these considerations become pointless upon realizing that, unlike the remaining specific symbols of QM, $\hbar$ has no referent: there is no such thing as the $\hbar$ *of* some physical system: $\hbar$ is in the same set of universal constants as R, k and $\varkappa$, not in the class to which c and e belong (see Ch. 1, 2.2.1). (*n*) In the usual presentations of QM most of the preceding postulates are missing — which suggests that they are p-incomplete (they do not characterize the primitives sufficiently). On the other hand in the usual presentations several axioms occur which we shall either discard altogether or deduce from the previous assumptions. Our most important exile is von Neumann's projection postulate, according to which a measurement of Q projects ψ onto an eigenstate φ_k of $\mathscr{Q} \triangleq Q$. The reasons for discarding it are the following. Firstly, it restricts without any reason the generality of the Schrödinger equation (recall comment *g*). Secondly, it is the only hypothesis that cannot be restated in a general way, one not involving measurement — and when it is stated in this way it becomes inconsistent with the basic philosophy of the orthodox interpretation for it enthrones the unobserved system (see comment *h*). Thirdly, it is inconsistent with the rest of QM (Margenau, 1936 and Sec. 7). On the other hand, we accept the stochastic interpretation of ψ and its components but we do not postulate it: it shall be deduced right away.

4. Comprehensive Theorems

We shall now deduce a few typical comprehensive theorems, i.e. formulas derived from the previous axioms without specifying the form of H.

Thm. 1: *probability amplitudes.* If $\sigma + \bar{\sigma}$ $(\sigma \in \Sigma, \bar{\sigma} \in \overline{\Sigma})$ is in the single (nondegenerate) state $\psi \in \mathfrak{H}$ and if $\mathscr{Q} \triangleq Q$ and if $c_k \in C$ is the component of ψ along the kth axis $\varphi_k = e i f \mathscr{Q}$ of the basis $\{\varphi_k\}$ in $\mathfrak{H}$ [recall (5.2)], then the probability that the value q_k of Q lies in the interval $[q_{k'}, q_{k''}]$ is

$$w(q_k \in [q_{k'}, q_{k''}]) = \sum_{k \in \Delta k} |c_k|^2, \qquad \Delta k = [k', k''], \qquad (5.6)$$

where the summation (or the integration if the spectrum of $\mathscr{Q}$ is continuous) is extended over those k for which $q_k'' \leq q_k \leq q_{k'}$. *Proof.* The proof that the function w satisfies Kolmogoroff's axioms for probability (Ch. 2, 2) is straightforward: (*a*) by QM 3.2c the set of k's on which w is "defined" is in fact nonempty; (*b*) every square of a real number (such as $|c_k|$) is nonnegative; (*c*) the sum of the $|c_k|^2$ for the union of two disjoint eigenvalue intervals equals the sum of the corresponding sums, and (*d*) for any normalized $\psi \in \mathfrak{H}$ and any $\mathscr{Q} \in \{\mathscr{Q}\}$, the

$|c_k|^2$ add up to unity as shown with the help of QM 3.2c:

$$1 = \|\psi\| = \sum_{kk'} c_k^* c_{k'} (\varphi_k, \varphi_{k'}) = \sum_{kk'} c_k^* c_{k'} \delta_{kk'} = \sum_k |c_k|^2.$$

Remarks. (1) Ordinarily Thm. 1 must be postulated because it is stated with reference to measurements. Sometimes it is derived from the statement regarding the average of $\mathcal{Q}$ (our Thm. 2) — which is in turn sometimes postulated and at other times regarded as a definition. But such derivations are not rigorous: indeed while given a probability distribution the average of a random variable is uniquely determined, the converse is not true: one and the same average can be shared by any number of distributions. (2) Proving that the c_k are probability amplitudes is important because it shows that the probability interpretation of QM is inescapable not *ad hoc* hence dispensable. This does not entail that QM has espoused the whole of mathematical statistics: the mere fact that covariances (equivalently, joint probability distributions) cannot always be defined (see Thm. 10 in Sec. 6) suggests that the marriage is not a happy one (SUPPES, 1961). (3) The long-range experimental implication of Thm. 1 is clear: if Q is measured on a large collection of copies of σ–$\bar\sigma$ pairs all in the same state ψ, then a histogram closely fitting the $|c_k|^2$ curve will be obtained provided QM is true. This theorem is therefore the chief predictive tool of QM. A knowledge of the (infinite!) set of probability amplitudes enables us to compute the statistics of the corresponding operator $\mathcal{Q} \triangleq Q$. But working the other way around, i.e. starting from a (finite!) number of empirical data concerning this statistics does not determine the c_k — the more so since these are complex, i.e. $c_k = |c_k| e^{i \alpha_k}$. (4) By virtue of Thm. 1 all physical variables of QM except the constants (such as $\hbar$, μ and e) are *random variables:* the stochastic character is not, as sometimes claimed, introduced as an after-thought but is deeply ingrained in QM. It is also seen that this stochastic character is not inherent in ψ but in the relation between ψ and *eiv* $\mathcal{Q}$: indeed ψ *induces* a probability distribution on every (otherwise stochastically amorphous) set $\{q_k\}$ of *eiv* $\mathcal{Q}$. Therefore ψ by itself does not represent a field of possibilities — nor does φ_k. Every set $\{q_k\}$ of eigenvalues of an operator $\mathcal{Q}$ is a set of possibilities to which ψ assigns definite weights. (5) The probabilities occurring in QM cannot be assigned a subjective interpretation (as credibility, degree of certainty, subjective utility, and so forth) because both ψ and $\mathcal{Q}$ refer, *ex hypothesi*, to a quanton in an environment, not to The Observer. In other words, the set $\{c_k\}$ does not represent the state of our knowledge concerning a σ–$\bar\sigma$ pair, much less an experimental arrangement, but is an *objective* (but potential) *property* of such a σ–$\bar\sigma$ pair — just as much as the Jacobi characteristic function S.

The claim (by BOHR, HEISENBERG and BORN) that ψ and its law of evolution concern our knowledge rather than nature is inconsistent with the very way the state of the ith quanton in an aggregate is symbolized, i.e. ψ_i: the index i names the object, not The Observer. Consequently, the subjectivist interpretation of probability, an inescapable ingredient of the orthodox interpretation of QM, does not square with the way one handles it in everyday work. (6) The interpretations of probability entering QM, like every other physical theory, are the ones yielding physical models of CP (Ch. 2, 2): the propensity and the randomness interpretations for theoretical work, the frequency interpretation for experimental work. When dealing with a single quanton σ in an environment $\bar{\sigma}$, we shall speak of the probability of the pair of things $\langle \sigma, \bar{\sigma} \rangle$ as regards the state-property pair $\langle \psi, Q \rangle$. As long as the formulas are not subjected to test, the $|c_k|^2$ are intended to be objective (yet prior and possibly false) probabilities, not observed relative frequencies. The fact that they are calculated does not render them subjective. (7) The fact that in the usual discussions of QM only two interpretations of CP are mentioned, the subjectivist and the statistical one, as if they were the only possible ones, shows that those discussions remain in the stage prior to the modern developments in the foundations of mathematics — particularly model theory — which have shown that, *per se*, the calculus of probability is bound to no particular interpretation — this being the reason it can be interpreted in so many ways (BUNGE, 1967b).

Corollary 1. Under the conditions of Thm. 1, the probability w_k that Q has precisely the value q_k is

$$w_k = |(\varphi_k, \psi)|^2. \tag{5.7}$$

Proof. From Thm. 1 and the theorem of analysis concerning the inversion of c_k.

Corollary 2. If $\sigma \dot{+} \bar{\sigma} (\sigma \in \Sigma, \bar{\sigma} \in \bar{\Sigma})$ is in the kth eigenstate of $\mathcal{Q}$ [i.e. if $\psi = \varphi_k$] then Q has the definite value q_k [i.e. $w_k = 1$].

Remarks. (1) Restatement: unless $\sigma \dot{+} \bar{\sigma}$ is in an eigenstate of $\mathcal{Q}$, the set of values of Q will exhibit a spread. (2) The long-range experimental implication is this. If a system is prepared to attain the clear-cut state φ_k, then the measured value of Q will be close to q_k — provided QM is true. This last statement comes close to the usual operationalist version of our Corollary 2, namely: "If the observable corresponding to $\mathcal{Q}$ is measured on a system, then the value q_k will result with certainty". In the strict Copenhagen doctrine the property Q has no definite value even if $\psi = eif \mathcal{Q}$, unless that property is being measured (see e.g. BOHR, 1932 and 1958; FRANK, 1946; ROSENFELD, 1953). In our version of QM a property Q will always have some *distribution* and it will have a particular value q_k provided $\psi = eif\, \mathscr{Q}$, even if no measurement is performed. It is only when we want to evaluate that hypothesis that we must subject it to experimental test (see Ch. 1, 4.1.4, 4.2.6 and 4.3.3). Nor does the concept of certainty enter our version of QM: while it is all right when referring to human expectations — concerning, e.g., measurement results — it is out of place in a physical theory.

Thm. 2: *averages*. Under the conditions of Thm. 1, the average of Q equals

$$\langle \mathscr{Q} \rangle_{\psi} = \sum |c_k|^2 q_k = (\psi, \mathscr{Q}\psi). \tag{5.8}$$

Proof. The first equality follows from Thm. 1 and Df. 4 (in Sec. 2). The second follows upon expanding ψ in eigenfunctions of $\mathscr{Q}$ and introducing the eigenvalue equations $\mathscr{Q}\varphi_k = q_k \varphi_k$ and the orthonormalization condition QM 3.2c. For continuous spectra replace summations by integrations — and be prepared to find that many averages simply do not exist.

Remarks. (1) We employ throughout the term 'average' not the expression 'expectation value', which has a psychological connotation and is a relic of the time when probabilities were associated with ignorance. (2) Thm. 2 is usually interpreted in terms of measurement results, which is absurd since in the computation of averages by means of (5.8) the whole infinite spectrum of $\mathscr{Q}$ occurs. Again, the most we can say is that, *if* QM is true, then the average of Q measured on a large number of systems all in a state represented by ψ will be close to the value predicted by Thm. 2. But this statement is tautological. (3) The first equality in (5.8) has a clear meaning by virtue of Thm. 1, but the second is the more convenient for computation purposes as it does not require the expansion of ψ in $eif\, \mathscr{Q}$.

Corollary: spreads. Under the conditions of Thm. 1, the standard deviation (*rms*) of the distribution of Q equals

$$\Delta_{\psi}\mathscr{Q} = \langle (\mathscr{Q} - \langle \mathscr{Q} \rangle_{\psi})^2 \rangle_{\psi}^{\frac{1}{2}}. \tag{5.9}$$

Proof. By Thm. 2 and Df. 5 (in Sec. 2).

Remarks. (1) Thm. 2 and its corollary cover the so-called statistics of any Q relative to any state of a pair $\sigma - \bar{\sigma}$. This theoretical or calculated statistics takes into account the infinitely many possibilities of a physical system in a given environment and should therefore be contrasted rather than identified with the finite sample of actualized possibilities a run of measurements can yield. (2) Equation (5.9) says nothing about the spread of a set of measurements of Q on $\sigma - \bar{\sigma}$; it is an *objective scatter relation* (POPPER, 1935). But of course it follows that, *if* QM is true, then any such run of measurements will satisfy the above corollary — which is again a tautology. Unless $\sigma \dotplus \bar{\sigma}$ happens to be in an eigenstate of $\mathcal{Q}$ (Corollary 2 of Thm. 1), even ideally accurate measurements should exhibit a spread. In other words, the q. m. latitudes or spreads are inherent in the physical systems not in their measurements. What measurements do is to let us have a glimpse at such spreads. If anything, by manipulating the system one can occasionally place it on a state φ_k where Q will exhibit no spread. (3) It follows that the names 'inaccuracy' and 'uncertainty' for the spread $\Delta \mathcal{Q}$ are inaccurate, the first for suggesting that the scatter is due to inaccurate measurement, the second for suggesting that it refers to our state of mind — "uncertainty" being a psychological not a physical predicate. (4) The interpretation of $\Delta \mathcal{Q}$ as a *subjective indeterminacy* (uncertainty) is rooted to the subjectivist interpretation of the probability theory (as is obvious in v. NEUMANN, 1931). And the interpretation of $\Delta \mathcal{Q}$ as *objective indeterminacy* or haziness is unavoidable if the classical model of the point particle is kept (as is obvious in BORN, 1956 and 1961 and LANDÉ, 1965). In our version of QM the subjective interpretation of probability occurs as little as the concept of point particle, whence none of the preceding interpretations of $\Delta \mathcal{Q}$ — as uncertainty and as indeterminacy — is accepted.

Thm. 3: *spreads of canonically conjugate variables.* For every $\langle \sigma, \bar{\sigma} \rangle \in \Sigma \times \bar{\Sigma}$, every $\psi \in \mathfrak{H}$ that solves the Schrödinger equation for $\sigma \dotplus \bar{\sigma}$, and every trio $\mathcal{A}, \mathcal{B}, \mathcal{C} \in \{\mathcal{Q}\}$ such that $\mathcal{A} \triangleq A, \mathcal{B} \triangleq B, \mathcal{C} \triangleq C$, with $A, B, C \in \{Q\}$ and $\mathcal{A}\mathcal{B} - \mathcal{B}\mathcal{A} = i\mathcal{C}$, the standard deviations of the A and B distributions in the state ψ are related by

$$\Delta_\psi \mathcal{A} \cdot \Delta_\psi \mathcal{B} \geq \tfrac{1}{2} |\langle \mathcal{C} \rangle_\psi|. \qquad (5.10)$$

Proof. By the Corollary of Thm. 2 and the Schwarz inequality.

Remarks. (1) This theorem, a generalization of HEISENBERG'S "principle", places a lower bound on the scatters of canonically conjugate properties. An equality can in principle be obtained upon specifying $\mathcal{A}, \mathcal{B}, \mathcal{C}$, and ψ. (2) Once again, nothing has been assumed about measuring devices. Consequently the preceding theorem is supposed to be satisfied by every $\langle \sigma, \bar{\sigma} \rangle$, whether or not $\bar{\sigma}$ happens to idealize an apparatus, and even when there is no $\bar{\sigma}$ — e.g., for a free electron. How could we specify anything about the experimental set-up if we have not even specified the properties A and B? (3) The usual interpretation of the random fluctuations as uncontrollable disturbances caused by the apparatus is *ad hoc* since (*a*) no apparatus variable occurs in the theorem (not even when deduced in the orthodox textbooks) and (*b*) in general there exists no joint probability distribution for pairs of noncommuting variables, whence no conclusion about the spreads of measurement results can be drawn even applying (5.10) to an experimental situation (SUPPES, 1961). (4) The widespread belief that the scatter relations (5.10) can be inferred from an analysis of measurements and then the corresponding commutation relations deduced from the scatter relations (HEISENBERG, 1927) is plainly false: (a) Thm. 3 is a universal statement involving unobservables, and no such statement can be gotten by induction from observed instances; (b) every analysis of a measurement procedure is made in some context or other (see Ch. 1, 5.1.4), and if QM is not used to this end then some classical theory will be employed (as is usual in HEISENBERG's and BOHR's analyses of gedankenexperiments) — but then no quantal relations will come out; (c) if the scatters of $\mathscr{A}$ and $\mathscr{B}$ satisfy (5.10), then its commutator can be any of the infinitely many relations $[\mathscr{A}, \mathscr{B}] = i(\mathscr{C} + \mathscr{D})$ with an arbitrary $\mathscr{D}$ subject to the sole condition $\langle \mathscr{D} \rangle = 0$. (5) Similar relations occur in CEM and other field theories, also independently of considerations on measurement. This is one more reason for not interpreting them as either indeterminacy or uncertainty relations caused by either The Observer's ignorance or His experimental activity.

Thm. 4: *rate of change of averages.* For a $\sigma + \bar{\sigma} (\sigma \in \Sigma, \bar{\sigma} \in \bar{\Sigma})$ in a state ψ at a time $t \in T$, and any $\mathscr{Q} \in \{\mathscr{Q}\}$,

$$\frac{\mathrm{d}}{\mathrm{d}t} \langle \mathscr{Q} \rangle_{\psi} = \frac{\partial}{\partial t} \langle \mathscr{Q} \rangle_{\psi} + \frac{i}{\hbar} \langle H\mathscr{Q} - \mathscr{Q}H \rangle_{\psi}. \tag{5.11}$$

Proof. Differentiate (5.8).

Corollary: conservation laws. If $\mathscr{Q} \triangleq Q$ does not depend on t explicitly and commutes with H, then Q is conserved (time-independent).

Remark. If $\mathscr{Q}$ satisfies the above corollary, Q is said to be a constant of the motion of σ (or rather of $\sigma + \bar{\sigma}$). According to the usual interpretation of QM, the phrase 'Q is a constant of the motion' means "Measuring Q now or at a future time gives the same result" (see e.g. FERMI, 1961). Yet the theorem says nothing about measurement: indeed, no variable occurs in it which represents an experimental arrangement. Moreover the operationalist interpretation fails for any measurement disturbing the Q-distribution.

Thm. 5: *time-independent Schrödinger equation.* If H does not depend on t, then ψ can be separated in the form

$$\psi(x, t) = u(x) \cdot \exp\left(-\frac{i}{\hbar} E t\right) \tag{5.12a}$$

where $E \in R$ and u satisfies

$$H u = E u. \tag{5.12b}$$

Proof. Substitute (5.12a) into QM 5a.

Df. 11. A state represented by a function of the class (5.12a) is called a *stationary state*.

Df. 12. The discipline dealing with problems fitting equations (5.12) is called *quantum statics*.

Corollary. If $\sigma \dotplus \bar{\sigma}\,(\sigma \in \Sigma, \bar{\sigma} \in \bar{\Sigma})$ is in a stationary state then its energy has a single ["well-defined"] value E.

Remarks. (1) Stationary states are, if anything, even more hidden than any other quantum states: as long as a system remains in a stationary state nothing happens to it that can leave a trace on its environment. In particular, the energy of a bound system, such as an atom in its ground level, is unobservable. Therefore stationary states should be banned from the Copenhagen interpretation of QM and accordingly also the transitions among such states — which transitions are occasionally observable. Unless such a surgical operation is performed no operational interpretation of QM is possible. If performed not much of QM is left alive. (2) Most problems so far solved, whether in closed form or with the help of approximation (e.g., perturbation) techniques, are problems in quantum statics.

Thm. 6: *superposition of stationary states.* If, for a $\langle \sigma, \bar{\sigma} \rangle \in \Sigma \times \bar{\Sigma}$, and for every k, $E_k \overset{\text{df}}{=} eiv\,H$, $u_k \overset{\text{df}}{=} eif\,H$, and $c_k \in C$, then

$$\psi(x, t) = \sum_k c_k u_k(x) \exp\left(-\frac{i}{\hbar} E_k t\right) \qquad (5.13)$$

satisfies the Schrödinger equation QM 5 a.

Remark. On the usual interpretation, as long as the energy is not measured, the system is in no particular state. But this contradicts the postulate QM 5 b that, as long as ψ is a definite function and as long as it exists, it represents a definite state of a physical thing. And a run of measurements on one and the same pair $\sigma - \bar{\sigma}$ initially in a state (5.13) will yield a distribution which, if QM is true, will have nearly the profile $|c_k|^2$.

So far the main theorems of the generic part of QM. This part is so comprehensive (noncommittal) that we do not know yet whether it refers to matter or to fields — indeed, the mass concept has not yet been introduced. And it refers to no special physical situation, much less to experimental situations. We now proceed to introduce further assumptions that will enable us to give the generic Schrödinger equations (5.3) and (5.12) specific forms, and therefore to pose and solve definite problems. This will amount to introducing quantum analogs for position and momentum, and specifying the still unused primitives of our initial list: V, M, $\mathscr{E}$, A_0, A, c, and H.

5. Specific Postulates

We now add a set of postulates concerning the q.m. "position" and "momentum", and three of the remaining primitives: V, M, and H. We need not add postulates regarding the electric charge, the e.m. potentials and the velocity of e.m. signals, because CEM (Ch. 4, 1) takes care of them: indeed, we counted CEM among the presuppositions of QM when we scanned the material background of the theory (see Sec. 2). Here go the remaining postulates.

$QM\ 6$: *position and momentum densities.* If $\sigma \in \Sigma$, when associated with $\bar{\sigma} \in \bar{\Sigma}$, is in the state represented by $\psi \in \mathfrak{H}$, and if $x \in E^3$, then

(a) The position density of σ in $\bar{\sigma}$ at any $t \in T$ is $\psi^* x \psi$.

(b) The momentum density of σ in $\bar{\sigma}$ at any $t \in T$ is $\psi^* (\hbar/i) \nabla \psi$.

Remarks. (1) The word 'density' is to be understood here in the sense of mathematical statistics not of classical physics. (2) $QM\ 6a$ shows once again that, from a mathematical point of view, QM in the Schrödinger version is a *field theory*. Indeed, the coordinate "operator" x (sometimes curiously written '$\cdot x$', which is just bad mathematics) coincides with the space label x represented by a triple of real numbers, whereas in classical mechanics the position coordinate is a time-dependent vector differing from zero only when its tip falls inside the body. Therefore analogies between the q.m. x (which is a "c-number" all right although it is usually called a "q-number") with the classical material or physical coordinate X (a function on $\Sigma \times K \times T$) are misleading. A good deal of confusion would be spared if instead of speaking of position in the context of QM we spoke of, say, *quosition.* (3) $QM\ 6b$ was "found" neither by induction nor by trial and error but by analogy between the term $-(\hbar^2/2\mu) \nabla^2$ in the de Broglie and Schrödinger equations, on the one hand, and the classical kinetic energy in hamiltonian PM. This analogy between QM and PM, so fertile as misleading, breaks down for curvilinear coordinates. Thus in spherical coordinates the radial momentum operator is not $(\hbar/i) \partial/\partial r$ but $(\hbar/i) (\partial/\partial r + 1/r)$. Due to the limitations of the analogy with PM it might be convenient to call the operator $(\hbar/i) \nabla$ the *quomentum.* (4) If, as is customary, we had said that x represents the position and $(\hbar/i)\nabla$ the momentum of σ, then we would have presupposed that quantons are point particles and this would have invited the questions 'How is it possible for a particle to be at a given place with no definite velocity?' and 'How come that a particle moving with a given velocity is nowhere?' (These baffling questions, inevitable in the usual formulations of QM but senseless in ours, have given rise to the so-called quantum logic: see Sec. 8.) In our version of QM there are only position and momentum densities and averages. (Something similar holds for the remaining dynamical variables according to QM 3.2f.) Incidentally, to find out whether the position concept "makes sense" for the quantons QM refers to, one should not discuss the diffraction and polarization of photons (as is done e.g. in DIRAC, 1958), if what one wants to know is what meaning should be assigned to ' x ' in elementary QM, which is not about photons. No such manoeuvres are needed if the classical metaphors of particles and waves are left at the gate of QM.

Df. 13: orbital angular momentum operator: $L \overset{\text{df}}{=} x \times (\hbar/i) V$.

Remark. Just as in CM, the concept of angular momentum is here defined, hence logically redundant. But, just as in the classical case, L represents an important physical property (one occurring in a number of law statements) and moreover one that, in QM, cannot be entirely reduced to its ingredients x and $\mathscr{P}$: in fact the *eif's* and *eiv's* of L cannot be constructed out of those of x (which are anyhow ill-behaved) and $\mathscr{P}$ separately.

QM 7: nonelectromagnetic forces.

(a) V is a function of σ, $\bar{\sigma}$, x, and V.

(b) For any given pair $\langle \sigma, \bar{\sigma} \rangle \in \Sigma \times \bar{\Sigma}$, the value of $\psi^* V \psi$ at $\langle \sigma, \bar{\sigma} \rangle$ represents the density of the nonelectromagnetic action of $\bar{\sigma}$ on σ.

Remark. The potential V represents the nonelectromagnetic part of the action of the environment $\bar{\sigma}$ on our system of interest σ. The partner $\bar{\sigma}$ of σ is usually not mentioned. Thus one speaks of a system in a potential well without indicating the source of this potential — which is mathematically correct but physically mysterious. Some physical thing (body or field) must be there in order to act on the given σ; we may well be uninterested in the details of $\bar{\sigma}$ but we cannot eliminate it from the picture.

QM 8: mass.

(a) $\mathscr{M}$ is a function from Σ to R^+.

(b) The value μ of $\mathscr{M}$ at $\sigma \in \Sigma$ represents the mass (inertia) of σ.

Remarks. (1) This postulate is insufficient to get a "feel" of the meaning of '$\mathscr{M}$'. In this case we cannot resort to classical mechanics because it is a rival theory. But we can wait until some formulas involving μ are introduced: they, and particularly EHRENFEST's theorems, will jointly delineate the meaning of '$\mathscr{M}$'. (2) Mass and charge are, like position and time, classical properties in QM, in the sense that they are not represented by hermitian operators (save trivially). In more sophisticated theories they are represented by operators in some space. (3) By means of $\mathscr{P}$ and μ we may, but need not, introduce the linear velocity operator $\mathscr{V} \overset{\text{df}}{=} \mathscr{P}/\mu$ and several others. All the terms without a classical analog occurring in the case of $\mathscr{P}$ when using curvilinear coordinates will have their ugly velocity partners (BUNGE, 1960). But nothing as puzzling as the velocity operators in relativistic QM, which are logically independent of the momentum operators (BUNGE, 1955 a and 1960).

QM 9: *hamiltonian of quanton with mass and charge in external field.* If $\sigma \in \Sigma$ and $M(\sigma) = \mu \neq 0$ and $\mathscr{E}(\sigma) = e$ and if the action of $\bar{\sigma} \in \bar{\Sigma}$ on σ is represented by the nonelectromagnetic potential V and the e.m. potentials A_0, A, then the hamiltonian of σ in the company of $\bar{\sigma}$ is

$$H = \frac{1}{2\mu} \left(\frac{\hbar}{i} \, \nabla - \frac{e}{c} \, A \right)^2 + \frac{e}{c} \, A_0 + V. \tag{5.14}$$

Remarks. (1) The hypothesizing of H independently of the Schrödinger equation leaves us the freedom to change H whenever necessary — e.g., to account for spin. (2) Since the e.m. field represented by A_0, A is here treated classically (by CEM), the present theory is semiquantal. For small field energies the field fluctuations (or the *rms* of the field occupation numbers) become important and elementary QM breaks down. Then it becomes necessary to borrow the e.m. potentials from quantum electrodynamics; in such a case CEM should cease to be part of the background of QM. (3) Even if the e.m. potentials occurring in *QM* 9 are classical, they have quantal effects: they produce phase differences that can lead to interferences (AHARONOV and BOHM, 1959, 1961).

The foregoing axioms *QM* 1—9 constitute the basis of our formulation of QM. It is p-complete insofar as it specifies, with the help of protophysics and CEM, all the primitives employed; and it is d-complete in the sense that it yields all the known general theorems — which is not surprising as ours is just a reordering and reinterpretation of the available body. As usual in our version of physical axiomatics (see Ch. 1, 4.2), we have introduced axioms of two kinds: mathematical (e.g., "$\mathscr{Q}$ is a hermitian operator in $\mathfrak{H}$") and physical ones (e.g., "If Q is a property of $\sigma \in \Sigma$, then there exists a hermitian $\mathscr{Q}$ such that $\mathscr{Q} \triangleq Q$"). In this way both the mathematical status and the physical meaning of the basic symbols are specified. How meanings came to be assigned is a historical not a semantical question. In some cases they were suggested by analogy with classical theories, either by noticing formal similarities with them, or by averaging, or by taking a correspondence limit (e.g., $\hbar = 0$). But none of these procedures tells us anything about the meaning of the boundary conditions on ψ and on the operator eigenfunctions (NISHIYAMA, 1963). At other times symbols and entire formulas had to be interpreted in an entirely new way — as in the case of the components of ψ along the axes φ_k. In no case did experimental situations really help, for the basic symbols of QM are too far from them. And gedankenexperiments may illustrate some theorems but they can hardly suggest the meaning of symbols that do not concern experimental situations. In any case the interpretation process of QM has been a long and twisted one and it should not be regarded as finished for, after all, meaning postulates determine only the semantical profile of a theory (see Ch. 1, 4.2.5 and 4.2.6).

We shall next give a few typical and specific theorems of QM in order to discuss some of their philosophical implications.

6. Specific Theorems

The most famous among the theorems of QM is

Thm. 7: position distribution. For any $\sigma \dotplus \bar{\sigma} (\sigma \in \Sigma, \bar{\sigma} \in \bar{\Sigma})$ in a state $\psi \in \mathfrak{H}$, the probability that σ lies in a region $V \subseteq E^3$ equals

$$\int_V d^3 x |\psi(x)|^2. \tag{5.15}$$

Proof. By Thm. 1 (5.6) and its Corollary 1 (5.7) upon the specification $\mathcal{Q} = x$ and the assumption *QM* 6a.

Remarks. (1) Thm. 7 allows us to compute the probability that any given $\sigma \in \Sigma$ be contained in V (which may or may not be hollow); it does not give the probability that the *position* of σ lies on a certain interval, because this presupposes that quantons are point particles with either a hazy or an unknown position. We are taking seriously the usual contention that quantons have no definite trajectories — which claim is contradicted when speaking of definite positions. (2) For $V = E^3$ and upon suitably normalizing ψ, the probability becomes unity. This result presupposes, of course, *QM* 1e (ψ is square-integrable), which is the case only when H has a discrete spectrum. In the case of continuous spectra a different normalization must be adopted (see e.g. BUNGE, 1960). (3) The preceding theorem, originally postulated by M. BORN (in a different form) gave wave mechanics — which is neither undulatory nor mechanical — some of its physical content; the rest of the interpretation was contrived with the help of classical analogies, particularly those of particle and wave. Thanks to Thm. 1, Thm. 7 need no longer be postulated and BORN's statistical interpretation is again seen to be natural not *ad hoc* — although it was suggested by the assumption that quantons are point particles, only endowed with an Epicurean swerve motion. (4) In elementary treatments Thm. 7 is formulated as a semantic postulate or correspondence rule attaching ψ a physical meaning. While this procedure is not objectionable it leaves the doubt that ψ might, within current QM, be interpreted differently. (5) BORN's hypothesis (5.15) was not subjected to experimental test for single particles until the 1950's with the help of photomultipliers. Even now only its generalization for aggregates of quantons has been confirmed, particularly with scattering experiments. This has contributed to the belief that QM is a theory of aggregates. (6) In our version of QM, Thm. 7 gives the probability of the objective presence of σ in V or, if preferred, the propensity of σ to dwell in V at a given moment (DE BROGLIE'S *probabilité de présence*). The usual operationalist interpretation is in terms of the probability of *finding* σ in V when a position *measurement* is actually performed. This interpretation is *ad hoc* and wrong. It is unwarranted because "finding" is a pragmatic predicate absent from the primitives of QM. And it is false because the probability of finding a σ in a given region depends not only on the state of σ but also on the search technique. The probability of *finding* σ in V differs from the probability that σ *be* in V: the former depends not only on σ and $\bar{\sigma}$ but also on the parameters characterizing the search, so much so that $w(\sigma \dot{\in} V) \neq 0$ is consistent with a zero probability of finding σ in V, due to inadequate technique or sloppy application of a good technique. (7) It is legitimate to introduce the nonphysical concepts of *search* and *finding* in the context of operations research (see e.g. GUENIN, 1961). To this end two primitive nonphysical concepts must be introduced in addition to our physical concepts: (*a*) the *search effort* devoted to locating σ within $d^3 x$, namely $S(\sigma, x) d^3 x$, where S is a nonnegative function, continuous and integrable w.r.t. x, and (*b*) the *probability of detecting* σ upon spending an effort S when σ is assumed to be in fact within $d^3 x$, i.e. $P[S(\sigma, x) d^3 x | \sigma \dot{\in} [x, x + d x]]$. This *operations research* probability is obviously different from the *physical* probability $w(\sigma \dot{\in} V)$. Thus σ may *be* within a certain region yet we may fail to detect it — even if σ is as big as a galaxy. In short, the probability that something lies somewhere depends on that something and its environment whereas the probability of detecting it depends both on that probability and on the detection procedure. To miss this difference is to mistake theoretical physics for experimental physics. This is not just a question of division of labor: there is a conceptual and a quantitative difference between the two probabilities, and the experimenter who overlooks it may draw the wrong inferences concerning the actual position distribution of an object. (8) A corollary of Thm. 7 is that the probability that σ be at some *point* in space is zero, for then the two integration limits coincide. Yet if measurements are performed quantons will actually be found at certain "points" — actually small regions. In general, a zero probability is consistent with a nonzero frequency. Morals: (*a*) the frequency theories of probability, of VENN, v. MISES, etc., which equate probability with frequency, are false; (*b*) theoretical predictions cannot be translated into an experimental language without further ado (BUNGE, 1967b).

Thm. 8: *momentum distribution.* For any pair $\langle \sigma, \bar{\sigma} \rangle \in \Sigma \times \bar{\Sigma}$ in a spatially unbounded state $\psi \in \mathfrak{H}$, the linear momentum distribution is given by the square of the Fourier transform of ψ:

$$|c_k|^2 = (2\pi)^{-\frac{3}{2}} |(e^{ikx}, \psi)|^2, \quad \text{with} \quad k \overset{\mathrm{df}}{=} eiv \, V/i. \qquad (5.16)$$

Proof. Solve the eigenvalue equation for V/i and use Thm. 1.

Remark. The Fourier transform involved in this theorem enables one to compute momentum distributions directly, without using *QM* 6*b*. It also constitutes the bridge between the coordinate and the momentum representations of QM. The relations among the operators themselves are stated in

Lemma 1: *basic commutation relations.* If $\mathscr{P} \equiv (\hbar/i) V$, then

$$x_i \mathscr{P}_j - \mathscr{P}_j x_i = i\hbar \delta_{ij}. \qquad (5.17)$$

Remark. We have not assumed that x and $\mathscr{P}$ *refer* to anything: the preceding is a purely mathematical statement; only position and momentum densities refer to a physical system and have therefore a physical meaning (recall *QM* 6). On the other hand in the matrix formulation of QM the above commutation relations are physically meaningful and are not deduced but postulated: they must, because there x and $\mathscr{P}$ are primitives, which is not the case with wave mechanics.

Lemma 2: *derived commutation relations.*

$$[L_i, L_j] = i\hbar \varepsilon_{ijk} L_k, \quad i, j, k = 3 \qquad (5.18)$$

where ε is the Levi-Civita 3-tensor. *Proof.* By Df. 13 and Lemma 1.

Remarks. (1) In alternative theories L may be taken as a primitive, in which case the (5.18) must be postulated. If this course is taken, a more comprehensive concept of angular momentum is obtained, one covering both the orbital and the spin angular momenta. A whole theory of angular momenta can then be built on the basis of the above commutation relations and some algebra. But in the Schrödinger theory these relations are purely formal because only the angular momentum densities, not their operators, are physically meaningful. (2) In spherical coordinates the azimuthal component $\mathscr{P}_\varphi$ of the linear momentum coincides with the z-component of the angular momentum, and their eigenvalues are: $eiv \, \mathscr{P}_\varphi \equiv eiv \, L_z = m\hbar$ with $m \in I$. This theorem shows that, except for solving problems characterized by special symmetry properties, it is preferable to use either Cartesian or nonspecific curvilinear coordinates. It also teaches that quantization is not totally inherent in physical systems: sometimes it comes with a certain representation. One should therefore distinguish *genuine quantization*, invariant under representation changes, from *pseudoquantization*, or quantization depending on the representation (e.g., the kind of coordinates).

Thm. 9: *Heisenberg relations*. For every $\langle \sigma, \bar{\sigma} \rangle \in \Sigma \times \bar{\Sigma}$ in a state $\psi \in \mathfrak{H}$, the scatters in the position and momentum distribution at any given $t \in T$ are reciprocal:

$$\Delta_\psi x \cdot \Delta_\psi \mathscr{P} \geq \hbar/2. \tag{5.19}$$

Proof. By axiom *QM* 6, Thm. 3 and Lemma 1.

Remarks. (1) Since this is a very special theorem, to start philosophical discussions with it is misleading — as much as calling it a "principle". (2) Read in terms of particle mechanics, these relations mean that the dynamical state of a system — another classical concept — is not sharply determined since the better the position of a *particle* is "defined" the worst its momentum is "defined". No such indeterministic interpretation is possible in our version of QM because the concept of classical particle is alien to it. (3) Similar relations occur in classical field theories and for the same formal reason — namely that the two distributions are related by a Fourier transformation. Thus in communications theory the relation "$\Delta x \cdot \Delta k \cong 1$" between the size Δx of a wave packet and its band width Δk holds. But this is not a stochastic relation. Consequently the wave interpretation of (5.19) is as *ad hoc* as its particle interpretation. (4) Thm. 9 is sometimes "deduced" by reasoning on ideal measurements, e.g. by means of HEISENBERG'S microscope. It is even claimed that the statement was originally inferred from a detailed analysis of measurement procedures. But no such deduction is possible, for the reasons given in Remark 4 to Thm. 3 (Sec. 4) and because (*a*) experimental arrangements are classically describable (as BOHR himself has untiringly emphasized), (*b*) the Heisenberg relations happen to be probability statements, and (*c*) the theoretical spreads require the knowledge of the unobservable ψ. What happens is (*a*) that, by interpreting x and $\mathscr{P}$ as *classical* variables referring to a point particle, relations *similar* to HEISENBERG'S can be obtained for some examples (but then $\hbar$ is missed); (*b*) by interpreting ψ as a *classical* wave field, relations *similar* to HEISENBERG'S can be obtained (see Remark 3 above); (*c*) classical gedankenexperiments can always be imagined to illustrate (5.19) and other relations, which is no wonder as they are tailored to that task but not every actual calculation for such idealized situations happens to confirm HEISENBERG'S relations (BECK and NUSSENZVEIG, 1958); (*d*) many real experiments are likely to confirm Thm. 9, which is so general (by hypothesis) that it should hold, in particular, for a quanton-apparatus pair — yet no such experiments seem to have been performed. But one thing is to put the theoretical formula (5.19) to the test and quite another to claim that any such high-powered hypotheses can be extracted from experiment: if the latter were true, logic would be false. (5) The Heisenberg relations pass for being an illustration of the *wave-particle duality*, which would in turn be a case of the *complementarity principle*. There is no such duality in our version of QM, because it contains neither the concept of wave nor that of particle. Being a stochastic theory, it is only natural that spreads should occur in it alongside averages. And Thm. 9 states only that, the narrower the x-distribution, the broader the p-distribution and conversely: this has nothing to do with a complementarity between experimental set-ups (or alternatively modes of description). So much the better, because the complementarity principle is subject to grave philosophical difficulties (BUNGE, 1955 b). (6) A sharp position measurement requires "illuminating" the quanton with either short-wave radiation or high speed quantons. If the energy of either surpasses the rest energy of the quanton to be located, the latter may disappear as such and a whole shower of new things may emerge. That is, the very object we wish to locate will turn into something else if we try too hard. From this it has been concluded that space is inscrutable and must be eliminated from physics (CHEW, 1963). Of course the assumption of a continuous spacetime is not directly testable — but then most assumptions are in the same predicament. The problem with the spacetime continuum is whether (a) it is necessary to build deep and fertile theories and (b) the assumption does show through some testable logical consequence. Now in all current microphysical theories, particularly in the relativistic ones, the assumption is made that space and time are continua. For example, q.m. averages are computed by integrating over space, and space and time variables are needed to state any formula about which it can be said that it is Lorentz-covariant. What is true is that QM, in the Schrödinger representation, does not use material or physical coordinates such as those occurring in CM — but this is a far cry from claiming that space is dispensable. (7) A frequent reading of HEISEN-BERG's relations is this: "One can never know exactly the simultaneous values of the position and the momentum of a system". This presupposes that a quanton has a definite position and a definite momentum at any time, only for some reason (incompleteness of the theory, or else disturbance by a measuring device) we cannot get to know them accurately. This is not what our version of Thm. 9 states: it speaks not about human knowledge and ignorance but about a chunk of reality that has no simultaneous sharp position and momentum, for the excellent reason that it has neither a position nor a momentum *tout court* but position and momentum distributions (by *QM* 6). For having such distributions it has the *possibility* of going exceptionally either to a sharply localized state or to a state of sharp momentum. But in general it will be in a state in between which, from a classical viewpoint, is an intolerable wishy-washy situation. (8) That *measurements* of position and momentum give no simultaneous sharp values follows from the preceding in conjunction with the truism "What does not exist cannot be measured". On the other hand, as we shall see later on (Thm. 14), for some exceptional states one can ask for the joint probability that x and p have definite values. (9) The Heisenberg relations are supposed to hold for sets of measured values because they are assumed to hold for an autonomous system (no $\bar{\sigma}$) to begin with. They are not *due to measurement*, they are not effects of the famous "uncontrollable and unpredictable disturbance caused on the system by the measuring device" — as is so often claimed in an attempt to justify indeterminism with a causal argument. Since nothing has been assumed about measurements in deriving Thm. 9, we conclude that it refers to what it says it does, namely to any pair $\sigma - \bar{\sigma}$, whether or not $\bar{\sigma}$ happens to represent an observation device (see also SUPPES, 1961, 1963). The operational interpretation of HEISENBERG's relations, by slipping into it a measuring device whose existence is neither affirmed nor denied by either Thm. 9 or its premises, introduces a semantical inconsistency in QM. (10) The operational interpretation of HEISENBERG's relations — the very kernel of the usual interpretation of QM — is inconsistent with the two basic ontological assumptions of the Copenhagen interpretation itself, i.e. subjectivism and indeterminism. Indeed, if the apparatus *causes* uncontrollable disturbances in the state of the system, then the system *exists* in some state or other before being subject to "observation". (11) Just as one can invent gedankenexperiments to justify HEISENBERG's relations, so one can contrive ideal experiments that do not satisfy the operationalist interpretation of Thm. 9 because they involve no disturbance of the measured object (RENNINGER, 1960).

And now to a *pseudotheorem*. It is usually claimed that a relation similar to HEISENBERG's holds for energy and time, namely

$$\Delta_\psi t \cdot \Delta_\psi E \geq \hbar/2. \tag{5.20}$$

It could be that, properly interpreted, this formula were true. But it does not belong to QM; in particular, it cannot be proved along the lines of HEISENBERG's relations. The reasons for this are: (a) t is not an operator in Hilbert space, hence it exhibits no scatter; (b) although $i\hbar \partial/\partial t$ is sometimes said to be the energy operator, only H plays this role, and for a stationary state H has no spreads either (indeed, $\langle H \rangle = E$ and $\langle H^2 \rangle - E^2$), whereas (5.20) is alleged to be completely general. So far there are only heuristic reasons for the so-called Heisenberg 4th relation (actually proposed by N. BOHR). But in any case it cannot be interpreted similarly to HEISENBERG's relations if only because 'Δt'

cannot be a standard deviation as long as t is a "c-number". If one wants to have (5.20) one must modify QM by introducing a suitable time operator — and then destroy the equivalence between the Schrödinger and the Heisenberg "pictures", as in the latter the dynamical variables evolve *in* time. Contrary to a widespread belief, the same holds for relativistic QM: in this theory, too, t is a classical variable.

The elimination of formula (5.20) solves several puzzles. Thus it is sometimes said that it entails that it is unreasonable to require that the energy of a system be conserved to an accuracy greater than ΔE.

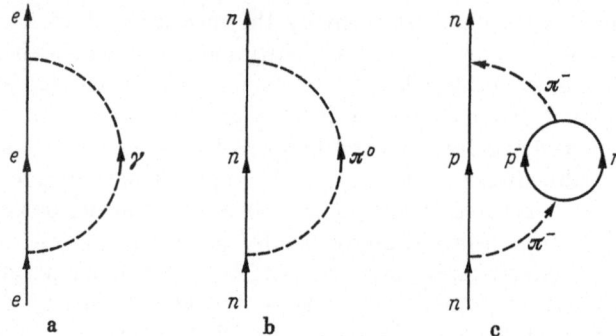

Fig. 5.1. Virtual (unobservable) self-energy processes: (a) second order process for electrons; (b) one of the possible 2th order processes for neutrons; (c) one of the possible 4nd order processes for neutrons. There are as many diagrams as terms in the expression for the total self-energy of every "particle" — i.e. infinitely many. And none of them represents an observable process

This is in turn justified by saying that anyhow such a limited violation of energy conservation could not be observed provided Δt were large enough. Whereupon decay schemata are proposed which involve a temporary gain or loss of energy that is paid for or recovered within the interval Δt — before someone could notice them. Quite apart from the fact that 'Δt' is stochastically meaningless, 'ΔE' cannot be interpreted as an energy change, particularly if H is a constant of the motion. One muddle invites another: if energy need not be conserved even for a closed system (in which case H is conserved), then why not introduce *virtual* processes accompanied by the emission and absorption of *virtual* quanta? Thus it is said that an electron can emit and reabsorb a virtual photon provided this whole virtual process takes place in a time less than $\Delta t \cong \hbar/2\Delta E = 1/4\pi\nu$ [according to (5.20) and to $\Delta E = h\nu$]. Similar virtual processes are invented for nucleons and other "particles": see Fig. 5.1. In this way the weird idea of self-interaction is introduced and the corresponding self-energy calculated; and if, as so often happens, the latter does not want to converge, then no mathematical unfaithfulness will be spared to force it to converge. For want of a theorem a kingdom of ghosts was gained. (Another source is of course the Pythagorean faith in the physical meaningfulness of every term in a perturbation computation: see Ch. 1, 4.3.2 and BUNGE, 1955 c). Let us now turn to some genuine theorems.

Thm. 10. For every $\langle \sigma, \bar{\sigma} \rangle \in \Sigma \times \bar{\Sigma}$ in a state $\psi \in \mathfrak{H}$, the scatters in the angular momentum components at any $t \in T$ are related by

$$\Delta_\psi L_i \cdot \Delta_\psi L_j \geq \varepsilon_{ijk} \tfrac{1}{2} \hbar \left| \langle L_k \rangle_\psi \right|. \tag{5.21}$$

Proof. By Axiom *QM* 6, Thm. 3 and Lemma 2.

Remarks. At any given moment there exists always a definite pair $(eiv |L|, eiv\, L_i)$ for any i, because $|L|$ commutes with every one of its components. But since the various components of L do not commute with each other, if $eiv |L| \neq 0$ then there is no triple $\langle l_x, l_y, l_z \rangle$ of definite numbers. Consequently L is not a vector proper, whence its usual visualization, though often handy, is misleading. Moreover, since for $eiv |L| \neq 0$ there exists no state of $\sigma \dotplus \bar{\sigma}$ for which L *has* all three components, it is impossible to measure them simultaneously. This is not an operational interpretation of Thm. 10 but a trivial corollary of it.

Thm. 11: *definite Schrödinger equation.* Under the conditions of *QM* 9, any state $\psi \in \mathfrak{H}$ of σ when associated with $\bar{\sigma}$ evolves according to

$$i\hbar\, \partial \psi / \partial t = \left[\frac{1}{2\mu} \left(\frac{\hbar}{i} \, V - \frac{e}{c} A \right)^2 + \frac{e}{c} A_0 + V \right] \psi. \tag{5.22}$$

Proof. Introduce *QM* 9 into *QM* 5 a.

Remarks. (1) It may sound queer to call this a theorem, as it was originally postulated. But, again, logical order bears a low correlation with historical order. In any case, Thm. 11 is the central hypothesis of QM in the position representation, because (*a*) all logically previous assumptions prepare its entry and (*b*) every specific problem in QM consist in specifying the potentials occurring in (5.22) and the boundary conditions. (2) We shall waste no time in trying to interpret (5.22) either as a real wave equation or as a diffusion equation or as the pattern of the spreading of information. Every one of these interpretations is analogical, it conflicts with some other analogical (semiclassical) interpretation, and it is unwarranted by the axioms. QM is not properly interpreted by borrowing images from other theories but by recognizing that it is a radically new theory involving peculiar semantic hypotheses. Those images are therefore didactically as misleading as the hydrodynamical analogies to explain heat and electricity "flows": good teaching does not consist in feeding easy but mistaken analogies.

Thm. 12: *conservation of probability.* For any $\langle \sigma, \bar{\sigma} \rangle \in \Sigma \times \bar{\Sigma}$ in any $\psi \in \mathfrak{H}$ and at any $x \in E^3$ and every $t \in T$,

$$\frac{\partial \varrho}{\partial t} + \nabla \cdot j = 0, \tag{5.23}$$

where

Df. 14 $$\varrho \overset{\mathrm{df}}{=} |\psi|^2. \tag{5.24}$$

Df. 15 $$j \overset{\mathrm{df}}{=} \frac{\hbar}{i\,2\,\mu} (\psi^* \cdot \nabla \psi - \psi \cdot \nabla \psi^*). \tag{5.25}$$

Proof. Compute on (5.22) and its complex conjugate.

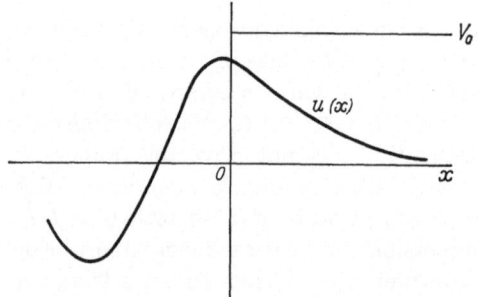

Fig. 5.2. Scattering of a quanton by a potential step as example of ghost waves

Corollary. For any $\langle \sigma, \bar{\sigma} \rangle \in \Sigma \times \bar{\Sigma}$ in any $\psi \in \mathfrak{H}$, at every $t \in T$ and every $V \subseteq E^3$,

$$\frac{\partial}{\partial t} \int_V \mathrm{d}^3 x \varrho = 0. \tag{5.26}$$

Remarks. (1) The derived laws (5.23) and (5.26) have suggested hydrodynamical interpretations of QM, in which the "probability fluid" diffuses and even, for spinning systems, has vortices. But these analogies do not say what a quanton really is. It general it is not advisable to attempt to squeeze the new into the old. (2) When interpreted *classically* the continuity equation says that ϱ represents the density and j the current density of matter, and that the latter is conserved. But this interpretation, encouraged by an analogy, is not warranted by our axioms. Moreover it meets a number of counterexamples. Think of the familiar potential step — some obstacle to the referent of σ created by the referent of $\bar{\sigma}$ (see Fig. 5.2). If the quanton energy is smaller than the barrier height, the "current" vanishes to the right of the discontinuity although the "field" does not: in this region the "current" does not accompany the de Broglie "waves". On the other hand if the quanton energy surpasses the potential barrier, the "current" goes through unabated while the "waves" are partially reflected at the discontinuity. From a corpuscular point of view, unaccompanied waves are ghosts — but set their amplitudes equal to 0 and you destroy probability conservation. Similarly with the "waves" reflected at the barrier even though they are unaccompanied by bouncing-off "corpuscles". Once again we see that neither the corpuscular nor the undulatory interpretations of QM are possible. (3) Mathematically, the conservation of the distances $\|\psi\|$ in $\mathfrak{H}$ is very restrictive. Should a more flexible theory be proposed, the probability interpretation would have to be deeply modified or even given up.

Metatheorem 1: gauge invariance. The Schrödinger equation (5.22) is preserved if the e.m. potentials are subjected to a gauge transformation of the 2nd kind (see Ch. 3, 1.2.2) provided ψ is subjected to a gauge transformation of the 1st kind, i.e.

$$\psi \to \psi \cdot e^{(ie/\hbar c)\Lambda}. \tag{5.27}$$

Proof. Compute.

Remark. Like any other statement concerning a property of a given statement, this one is a metastatement — in particular a metanomological statement (see Ch. 1, 1.3.1).

Thm. 13: *free nonlocalized quanton.* For any $\sigma \in \Sigma$ in the absence of every $\bar{\sigma} \in \bar{\Sigma}$ and in a spatially unbounded state,

$$\psi(x, t) = A\, e^{i(k \cdot x - \omega t)} \tag{5.28}$$

with $A \in C$, $k_i \in R$, and

Df. 16 $\qquad\qquad \omega \overset{\mathrm{df}}{=} \hbar k^2/2.$

Proof. Check. (To avoid integrability difficulties take eigendifferentials, i.e. "wave" packets rather than monochromatic "waves".)

Remarks. (1) The arbitrary integration constants A and k are so far physically meaningless. If (5.28) were square-integrable it would be possible to compute $\langle (\hbar/i)\nabla \rangle$ to obtain $\hbar k$. In this case we would say that k equals the average (or the classical) momentum of σ divided by $\hbar$. But this is not the case with a free unlocalized system; and if it is bounded, its momentum is quantized. (2) A free system is, by definition, one which interacts with no other system — in particular with no apparatus. Hence it is an unobservable entity — a nonentity for subjectivism. Yet the concept of free system is essential in QM since every solution of the Schrödinger equation can be represented as a linear combination of free solutions. (3) By analogy with classical waves we could say that ω in (5.28) is the circular frequency and

Df. 17 $\qquad\qquad \lambda \overset{\mathrm{df}}{=} 2\pi/k$

the wavelength of σ. Moreover, the distributions of fringes in electron diffraction experiments would seem to lend support to this interpretation. Still, that is nothing but a classical analogy — a fertile one in some cases (e.g., for conceiving the electron microscope) but a misleading one in many others. For example, a stationary state of a free quanton in spherical coordinates is $\psi_{klm} = j_l(kr) P_l^m (\cos \vartheta) \cdot e^{im\varphi}$. Not only does this solution exhibit pseudoquantization (Lemma 2, Remark 2) but, since for $l > 0$ the zeros of the Bessel spherical functions j_l are not equally spaced, k cannot be interpreted as a reciprocal wavelength. (4) The states represented by (5.28) are not spatially localized: the position distribution is homogeneous. But by suitably superposing slightly out of phase nonlocalized states, the x-distribution can be narrowed as much as desired. Which shows that a sharp localization is not a permanent attribute of quantons but rather a state a quanton can occasionally indulge in — especially if forced by other quantons, as in the case of the atoms in a crystal. (5) It follows that in principle QM permits the preparation of a system with any given localization. This does not mean that a fairly sharp position is a property a quanton can aquire only upon a position measurement: the solutions (5.28) we can superpose to get a narrow "wave" packet contain no apparatus variable. If a σ can be squeezed or blown up it is because it can exist in either state. (Moreover, when forming a "wave" packet out of free "waves", it is taken for granted that no other system is in the vicinity of σ — for otherwise this would cease to be free.) (6) In particular, a quanton can exist within a closed box — an elementary example even the most orthodox books on QM carry. The only way of testing this hypothesis — i.e. of finding out whether there is anything inside the box — is by opening it and watching whether anything flies out of it. But by so doing the original situation changes radically: the new boundary conditions are time-dependent and the standard solution is useless. Hence the exercise concerning the particle in the box presupposes the existence of a system we cannot observe. Yet no operationalist seems to feel qualms about this.

Thm. 14: *joint position and momentum distribution.* If, at $t = 0$, $\sigma + \bar{\sigma}(\sigma \in \Sigma, \bar{\sigma} \in \bar{\Sigma})$ is in a state represented by

$$\psi(x, 0) = A \exp \left[-ax^2 + bx - (\text{Re } b)^2/4a \right] \qquad (5.29)$$

with $A, b \in C$ and $a > 0$, then

(*a*) the equality holds in Heisenberg's relations (minimum scatter state), and

(*b*) the covariance of x and $\mathscr{P}$ is zero — i.e. the quosition and the quomentum have a joint probability distribution which is multiplicative. *Proof.* See Suppes, 1961 and Urbanik, 1961.

Remarks. (1) Every pair of noncommuting dynamical variables lacks a joint distribution for every state, but certain pairs do have a joint distribution for certain exceptional states. (2) Thm. 14 is another nail in the coffin of the operationalist interpretation of QM: not only nothing is said in it about an apparatus and its bad influences on the microsystem, but it shows that, although x and $\mathscr{P}$ do not commute, their distributions are *on occasion statistically independent.* Consequently, if the corresponding properties are measured when $\sigma + \bar{\sigma}$ is in a state like (5.29), the measurement of one of them should not affect the statistics of the set of values of the conjugate variable. (3) The nonexistence of joint probability distributions in general has been said to conflict with the calculus of probability (CP). There is no conflict but simply a part of CP becomes pointless or inapplicable in QM. There are three possibilities: (*a*) leaving QM and CP as they are; (*b*) keeping QM but supplying whenever needed the clause "If X and Y are commuting random variables, then..."; (*c*) generalize CP to noncommuting variables — or rather to distributions generated by such.

Thm. 15: correspondence with PM (EHRENFEST). If a $\bar{\sigma} \in \bar{\Sigma}$ acts on a $\sigma \in \Sigma$ through a scalar field V and if $\sigma + \bar{\sigma}$ is in a state $\psi \in \mathfrak{H}$ then, at any $t \in T$,

(a)
$$\mu \frac{\mathrm{d}}{\mathrm{d}t} \langle x \rangle_\psi = \langle \mathscr{P} \rangle_\psi \qquad (5.30)$$

(b)
$$\frac{\mathrm{d}}{\mathrm{d}t} \langle \mathscr{P} \rangle_\psi = - \langle \nabla V \rangle_\psi. \qquad (5.31)$$

Proof. By *QM* 6, Thm. 2 and Thm. 11.

Remarks. (1) This theorem establishes a correspondence between QM and PM in the sense that the averages of the x and $\mathscr{P}$ distributions move classically as required by the correspondence principle (see Sec. 1). The correspondence is *global* not local: indeed, in the Schrödinger "picture" x is not t-dependent hence it cannot represent the position of a particle. Time enters through ψ; hence only the bilinear forms $\psi^* \mathscr{Q} \psi$ and their space integrals can "move". (2) We have not cared to introduce the definition: $F \overset{\mathrm{df}}{=} - \nabla V$ because in a Hamiltonian theory the force concept is dispensable and, more important, because we would not quite know how to interpret the symbol 'F' when V contains operators proper. But there is no question that a classical analogy guides very often the choice of the proper V.

Let the above theorems suffice as a sample of the kind of thing to expect from our formulation of QM. Any further development of the theory requires the introduction of specific assumptions about the nature of the system and the fields in which it is immersed, as well as its extension to quanton aggregates.

Whatever defects our formulation of QM may have, it seems to have certain important properties characterizing none of the semioperationalist formulations. In fact our formulation is (a) *logically organized* (explicit enumeration of the basic concepts and the basic assumptions, and indication of the status of the statements — postulates, theorems, metatheorems, definitions); (b) *formally consistent* (contradiction-free); (c) *semantically consistent* (no formula is interpreted *ad hoc*, i.e. in a way unwarranted by the interpretive axioms, because no predicate is introduced surreptitiously); (d) *thoroughly quantal and unitary* (no corpuscular and undulatory analogies and no wavering between them); (e) *thoroughly physical* — no psychological concepts such as "observer", "mind", "subjective probability", "expectation", "uncertainty", or "finding", and no fictions such as "ideal measurement" and extra "hidden variables" with no effects. Moreover this consistent and ghost-free formulation of QM applies to the *individual* microsystem — whether isolated or in an environment taken as a unit — and it preserves BORN's stochastic interpretation, which it proves and frees from subjectivism.

7. Measurement Theory

Like every other factual theory QM must be subjected to tests of two kinds: conceptual (internal and semantical consistency as well as compatibility with other theories) and empirical (adequacy to old and new empirical information). The conceptual tests of QM require a number of metatheoretical techniques developed in the course of this century (see Ch. 1, 4.1.6 and 4.2); they do not seem to have been used before because they require prior axiomatization, and QM has never been properly axiomatized (see the introduction to this Ch.). The empirical tests involve exact observations (measurements) such as the spectroscopic ones, and experiments — e.g. the scattering of quantons by targets. To count as evidence, these operations must be real not phony: gedankenexperiments, so useful — and also misleading — have no test value (POPPER, 1935).

Now every empirical test calls for measurement theories even if sketchy: thus some tests of CM employ the theory of the balance, itself an application of CM, and others require the theory of the transistor, which is an application of QM. In short a general physical theory, if applied to measurements of a class, yields a *measurement theory* (see Ch. 1, 5.1.4); this in turn backs all the measurements of that class: it renders their design and interpretation possible and it explains their *modi operandi*.

By itself no measurement makes sense: in the absence of every theory the most that can be inferred from an apparatus reading is that a certain macroeffect — a change in the state of an experimental device — was observed in given circumstances. To go any further, unaided by theory, is just like reading the future in tea leaves. Only a theory, by hypothesizing object-index, and cause-effect connections, enables us to conclude something about the object and the cause out of instrument readings — the more so if the object is imperceptible and therefore accessible in an indirect way (see Ch. 1, 5.1.1). Moreover experiments can check only relatively low-level statements — say about q.m. averages and scatters about the mean. Without the logical relations existing within a theoretical body between such statements and the basic statements or axioms, most of which are experimentally inaccessible, we would be unable to test such high-level statements.

Yet no basic theory refers to experimental situations — e.g. the tracks in a bubble chamber; these are all extremely specific and complex: they involve many variables of different kinds and therefore require the joint intervention of several theories. Consequently the checking of any theorem in a high-brow theory like CM, CEM, GR or QM, is extremely roundabout: it calls for a whole set of different theories by means of which the empirical operations can be designed, carried out and interpreted. These additional theories which are used (hence not tested) in the course of an experimental test of a substantive theory, are auxiliary in this context although they may become substantive in alternative contexts (see Ch. 1, 5.1.4). In brief, every measurement in physics employs two sets of theories: those which are being tried out — say QM — and those which help design, operate and read the experimental arrangement — e.g., CM, CEM and the theory of the galvanometer. Pure experiential data, uninterpreted by theory, make no sense to a theory.

This holds for classical as well as for quantum physics. The usual claim that no such theoretical background is necessary in classical physics because classically observation errors can be made as small as desired, is false: no delicate measurement is error-free, if only because of the spontaneous Brownian motions. And even if error-free measurements were feasible we should know what is it they confirm or disprove: we must make inferences, and inferences are possible only within a body of ideas. And the claim — of part of the orthodox school — that QM needs no special measurement theory because it is already a theory concerning experimental results, is no less mistaken: if basic QM referred to an unanalyzed microsystem-apparatus unit (BOHR, 1932), this should be seen through an analysis of the basic formulas of the theory. Since measuring devices are so far describable largely in classical terms, every q.m. formula should contain two sets of variables, one quantal referring to the microlevel of the situation, the other classical referring to its macrolevel. Yet this is not the case with the basic formulas of QM. Therefore the usual references to measuring devices and observers in relation to the fundamental hypotheses of QM are as pertinent as the phrase 'If God wills it, then $2 + 2 = 4$'. If the supernumerary apparatus were there, it would drag the whole of classical physics into every q. m. formula; and if the equally supernumerary observer were involved in every q. m. formula then the whole of psychology would be dragged into it as well.

The experimental devices occurring in QM textbooks are cozily ideal and moreover most of them are universal: usually nothing is specified about them save their name — *the apparatus*. Yet there is no such thing as the general purpose apparatus, and consequently there is no point in trying to build a general theory of it. General electron theories are possible because electrons are all alike. And a generic theory of quantons disregarding specific differences is possible as well (see Secs. 2 to 4) — but it can make no definite prediction that could be checked by experiment: specific predictions call for specifications. For the same reason there can be no *general* theory of measurement yielding *specific* predictions concerning the outcome of a *generic* measurement by means of a *general purpose* device. There is no generic measurement save in QM textbooks and consequently no *general* physical theory of *real* measurements is available — nor is it possible. This is not prophesying but a matter of logic. There can be at most a *broad approach* to the measurement problem, with the hope of seizing on some key traits of measurements in a given field; as soon as real measurements are envisaged very specific theories using fragments of a host of fundamental theories come up. Generic theories of measurement (e.g., v. NEUMANN, 1931; LONDON and BAUER, 1939; BOHM, 1951; LUDWIG, 1961 and 1964) can produce no single numerical statement concerning a real measurement, for the excellent reason that there do not exist universal measurement instruments.

There are conflicting views about the status of a comprehensive QMM (quantum mechanics of measurement). They are: (*a*) QMM is a full-fledged theory used for testing QM (widespread opinion); (*b*) QMM does not exist as a separate theory: QM is already the theory of experiment (v. NEUMANN and DIRAC in places); (*c*) QMM is possible but to build it would be misleading (ROSENFELD in INFELD, 1964); (*d*) QMM is possible and even necessary but it must take the observer's consciousness into account (LONDON and BAUER, 1939; WIGNER, 1962, 1963); (*e*) QMM is a forlorn embryo consisting in an application of QM to roughly sketched experimental (micro-macro) situations, and it should contribute to understanding the measurement process — but as it stands it has hardly a predictive power: it ought to become a more realistic theory, which it won't unless all preceding views are discarded. We take the last stand.

Let us insist: QMM is not indispensable for testing QM since the observations and experiments that have corroborated QM have macrophysical upshots and classical physics is supposed to take care of their macrophysical level (until superconductores and superfluids become important in instrumentation — which they will). Think of typical empirical operations in microphysics such as the display and measurement of spectral lines, the preparation and measurement of atomic beams, and the exposure and analysis of nuclear plates. These count as tests not only because they involve observable effects but also because there are theories which tell us — if only in outline — that these macroeffects *are* effects of microevents. That is, atomic and subatomic experimental physics are at all possible because there are theories that, on hypothesizing a hidden reality, enable us to infer back from effects to causes. In all cases the explanation of how an arrangement works, and therefore the justification for the experiment, makes some use of QM — e.g., the semiclassical theory of the tracks left by ionizing "particles". In no case are data gathered and read without using the very theory that is being tested; in particular QM (but usually not QMM) is involved in its own test. In general, empirical tests are partly theoretical — therefore rarely final.

In any case every experiment involves specific physical processes: there are no all-purpose experiments, let alone ideal experimental arrangements such as an infinite and perfectly smooth edged slit. Gedankenexperiments illustrate theories without testing them; and they alone can do it in a simple way, for no rough, complex and dirty chunk of matter could ever climb as high up the theoretical pyramid. Now current QMM, though an application of QM, is not an application to real situations but is still a high-brow and generic theory and therefore it can be used to *sketch the analysis of highly idealized typical* experimental situations but can predict accurately no outcome of a single real experiment: it is a semitheory of pseudoexperiments. Let us outline such an analysis (see LONDON and BAUER, 1939; BOHM, 1951; MERZBACHER, 1961; MESSIAH, 1961 and MARGENAU, 1963 for different though related treatments, and BOHM and BUB, 1966, for a novel approach.)

We start out by adding a single postulate to the theory of quantons developed in the previous sections:

QM 10 Every measuring device is an aggregate of quantons with macroproperties.

This assumption entails: (*a*) that every system consisting of a quanton and a measuring device is a physical sum $\sigma + \bar{\sigma}$ where, as before, $\bar{\sigma}$ represents the environment of its partner σ; (b) that measurement processes are physical processes occurring, like all those with which physics is concerned, outside our skulls; (c) that the human mind has no direct influence on measurement processes although it is indispensable to imagine, carry out and interpret them; (d) that no quantal law is suspended during a measurement process — in particular the Schrödinger law applies to it contrary to v. NEUMANN'S assumption.

We next add two linguistic conventions (MARGENAU, 1936 and LUDWIG, 1964):

Df. 18. A *preparation* of a state ψ [or rather of a whole state ray] of a σ associated with $\bar{\sigma}$ consists in the action of a macrosystem $\bar{\sigma} \in \overline{\Sigma}$ on a quanton $\sigma \in \Sigma$ such that $\sigma \dotplus \bar{\sigma}$ falls in a preassigned state. [This state will be pure not mixed but it need not be sharp: indeed, in general it will not be an eigenstate of the operator representative of the property that is being measured.]

Df. 19. A *measurement* of Q on a $\sigma \in \Sigma$ by means of a macrosystem $\bar{\sigma} \in \overline{\Sigma}$ is a linkage of a property Q of σ and a property $\bar{Q}$ of $\bar{\sigma}$ brought about by a $\sigma - \bar{\sigma}$ interaction and such that the values of $\bar{Q}$ become correlated to those of Q through a known law.

These definitions are consistent with the assumption QM 10 that measuring devices are purely physical and particularly macrophysical systems — a necessary condition for being manipulable by an observer other than an amoeba. And Df. 19 embodies the idea that there is no significant measurement without some knowledge of the pertinent laws — in turn items of theories. Notice that the observer, a psychophysical entity, lurks behind QM 10 and Dfs. 18—19: there can be quantons without observers — as we have assumed in our formulation of the basic theory — but as soon as we propose to "observe" them, we ourselves must step in and design an arrangement that, though purely physical, will embody our knowledge and our purpose: it is an artifact not a natural system like the one fundamental QM is concerned with. In short the observer — but not The Observer — makes his appearance in the test stage though not before it. But even then he carefully draws a line between the physical processes he studies and himself — unless he happens to be more interested in himself than in nature, which is all right but is not physics. In other words: a physical system $\bar{\sigma}$, not a psychophysical entity, much less a disembodied Mind, is the partner of the quanton σ involved in an "observation". This holds in quantum as well as in classical physics.

Now any measurement of a microproperty Q on a quanton σ requires coupling σ with a macrosystem $\bar{\sigma}$ in such a way that the interaction will leave an observable trace on $\bar{\sigma}$ — i.e. will make some impact on some macroproperty $\overline{Q}$ of $\overline{\sigma}$ while leaving Q unaltered or nearly so. (If the measurement act did disturb Q appreciably while the interaction is on we would not be measuring Q unless we knew how to correct for the disturbance — as we usually do in classical physics. This shows that when performing atomic measurements one assumes the objective existence of microsystems and moreover the constancy of some of their properties.)

Furthermore the Q–$\overline{Q}$ linkage must be such that the values q_k of $\mathcal{Q} \triangleq Q$ become closely correlated with the values $\overline{q}_{\overline{k}}$ of the $\overline{\mathcal{Q}} \triangleq \overline{Q}$ that will be read on the apparatus, so that an observer can infer the unobservable q_k from the observed $\overline{q}_{\overline{k}}$ — without such an inference (a purely conceptual operation) introducing any changes in either q_k or $\overline{q}_{\overline{k}}$. (No reduction of the "wave" packet upon reading the instrument: what may happen is that, on taking cognizance of the experimental result, we are forced to modify the originally hypothesized state of the system: a mere hypothesis correction not a physical effect of the reading act. In short if anything happens upon reading a gauge it happens within our skulls.) This requires designing an apparatus operating on the basis of a σ–$\overline{\sigma}$ interaction represented in the total hamiltonian for the compound system $\sigma + \overline{\sigma}$. The interaction will have the form $H_{\sigma\overline{\sigma}} = f(\mathcal{Q}, \overline{\mathcal{Q}}, t)$, where $\overline{\mathcal{Q}} \triangleq \overline{Q}$ and $\mathcal{Q} \triangleq Q$ are not additively separable. Indeed, a term $H_{\sigma\overline{\sigma}}$ with these properties represents a σ–$\overline{\sigma}$ interaction since both sets of variables are present and coupled, so that the Q-distribution will become correlated with the $\overline{Q}$-distribution.

Before proceeding any further notice that the object-apparatus interaction is treated as a purely physical event: the compound system $\sigma + \overline{\sigma}$ is a physical entity. In other words, no observer coordinates occur in $H_{\sigma\overline{\sigma}}$. Should the observer himself act as a (qualitative) detector, he would have to be treated as a physical system satisfying QM. In other words, QM does not handle the psychological and epistemological problems of apperception: QM takes for granted that a subject-object interaction exists and it hopes other disciplines will take care of it. If, as is sometimes held, the observer's mind were decisive in the very process of measurement — not just in its planning and interpretation — then we would have to include psychical variables in $H_{\sigma\overline{\sigma}}$ — but who, among those who hold this view, knows how to proceed? In short QMM is as much of a physical theory as QM. Even those who invoke the omnipotence of The Observer fail to add psychical variables where they belong, whence their interpretation is *ad hoc*. When hypothesizing a given $H_{\sigma\overline{\sigma}}$ one assumes that both the object and the apparatus exist objectively and independently though coupled: a nonentity could not possibly be correlated with "its" partner. And their separate existence entails that each has its own properties, however much they may be

modified by its partner. (The orthodox school holds that atomic properties emerge during observations and do not exist between these, so
that they cannot be neatly separated from the macroproperties of the
apparatus — whence one could never write down a hamiltonian in
which the two sets of variables, $\mathscr{Q}$ and $\bar{\mathscr{Q}}$, where distinguished. For
that school there is no $\sigma-\bar{\sigma}$ interaction and therefore "it" cannot
be analyzed: atomic phenomena are unanalyzable units, irrational
wholes (BOHR). In particular, it is allegedly impossible to draw a dividing
line or cut between the microobject and the observation device, which
is said to include the observer: see e.g. HEISENBERG, 1930; v. NEUMANN,
1931; LONDON and BAUER, 1939; BOHR, 1958; WIGNER, 1963; HEITLER,
1964 — and for criticism POPPER, 1935; BUNGE, 1955b; and the SCHRÖ-
DINGER-EINSTEIN correspondence in PRZIBRAM, 1963). Let us now proceed
with the formulas.

Let us agree to associate $t=0$ with the onset of the interaction.
Before that instant the state of the combined system $\sigma + \bar{\sigma}$ evolves
according to

$$i\hbar \frac{\partial \Psi_0}{\partial t} = H_0 \Psi_0, \quad \text{with} \quad H_0 = H(x, \mathscr{P}) + H(\bar{x}, \bar{\mathscr{P}}), \quad t<0. \quad (5.32)$$

Due to the $\sigma-\bar{\sigma}$ independence, reflected in the additivity of the partial
hamiltonians, the state of $\sigma + \bar{\sigma}$ up to $t=0$ can be represented by the
product of two pure states and accordingly a pure state itself by Df. 9:

$$t<0, \quad \Psi_0(x, \bar{x}, t) = \psi_\sigma(x, t) \cdot \psi_{\bar{\sigma}}(\bar{x}, t). \quad (5.33)$$

At $t=0$ an interaction is turned on during the time Δt. As a result
the compound system abandons state (5.33) and goes over into a final
state in accordance with SCHRÖDINGER's equation. In the final state
some of the dynamical variables of the parts will get mixed up, for
otherwise there will be no measurement proper. While the interaction
lasts, $H_{\sigma\bar{\sigma}}$ overpowers H_0, whence

$$t \in \Delta t, \quad i\hbar \frac{\partial \Psi}{\partial t} = H_{\sigma\bar{\sigma}} \Psi. \quad (5.34)$$

(This treatment could be improved by adopting the interaction representation, in which (5.34) would be the exact equation of evolution for the
transformed state representative.)

As usual in QM, the solutions Ψ_0 and Ψ are required to join smoothly
at $t=0$. (We do not postulate the mysterious jump of state but assume
that states evolve lawfully all along.) If we knew the initial state Ψ_0 and
the interaction hamiltonian $H_{\sigma\bar{\sigma}}$, the problem of measurement would
be solved in principle, no matter how hard the mathematics were. But
even assuming that we could read Ψ_0 from the way σ and $\bar{\sigma}$ are prepared,

as textbooks instruct us to do, we know of no $H_{\sigma\bar{\sigma}}$ corresponding to a *real* measurement. (The solitary textbook case of the Stern-Gerlach split beam *experiment* is a poor example of a *measurement* because it dodges the micro-macro transition: in fact it pretends that the two beams are "observed" directly, while the problem lies precisely in correlating the unobservable microevents with the observable macro-events. And to insist on the two-slit experiment as an illustration of QMM is short of obscene because no exact solution to this problem is known even for classical waves. In general, too many experiments discussed in QM are not measurements proper: see DE BROGLIE, 1964). In short all measure-ments discussed in QMM are phony and so will be the class of measurements we shall now discuss on the basis of the preceding schema. The only difference between v. NEU-MANN's QMM and ours is that the latter is consistent with QM: both concern bogus measurements.

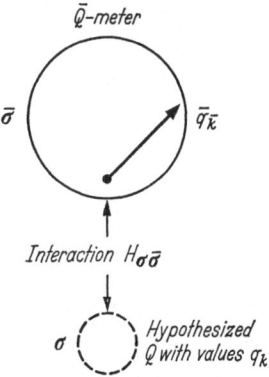

Fig. 5.3. Inferring an unob-servable $\bar{q}_k$ from reading $\bar{q}_k$ on a $\bar{Q}$-meter coupled to a microproperty Q

Pretend now we design a measuring device based on an interaction of the form

$$H_{\sigma\bar{\sigma}} = b\mathscr{2} \cdot \bar{\mathscr{2}}, \quad \text{with} \quad b \in R, \quad (5.35)$$

and $\mathscr{2} \triangleq Q$ (object property), and $\bar{\mathscr{2}} \triangleq \bar{Q}$ (in-strument property). For example, $\bar{Q}$ could be the position or the angular velocity of the pointer of a galvanometer connected to an electron multiplier. Needless to say Q will be a microproperty and $\bar{Q}$ a macroproperty and moreover a sufficiently ostensive one — yet we shall pretend that $\bar{Q}$ too be represented by a q.m. operator (see Fig. 5.3). Assume now that, before the interaction sets on, the apparatus has been put in an eigenstate of its indicator $\bar{Q}$, and precisely in its nondegenerate ground state $\bar{\varphi}_0$. If QM is true the pointer will show exactly $\bar{q}_0$, which may be taken as the zero of the scale. (But look attentively: the real pointer is fluctuating due to thermal and electrical noise. Let some time pass: the pointer will most likely have shifted.) In short, prepare $\bar{\sigma}$ so that

$$t < 0, \quad \psi_{\bar{\sigma}}(\bar{x}, t) = \bar{\varphi}_0(\bar{x}) \cdot e^{-i\bar{E}_0 t/\hbar}. \quad (5.36)$$

Having prepared the phony apparatus let us now prepare the object of observation. Imagine we build and operate a device that puts σ in a preassigned stationary state — not necessarily an eigenstate of $\mathscr{2}$. Moreover pretend we know exactly what we have done, i.e. that we can write

$$t < 0, \quad \psi_\sigma(x, t) = u(x) \cdot e^{-iEt/\hbar}. \quad (5.37)$$

Now leave the lab, retire to the office and expand u in $eif\,\mathcal{Q}$:

$$u(x) = \sum_k \overset{\circ}{c}_k \varphi_k(x), \quad \text{with} \quad \overset{\circ}{c}_k = (\varphi_k, u). \tag{5.38}$$

Since σ and $\bar{\sigma}$ are uncoupled for all $t < 0$, the state of the compound system is, by (5.33) and (5.36),

$$t < 0, \quad \Psi(x, \bar{x}, t) = \bar{\varphi}_0(\bar{x}) \cdot e^{-i(\bar{E}_0 + E)t/\hbar} \sum_k \overset{\circ}{c}_k \varphi_k(x), \tag{5.39}$$

which is a vector in the product space $\mathfrak{H}_\sigma \otimes \mathfrak{H}_{\bar\sigma}$ of the Hilbert spaces of σ and $\bar{\sigma}$. At $t = 0$ the interaction is turned on — e.g., an insulating partition is removed. (Needless to say, this is a purely physical operation that can be entrusted to an automaton.) Now expand the state function in products of $eif\,\mathcal{Q}$ and $eif\,\bar{\mathcal{Q}}$:

$$t \geqq 0, \quad \Psi(x, \bar{x}, t) = \sum_{k\bar{k}} c_{k\bar{k}}(t)\,\varphi_k(x)\,\bar{\varphi}_{\bar{k}}(\bar{x}), \tag{5.40}$$

still another vector in $\mathfrak{H}_\sigma \otimes \mathfrak{H}_{\bar\sigma}$. But the interaction will not constitute a measurement unless the apparatus indications $\bar{q}_{\bar{k}}$ are translatable into q_k values. In other words, $\bar{\mathcal{Q}}$ must be a faithful indicator or objectifier of $\mathcal{Q}$. For this it is necessary that there exists a function f from the set $\{\bar{q}_{\bar{k}}\}$ into the set $\{q_k\}$ or, what is equivalent, from the index $\bar{k}$ to the index k. It suffices that $\bar{k} = k$ (LONDON and BAUER's condition), a very strong requirement indeed. Pretending that our (nondescript) experimental arrangement fulfils this condition, (5.40) simplifies to

$$t \geqq 0, \quad \Psi(x, \bar{x}, t) = \sum_k c_k(t)\,\varphi_k(x)\,\bar{\varphi}_k(\bar{x}), \tag{5.41}$$

where $|c_k|^2$ is the joint probability that the unobservable Q has the value q_k while the pointer shows the value $\bar{q}_k$. To compute these probabilities replace (5.41) in the law of QM (5.34), multiply by φ_k^*, integrate over the region accessible to σ and use orthonormalization, to get

$$t \geqq 0, \quad i\hbar\,\dot{c}_k = b\,c_k\,q_k\,\bar{q}_k \quad \text{whence} \quad c_k(t) = c_k(0)\,e^{-i(b\,q_k\bar{q}_k t)/\hbar}, \tag{5.42}$$

and finally

$$t \geqq 0, \quad \Psi(x, \bar{x}, t) = \sum_k c_k(0)\,e^{-i(b q_k \bar{q}_k t)/\hbar}\,\varphi_k(x)\,\bar{\varphi}_k(\bar{x}). \tag{5.43}$$

The new state will join the state prior to the fall [i.e. (5.39)] at $t = 0$ provided

$$\text{for all } k \text{ and all } \bar{x}, \quad c_k(0) = \overset{\circ}{c}_k \cdot \frac{\bar{\varphi}_0(\bar{x})}{\bar{\varphi}_k(\bar{x})}. \tag{5.44}$$

Hence the upshot of the measurement interaction represented by (5.35) is

$$t \geqq 0, \quad \Psi(x, \bar{x}, t) = \bar{\varphi}_0(\bar{x}) \cdot \sum_k \overset{\circ}{c}_k\,e^{-i(b q_k \bar{q}_k t)/\hbar}\,\varphi_k(x). \tag{5.45}$$

(The generalization to $H_{\sigma\bar{\sigma}} = \sum b_{mn} \mathscr{Q}^m \bar{\mathscr{Q}}^n$ is straightforward: in (5.45) replace $b\, q_k\, q_k$ by $\sum b_{mn} q_k^m q_k^n$.) Interpretation: the q_k are now correlated with the $\bar{q}_k$. When the $\bar{Q}$-meter shows $\bar{q}_k$, we can infer that the unobservable Q of the object σ has the value q_k with probability $|c_k(0)|^2$. And this probability is t-independent: it is conserved no matter how violently the other properties of the observed object are altered by the interaction. In particular *two successive measurements* of Q, whether in rapid succession or not, have *the same probability* — which is not surprising as the hypothesized interaction is static except at $t = 0$. This does not mean that they yield the same *result:* this would be surprising. We are not saying that, if $\bar{q}_k$ is read on the $\bar{Q}$-meter, then q_k obtains both times but that, whenever $\bar{q}_k$ is read, the probability that q_k obtains is the same at all times after the interaction sets on. (In v. NEUMANN's theory it is *postulated* that two immediately successive measurements yield the same result no matter what the interaction may be. This assumption does not match the stochastic character of QM. Moreover, it cannot be "abstracted" from the Compton experiment as v. NEUMANN claimed, if only because no hypothesis can be more than suggested by experiment. Finally, the postulate makes repeated measurements unnecessary. One more proof that v. NEUMANN's interpretation, far from being faithful to experiment, is alien to it.)

Look again at (5.45). According to QM the system has *not* been thrown into an eigenstate of the unobservable represented by $\mathscr{Q}$. Indeed, the variables Q and $\bar{Q}$ have become inextricably mixed. In other words, v. NEUMANN's projection postulate is inconsistent with the remaining postulates of QM: it is a strange body in it. Moreover it is never used in actual calculations and it is unrealistic (MARGENAU, 1936, 1963, 1967 and WAKITA, 1962). Dropping the projection postulate is also sufficient to dissolve the Einstein-Podolsky-Rosen paradox (MARGENAU, 1936). Moreover, v. NEUMANN's postulate is not only inconsistent with QM but also with the operationalist philosophy underlying the orthodox interpretation of QM: indeed if on measurement states change abruptly — and in a lawless way — but evolve smoothly while no one is looking, then most statements in QM — all those related with the Schrödinger law — fail to refer to measurement. Therefore if one wishes to stick to what is observable then one must either discard v. NEUMANN's embryonic theory of measurement or dispense with the Schrödinger equation altogether (as LANDÉ, 1958 has proposed).

If the initial state of $\sigma + \bar{\sigma}$ is known, the probability amplitudes $\overset{\circ}{c}_k$ can be computed and the state of the compound system $\sigma + \bar{\sigma}$ can accordingly be determined and the unobservable q_k be inferred with probability $|c_k(0)|^2$ from the observed $\bar{q}_k$. The outcome is nonprobabilistic just in the exceptional case in which $\overset{\circ}{c}_k$ has a sharp peak somewhere, say at k_0, so that initially (not finally as in the v. Neumann theory) $u = \varphi_k$ or nearly so. This "reduction of the wave packet" is initial not final and it corresponds to the *preparation* of $\sigma + \bar{\sigma}$ in a state coinciding with a given eigenstate φ_{k_0} of $\mathcal{Q}$. In this exceptional case, during and after the measurement of Q

$$t \geqq 0, \quad \Psi(x, \bar{x}, t) = \bar{\varphi}_0(\bar{x}) \cdot e^{-i(b\,q_k{}^-kt)/\hbar} \varphi_{k_0}(x). \tag{5.46}$$

Interpretation: the reading $\bar{q}_{k_0}$ warrants the conclusion that $\mathcal{Q}$ has still the value q_{k_0} — as it should be since the measurement is not supposed to interfere with the measured property Q. But this can hardly be regarded as a measurement: the preparation stage has absorbed all the process. In the usual interpretation: The Observer said: 'Let the value q_{k_0} be'.

Let us now apply the preceding schema to (phony) position and momentum measurements, proceeding without great rigor.

Position measurement	*Momentum measurement*
$\mathcal{Q} = x, \bar{\mathcal{Q}} = \bar{\mathcal{P}}, H_{\sigma\bar{\sigma}} = b\,x\,\bar{\mathcal{P}}$	$\mathcal{Q} = \mathcal{P}, \bar{\mathcal{Q}} = \bar{x}, H_{\sigma\bar{\sigma}} = b\,\mathcal{P}\,\bar{x}$
$\varphi_k = \delta(x - x_k), \bar{\varphi}_k = (2\pi)^{-\frac{1}{2}} e^{i\bar{k}\bar{x}}$	$\varphi_k = (2\pi)^{-\frac{1}{2}} e^{ikx}, \bar{\varphi}_k = \delta(\bar{x} - \bar{x}_k)$
$q_k = x_k, \quad \bar{q}_k = \hbar\bar{k}$	$q_k = \hbar k, \quad \bar{q}_k = \bar{x}_k$

State during measurement

$\Psi(x, \bar{x}, t) = (2\pi)^{-\frac{1}{2}} \delta(x - x_k)$ $\cdot e^{i\bar{k}\bar{x}} \sum_{k''} \overset{\circ}{c}_{k''} e^{-ibx_kt k''}$	$\Psi(x, \bar{x}, t) = (2\pi)^{-\frac{1}{2}} \delta(\bar{x} - \bar{x}_k)$ $\cdot \sum_{k''} \overset{\circ}{c}_{k''} e^{-ib\bar{x}_kt k''} \cdot e^{ik''x}$

Interpretation

When the pointer shows $\hbar\bar{k}$, the object is at x_k with probability $|\overset{\circ}{c}_k|^2$.	When the pointer is at $\bar{x}_k$, the object has a momentum $\hbar k$ with probability $|\overset{\circ}{c}_k|^2$.

In particular

If σ has initially been placed at x_k (preparation), then	If σ has initially been imparted the momentum $\hbar k$ (preparation), then

$$\overset{\circ}{c}_{k''} = (2\pi)^{-\frac{1}{2}} \delta(k'' - k), \text{ whence}$$

$\Psi(x, \bar{x}, t) = \delta(x - x_k) \cdot e^{ik(\bar{x} - bx_kt)}$	$\Psi(x, \bar{x}, t) = \delta(\bar{x} - \bar{x}_k) \cdot e^{ik(x - b\bar{x}_kt)}$

But these states are not square integrable, therefore they are not physically possible states according to QM (see $QM\ 5d$). In other words, if QM in its puritanical version (no tampering with ∞) is true, then the position and momentum of quantons cannot be measured with arbitrary precision using devices materializing the above interaction hamiltonians.

In particular, no sharply localized states can be determined experimentally although they may exist for the free system. This is not too surprising since every measurement places an upper bound on the dimensions of the region accessible to the system. The theoretical possibility of point-like localization in the absence of interactions cannot be transferred to experimental situations. This worsens — or improves according to taste — in the relativistic theory of spinless quantons, where there is no hermitian operator with localized eigenstates (PHILIPS, 1964).

Let us insist that the preceding formulas do not concern real measurements: indeed we have no idea how to design devices embodying the above hypothesized interactions. In particular, we do not know how to switch them on instantaneously or even how to make an instantaneous momentum measurement. A realistic treatment should involve an analysis of the constitution of the apparatus — which is usually taken as a block — and would call for quantum statistics. Moreover, the preceding formulas are *theoretical statements* that cannot be contrasted with experimental ones without further ado. In order to compare the theoretical statements with the ones referring to a run of similar measurements, we should build up a dictionary like the following one, where '$[q_k]$' stands for a rational approximation to q_k:

Theoretical predictions	*Experimental findings*
The variable $\mathcal{Q} \triangleq Q$ concerning σ has the value q_k.	The property Q of the intended referent of σ, as measured on an aggregate of empirically identical quantons with the procedure t, has the value $[q_k]$ with standard error ε_{kNt} where N is the number of measurements in a single run.
The variable $\mathcal{Q} \triangleq Q$ concerning σ has the value q_k with probability w_k.	The property Q of the intended referent of σ, as measured on an aggregate of empirically identical quantons with the procedure t, has the value $[q_k]$ with relative frequency f_{kNt} and standard error ε_{kNt}.

Notice that, whereas the theoretical values q_k and w_k depend on σ, the corresponding empirical values $[q_k]$ and f_{kNt} depend on $\langle \sigma, \bar{\sigma} \rangle$ and t; in other words, the empirical magnitudes are functions on $\Sigma \times \bar{\Sigma} \times \mathcal{T}$, where $\mathcal{T}$ is the set of possible experimental procedures (techniques, apparatus, circumstances, and operations). For example, in the case of measurements of radioactive decay, what is measured is not the number N_0 of atoms at time $t = 0$ but the intensity I_0 of the source at that time, where $I_0 = N_0 f \Omega$, Ω being the solid angle subtended by the detector and f the latter's counting efficiency (JÁNOSSY, 1965). Only after such

a translation of a theoretical prediction to a semiempirical language
has been performed can we contrast it with an observational statement
couched in the same empirical language (BUNGE, 1967b, Vol. II). If for
a given property and a given procedure $|q_k - [q_k]| \leq \varepsilon_{kNt}$, the prediction
is confirmed, otherwise refuted relative to t — and similarly for the
probability of q_k. If the theoretical prediction is refuted, it may be
concluded that the actual Ψ differs from the calculated one — unless
the measurement itself has been defective. On the assumption that the
measurement has been performed correctly, the discrepancy with the
theoretical prediction leads us to modify the hypothesized state: it
prompts a change in our tentative beliefs (conjectures) not in nature.
(In v. NEUMANN's theory of QMM the mere act of reading the apparatus
makes Ψ jump, whence the state vector is declared to be a purely
epistemological item or at most only a half-physical one. Why not
extend this confusion to classical physics?)

To conclude. There exists an embryonic QMM: a chapter of QM which,
if consistent with it, has no use for the so-called reduction of the "wave
packet" (at superlight velocity) upon observation, much less for The
Observer's psychical coordinates. This theory is generic, it employs
unrealistic assumptions, and is badly underdeveloped — so much so
that it has failed to solve exactly even the single slit problem (not to
speak of the two slit one). The only solution known to the writer is
approximate and does not allow for the lower bound of the "indeterminacy principle". (See BECK and NUSSENZVEIG, 1958, who obtain $\Delta z \times$
$\Delta k_z \to \infty$ logarithmically for a plane wave incident on a narrow slit of
infinite length. Their proposal to change the standard deviation into
the half-width of the probability distribution would save HEISENBERG's
relations for this case but is too restrictive: QM cannot restrict a priori
all probability distributions to Gaussian-like ones.)

Therefore although QMM has some explanatory power it lacks predictive power. (For formalizations of these concepts see Ch. 1, 5.2.3.)
It won't acquire a predictive power until it handles, however schematically, real experiments done with real hence specific pieces of equipment.
Hence QMM has so far no test value: it illustrates QM without enhancing
its empirical support. Yet even though physically negligible to this
date, QMM has some philosophical interest. Firstly it exposes the operationalist claim that every q.m. formula has an experimental content
or reference: in v. NEUMANN's version because SCHRÖDINGER's equation
is explicitly postulated to refer to nonexperimental situations, in our
version because a measuring device can be read into a formula provided
the latter happens to involve apparatus variables in a non-supernumerary
fashion. Also, QMM refutes the claim that it is impossible to separate
the object from the subject in QM: that the observer's mind is inti-

mately intertwined with the object and consequently a realistic epistemology is untenable. Indeed, no psychical variables occur in QMM
(although all of them are constructs — but constructs with an intended
real referent). On the contrary only critical realism (more real than
ism), allied with logic, can dissipate the fog surrounding QM. Finally,
the nonexistence of a QMM specific enough to handle a single real measurement refutes the claim that QMM is indispensable for testing QM:
happily, QM was put to the test with the help of other theories, long
before QMM was quasiborn. But the underdevelopment of QM also
constitutes an indictment of those who, by insisting on the direct
experimental reference of QM, have hindered the recognition that QMM
is hardly born and that it should be given a chance to see a real measuring
device.

8. Debated Questions

Let us finally take up some tricky problems that were only glanced
at above. Firstly what are the *relations between* QM *and classical physics*,
particularly classical mechanics (CM)? The relations are of various
kinds: logical, semantical, heuristic, and methodological.

Logical relations. Certain classical concepts, such as those of mass,
momentum, and energy, recur in QM, often as the eigenvalues of the
corresponding operators; and so do the corresponding classical relations
among them, e.g. "$E = p^2/2\mu$" (or its relativistic extension) for a free
quanton. Moreover for certain limits (chiefly large quantum numbers,
$\hbar = 0$, and the ray or WKB approximation), QM gives back several
classical formulas while others disappear: they have no classical analogs.
Hence QM, in addition to presupposing CEM, contains a fragment of
CM. In other words, some of CM is built into QM. But this does not mean
that QM logically presupposes CM, as the Bohr-Heisenberg school holds.
In fact many q.m. formulas are inconsistent with their classical partners
and, from an axiomatic point of view (yet not heuristically), it is an
accident that we should recover some classical formulas. Nor does that
mean that QM is an extension of CM, in particular that PM is a subtheory of QM: this would require at least that every one of the primitives
of CM were either contained in the primitive base of QM or defined in
terms of them. But this is not the case: e.g., the classical particle
coordinate X, a function on $\Sigma \times T$, does not occur in QM as a primitive
although it does have a classical analog, namely $\langle x \rangle$ (recall Thm. 15a);
in the case of an aggregate of quantons, one recovers the classical c.m.
coordinate but not the body material coordinate X for every point in
a continuous body. Even if one were to enlarge the primitive base of
QM to include all the basic concepts of PM, the relation PM ⊂ QM would
hold only if every classical relation reappeared in QM — but this should

occur only in the various correspondence limits (as it does in Bohm's "hidden parameters" expansion of QM: Bohm, 1952). In short, QM and CM have a modest overlap and this intersection consists of a set of classical formulas that are either not modified by QM or are the correspondence limits of q.m. formulas. Were it not for this overlap there could be no semantical, heuristic and methodological relations between QM and CM.

Semantical relations. The meanings of certain q.m. symbols (e.g., "mass", but not "position") have been borrowed from CM and, whenever there is a nonvanishing correspondence limit, it is legitimate to assign those limiting formulas a classical interpretation. Thus a semiclassical theory of spinning particles (Bargmann, Michel and Telegdi, 1959; Nyborg, 1962 and Rafanelli and Schiller, 1964) not only enriches classical physics but also facilitates the interpretation of quantum theories of spinning particles. Yet in an axiomatic treatment of QM one pretends to make a fresh start: from this point of view, QM does not derive its content from CM. It is quite different in the orthodox interpretation of QM which, being in terms of phony measurements involving macrosystems, presupposes the whole of classical physics — only to contradict it piece-wise. One more inconsistency of the orthodox interpretation.

Heuristic relations. The builders of QM have used CM and CEM as heuristic crutches, e.g. in hypothesizing most (yet not all) of the non-relativistic hamiltonians and in discarding hypotheses with an utterly unusual correspondence limit. (Apparently it was said in Copenhagen that you could quantize your grandmother if only you were able to handle her classically.) But the use of classical analogies for stating and interpreting quantal formulas has been double-edged: at the same time indispensable (since one cannot start from scratch) and misleading. In our axiomatic reconstruction of QM we have adopted a most parsimonious attitude in this respect precisely because many difficulties in QM come from too strong an attachment to classical ideas.

Methodological relations. Since the empirical tests of QM are not direct but require the use of mechanical, optical, and electromagnetic systems satisfying classical physics (if only approximately), the latter is indispensable for testing QM. (From this it has been wrongly inferred that classical physics is also what gives QM its meaning — the usual confusion between test and reference.) Even the design, operation and interpretation of the simplest possible observational arrangement set up to produce data relevant to QM — e.g., spectroscopic ones — do involve various branches of classical physics, so that without them one would not know how to put QM to the test.

In sum QM, though radically new, is variously related to classical physics. And to advocate severing such ties — e.g. by renouncing altogether the concepts of space and time — is as wrong as hoping that QM will eventually be reduced to classical physics. A total divorce from classical physics would produce only an altogether untestable theory, and one incapable of entailing classical physics. This leads us to the verge of our second problem: what about classical particles and classical fields?

The *body-field duality* has been a stumbling block revered as a principle. The trouble with QM is that many of its formulas can be interpreted, when *out* of context, either in corpuscular or in undulatory terms, as well as both ways at a time — notwithstanding the complementarity "principle". Thus the phase shift δ_l in the ψ referring to a quanton scattered by a field of force can be interpreted in either of these ways: (a) the force field changes the *particle* angular momentum from l to $l' = l - (2\delta_l/\pi)$, whence $\Delta l \equiv l - l' = 2\delta_l/\pi$ and finally $\Delta|L| \cong 2\hbar\delta_l/\pi$, the additional rotation momentum; (b) the field acts as a refractive medium shifting the phase of the de Broglie *wave* by δ_l. (Note incidentally that both interpretations have a strong causal component.) But these are just metaphors.

The ease with which so many q.m. formulas can be given partial and *ad hoc* interpretations in classical or in semiclassical terms has contributed to both the growth and the muddling of QM. It has made it possible to argue with equal force — but in the loose and open context of the nonaxiomatic theory — in favor of a number of views every one is confirmed by as many refuting cases (BUNGE, 1956b). The chief views of this kind are: (a) material microsystems are particles which, due to their random motions, behave collectively as if they were waves (BORN); (b) they are particles which, due to the periodic structure of the bodies they interact with, exhibit a deceiving wave-like appearance (LANDÉ); (c) they are waves that can occasionally contract to point-like wave-packets (SCHRÖDINGER); (d) they are wave-particle compounds, i.e. they have at the same time corpuscular and field properties (DE BROGLIE). The difficulties inherent in each of these views (BUNGE, 1956b) have encouraged the alternative view, held by the most consistent among the upholders of the positivist interpretation of QM (e.g., FRANK, 1946), that material microsystems *are* nothing by themselves: that they do not exist except as words used in describing experimental situations, and that it is up to the experimenter to actualize or conjure up whatever situation He likes (within the bounds set by QM, just as according to LEIBNIZ God could do anything provided it was compatible with logic). This opinion is consistent but, like the previous one, it is not countenanced by the postulates of QM; in addition, it render physics unphysical.

As long as QM is not given a closed form one will be tempted to propose any number of *ad hoc* interpretations suggested by analogy with other disciplines — even information theory and decision theory. And as soon as QM is given a definite shape — as soon as it is axiomatized one way or the other — the only right answer to the question "What are the material microsystems?" is the sibylline one: "They are the hypothetical referents of QM". In particular, if our version of QM is adopted, the answer runs: "They are quantons — peculiar entities that cannot be described, except metaphorically, in classical terms such as 'particle' and 'wave'. For more information look at the axioms and theorems of QM." Similarly, the question "What are e.m. fields?" is best answered by saying that (classical) e.m. fields are the referents of MAXWELL's equations, not by mentioning analogies with elastic bodies or complex machines.

Within our formulation of QM the formulas "$p = h/\lambda$" (DE BROGLIE) and "$\Delta E = h\nu$" (BOHR) do not relate particle aspects (allegedly p and E) to wave aspects (supposedly λ and ν). After all, we have learnt that certain fields have momentum and energy, and that the motion of classical particles can be described in the Hamilton-Jacobi framework by means of waves in the phase space. (Moreover, it is possible to mimick QM within the Hamilton-Jacobi formulation of PM by associating with the particle a wave $\psi = \varrho^{\frac{1}{2}} \exp\,(iS/\hbar)$ where $\varrho = |\partial^2 S/\partial q\, \partial\alpha|$, α being a constant of the motion, S a solution of the Hamilton-Jacobi equation, and $\hbar$ an arbitrary constant of the proper dimension. Even second quantization can be classically mimicked: SCHILLER, 1967.) We adopt an ascetic attitude and recognize that the above formulas have no simple and literal interpretation in the present QM, and that it is only for their deceivingly simple appearance that they constitute the heart of the popular discussions of QM. This ascetic approach eludes paradoxes such as that an electron can interfere with itself but not with other electrons.

On the other hand we may, with some care, continue to talk about diffraction and interference of de Broglie waves and of particle trajectories and wave (or particle) scatterings as long as we recall that these are *analogies* (HEISENBERG, 1930) — as useful and as misleading as the heat flux and the tube of force and the observer riding a proton. In particular, the experimentalist will presumably continue to pretend that he produces, deflects and detects *particle* beams whose not quite corpuscular behavior he hopes the theoretician will explain (in the popular sense of rendering familiar not in the technical sense discussed in Ch. 1, 5.2.1) in terms of *waves*. Handling as he does macrosystems and macroevents, the experimenter cannot find out the nature of quantons by himself, without the assistance of theories. Only a theory of quantons, in part the current one, more fully some of the future theories, can tell us what quantons are like. (Descriptions of atomic and subatomic experiments are of no use to this purpose — particularly if they are gedankenexperiments.) And properly formulated QM is a unitary theory, not because it combines the concepts of particle and wave in some mystic way but because it uses neither of them. Therefore, to the extent that QM is true, there is no wave-particle duality in nature. Consequently there is no complementarity of the wave and the particle aspects either except at the classical level — e.g. for the description of some macroeffects. To keep complementarity at the quantum level is to perpetuate confusion. Which touches on the next question.

Indeterminacy and *uncertainty* vanish likewise for they are prompted by the classicist conception of quantons as point particles as well as by the subjective interpretation of probability. No point particles, no trajectory lines; no trajectories, nothing particularly indeterminate about them: only vaguely delineated strips remain in the particular case when the x-distribution is bell-shaped or nearly so. (The same holds for SCHRÖDINGER's purely undulatory interpretation of QM: Δx and $\Delta \mathscr{P}$ are there the spatial and the spectral widths respectively of a wave packet, which is in turn all there is. Unfortunately this view has difficulties of its own — e.g. the diffusion and the reduction of the "waves".) A definite probability distribution is nothing indeterminate for it satisfies a set of laws: only if the probability distributions of QM were to change in a lawless way would they be indeterminate and consequently QM indetermistic. True enough, according to v. NEUMANN's projection postulate there is no definite relation between the states of a system before a measurement (ψ) and after it (φ_k), because they are not tied up by the Schrödinger law. This is indeed indeterminism and even magic. But we have not adopted that postulate because it is inconsistent with QM (see Sec. 7). Laplacian determinism is not thereby saved: it is made pointless by QM at the quantum level, since the characterization of a quanton involves random variables and probability fields rather than a handful of numbers. QM is consistent with a flexible determinism allowing for stochastic regularities and espousing just two ontological hypotheses: lawfulness and nonmagic (BUNGE, 1959a, 1962b, 1962c).

As regards Laplacian determinism our version of QM is consistent with v. NEUMANN's theorem concerning the absence of "hidden" variables in QM. But this statement is, at its best, trivial hence redundant. In fact the famous theorem comes to say that nonstochastic (scatter-less, nonrandom) variables are absent from the *usual* version of QM. Now if the latter is properly axiomatized this goes without saying: the result is obtained by examining the list of primitives. Indeed, as axiomatized in Secs. 2—7, there are no more "hidden" variables than the quite hidden unobservables $\mathcal{Q}$ (miscalled "observables"). But if v. NEUMANN'S theorem is interpreted as asserting that no possible modification of the current theory could contain "hidden" (nonrandom) variables, then it ceases to be trivial: it is no longer a statement within QM but a piece of wishful thinking. This pious hope was destroyed (BOHM, 1952) with the addition to QM of a new primitive concept, namely a time-dependent material coordinate X allowing one to define a "hidden" momentum $P \overset{dt}{=} \mu \dot{X}$, related to the phase S of ψ by the new postulate: $P = \nabla S$. This enlargement of the primitive base of QM augments the explanatory power of the theory but not its predictive power. Since we adopt the strategy of maximizing the latter rather than the former (see Ch. 1, 5.2), we do not adopt it. The main role BOHM's version of QM has played is to refute v. NEUMANN's nonexistence "theorem" and to thaw the deeply frozen attitude towards QM. And this brings us to our next set of vexed questions: the epistemological ones.

The two main epistemological questions posed by QM are whether it really contains nothing but observables, and whether it has ended with objectivity. The first is easy to answer: QM has always contained *epistemologically hidden* (nonostensive) variables. None of the primitive concepts of QM concern direct observables if they refer to individual quantons. Macroproperties are derived only when the basic theory of QM is applied to aggregates of myriads of quantons; but the building blocks remain the Hilbert space associated with the single quanton in a nonspecific environment, and the operators in that space — none of which represent observable properties. The only ostensible thing about q.m. "observables" is their grossly inadequate name. So much so that several theoreticians have felt dissatisfied with the abundance of unobservables in QM and have proposed restricting the theory to really observable quantities. This was precisely the motivation of the S-matrix approach (HEISENBERG, 1943). It was supposed to be purely operational yet it retains ψ, which is unobservable and can be attached a physical meaning only in the wider context of QM, which contains the S-matrix theory (BUNGE, 1964).

A second epistemological question raised by QM is whether it has debunked *objectivity* as an ideal of scientific research endorsing instead subjectivism. The propounders of the not-too-homogeneous Copenhagen interpretation of QM oscillate between a subjectivism of the BERKELEY-MACH type and an amusing dualism — namely subjectivism as regards the microlevel and realism concerning the macrolevel (BUNGE, 1955b). Even when it refuses to count the mind of the physicist as part of the atomic event (HEISENBERG, 1958 but not BOHR, 1936 and *passim*), the school holds that the subject enters QM through two doors: probability and experiment. But probability drags the subject into QM only if a subjectivist (psychological) interpretation of the probability theory is adopted — an interpretation that has no place in physics (see Ch. 2, 2). The state vector and the coefficients of any of its series expansions have a purely physical referent and are therefore intended to be as objective as a classical position coordinate or a classical field strength. (Otherwise QM would explain human psychology, which unfortunately it does not.) That these are concepts and therefore the work of men and in *this* sense subjective (or rather intersubjective), is obvious; but this does not render QM any more subjective than CM or CEM. A dash of scientific semantics (Ch. 1, 1.3) would have spared much confusion in QM deriving from a lack of semantical analysis of physical formulas, in particular probabilistic ones. But unfortunately scientific semantics was born after the confusion had attained a gigantic size.

As to experiments, they introduce the subject into the physical situation under study only if they are mistaken for the psychological act of taking cognizance of experimental outcomes. Granted, the division of the world into the object of study and the rest is the product of a human act: it is up to us to pick the part of the world we want to study. But this partition is possible because the physical object is there — though not disconnected from the rest of the world. The part-whole relation is a physical one (recall Ch. 2, 5). Hence every analysis of an object-environment complex, whether conceptual or empirical, should be done in a purely physical manner, i.e. without projecting our own psychical events onto it — and this is done in QM as well as in classical physics and in actual scientific work it is performed by both adherents and critics of the Copenhagen interpretation.

A second reason often given for abandoning objectivism is the reduction of the wave packet on measurement: ψ is said to collapse to an eigenstate of the measured "observable" ($=$ unobservable) just because we happen to read the corresponding meter whence it is inferred that ψ must concern our knowledge. This is similar to the "reduction" of uncertainty when, having thrown a coin, we look at the outcome: would we then say that $P = \frac{1}{2}$ has suddenly become $P = 1$? Or would we rather suspect that two concepts of probability, objective propensity and subjective certainty, have been mixed up? Anyhow the famous reduction does not take place in our version of QM. In general, there is no place for the subject among the referents of QM. The observer comes in at the test stage but not as a disembodied ghost and he is most careful in distinguishing the thing observed from himself, for the empirical tests of QM involve physical processes not spiritual exercises. The task of the subject is much more important than that of counting himself among the referents of QM: he must invent and apply physics, which remains "an attempt conceptually to grasp reality as it is thought independently of its being observed" (EINSTEIN, 1949). That QM does in fact deal with an aspect of the external world is being increasingly recognized these days (BUNGE, ed., 1967d).

Finally what about *quantum logic*? We have seen (Sec. 2) that QM presupposes ordinary two-valued logic (PC=), as this is built into the mathematics employed by QM. Yet no less than BIRKHOFF and v. NEUMANN (1936) have claimed that QM does not conform to classical logic but to a logic of its own, and a few others have followed in their authoritative wake — practically all the German-speaking philosophers of physics. The central idea is this. Since properties represented by non-commuting "observables" are not simultaneously measurable with accuracy, statements concerning their exact values at one and the same instant cannot be jointly asserted. E.g., "σ is at x at time t" and "σ has the momentum p at time t" cannot both be true. Since every one of these statements can on occasion be true, it is concluded that ordinary logic collapses, or else that QM calls for an ontology radically different from the one associated with classical physics, in which such conjunctions can always be asserted.

But this conclusion does not tally with the fact that the formal structure of QM is given by ordinary logic and ordinary mathematics. Further, it is inspired by the classicist conception of quantons as point particles moving at random (see Remarks on Thm. 9, Sec. 6). Indeed, the question does not come up if probability distributions rather than classical but hazy or unknown properties are taken as the real thing. But even so the situation can be faced with the help of classical logic, just as in similar cases found in daily life. Thus "σ has now his hat on" is physically (but not logically) incompatible with "σ has now his fez on", whence we can assert truthfully either statement but not both of them. If one regards them as logically independent statements — i.e. if one takes them out of their context, namely hatmanship — then one may wonder whether their conjunction is always permissible. The puzzle is cleared up by recalling that the conjunction of those statements is false, not by chance but because of the additional premise "If σ has his hat on then σ does not have his fez on". This hidden conditional supplies the proper context by showing that the preceding statements are not logically independent. In the case of QM the missing conditional is of course a scatter relation — say "If the p-distribution is sharp then the q-distribution is not sharp". In either case, whether in relation to people or to quantons, the reasoning is this: $p \Rightarrow \neg q$, $p \vdash \neg q$, whence $p \wedge \neg q$, but not $p \wedge q$. The fallacy, then, originated in (a) taking statements about noncommensurable variables out of their theoretical context and (b) forgetting that logic and mathematics are not about things but about ideas, whence one should not ask (as BIRKHOFF and v. NEUMANN, 1936 did) about the "experimental meaning" of formal operations like conjunction and disjunction. What one can legitimately do is to investigate (with the help of ordinary logic) pairs of mutually incompatible situations and questions (MACKEY, 1963 and SCHEIBE, 1964). But this is not particularly relevant to QM; it is of interest to ontology and to the logic of problems.

Another change in the logic of QM was proposed for similar reasons by DESTOUCHES-FÉVRIER (1951) followed by REICHENBACH (1944). It was argued, in a nutshell, that an electron whose position is not measured has an indeterminate position, whence any statement concerning the value of the position in that state is indeterminate rather than true or false. If this were true it would require some system of three-valued logic. Objections: (a) if no such logic underlies the axiom basis then it should not leap up on the level of theorems; (b) this proposal presupposes again a corpuscular interpretation of QM, one which cannot be carried out consistently for all the formulas of QM (HEISENBERG, 1930); (c) this proposal, like the previous one of BIRKHOFF and v. NEUMANN, would render QM meaningless and untestable because it would isolate it from classical physics. In conclusion, QM is queer enough without an illogical change in logic. Moreover ordinary mathematical logic is, alongside common sense, the only light that can pierce through the fog currently enveloping QM.

To conclude. Our ghost-free formulation of elementary QM in its Schrödinger "picture" and coordinate representation has met and dissolved several paradoxes peculiar to the usual version of QM. To be sure not every problem has been solved nor, for that matter, will ever be solved. But at least there are no mysteries left if only because the pious attitude of revering the incomprehensible (GOETHE) has been given up. We have thereby parted company with those who, either merrily or gloomily, hold that QM is hopelessly irrational. What is irrational is this belief itself, which leads some to yearn for the good old 19th century (as if it had not been ridden by difficulties) and others to proclaim the Mussolinian-like credo *Believe, obey, cumpute!* The mysteries in the usual presentations of QM derive from a dusty philosophy: from the attempt to interpret every physical idea in terms of laboratory operations, and to read QM in classical terms. (The two are closely related, as experiments are macroevents hopefully describable in classical terms — at least according to BOHR and HEISENBERG though not to DIRAC and v. NEUMANN.) Eliminate operationalism and classicism, abide by logic and restore the forlorn physical object with the help of a bit of semantics: no mystery remains. (In this sense our version of QM is far more orthodox than the orthodox one.) What remain are genuine problems, both the intrasystematic ones (overhauling and developing QM) and the bigger problem of building new, finer conceptual microscopes to get at still more hidden recesses of reality.

Epilogue

When all was over — that is, when it became clear that nothing important is ever over in physics — in came the Duchess and drew her morals.

Logical moral 1. If one wishes to find out the structure of a theory then he has got to order it, disclosing its basic concepts and basic hypotheses.

Logical moral 2. Axiomatization does not eliminate controversy, because criticism is of the essence of research, but it can spare useless controversy by focusing on fundamentals, discarding nonconstitutive items (e.g., analogies), and clarifying meanings.

Semantical moral 1. If one wishes to determine the meaning of the key ideas of a theory one must start by recognizing which are those key ideas — i.e. one must axiomatize the theory if only in outline, for the meaning of a theory is sketched by the postulates that interpret the basic symbols of the system.

Semantical moral 2. When a meaning analysis of a logically organized body of physical ideas is carried out, one discovers that they are neither purely mathematical nor do they concern sense experiences but — guess what? — they are strictly physical, i.e. concerning supposedly autonomous entities of some sort or other.

Semantical moral 3. If one wishes to minimize misinterpretation and sterile controversy one should avoid (*a*) introducing supernumerary entities, i.e. things which are not really represented in the formulas, and (*b*) confusing the meaning (interpretation) of a symbol with the operations aiming at testing for the truth of the statements in which the symbol occurs — the more so in the case of high-level theories like CEM and QM, which are tried out in extremely indirect and incomplete ways.

Methodological moral 1. No physical theory is self-sufficient as regards its own test: the test of any such theory requires fragments of a number of other theories, which are sometimes partially incompatible with the theory concerned.

Methodological moral 2. There are three kinds of theory construction principles: (*a*) those which help mold the theory, leave their imprint on it but are not kept in the theory (e.g., covariance principles); (*b*) those which become absorbed by the theory either as axioms or as theorems (e.g., the equivalence "principles"), and (c) those which are wrongly supposed to play any role at all (e.g., MACH's "principle").

Methodological moral 3. No set of heuristic rules, however fertile, determines uniquely a rich theory: heuristic principles prune rather than build. In any case they rarely entail, whereas a theory is a set of formulas partially ordered by the entailment relation.

Philosophical moral 1. Physical research, if deep, poses philosophical problems of many sorts: in ontology — e.g., elucidating the category of time; in epistemology — e.g., clarifying the relations among invariance, degeneracy and objectivity (think of the spherical symmetry of the position density in a central field, where the arbitrary orientation of the polar axis does not show up — but then at the price of degeneracy or ambiguity); in the "logic" of nondeductive inference — e.g., the rejection of measurement results on the strength of a reliable theory; in semantics — e.g., an axiomatic definition of the relation $\triangleq$ of modelling; in the theory of theories — e.g., setting up semantical consistency criteria.

Philosophical moral 2. The philosophical analyses of physical theories are welcome as long as (a) they concern the theories proper not their popularizations, (b) they are carried out with the help of contemporary logical and semantical techniques, and (c) they are performed with an open mind rather than with the aim of illustrating some venerable philosophy.

Historical moral 1. Every partially true theory is as alive as the physicists who care to work it out, apply and analyze it: even classical theories continue to evolve and deserve being overhauled both for their own sake and as models of theory construction.

Historical moral 2. The infancy of a theory is usually so confuse that historical documents and testimonies are of little avail unless accompanied by penetrating analyses of the problem situation and of the theory itself. Very often the inventor himself is not aware of all the heuristic clues he used and of the very character of his creation; he may even have reconstructed the whole story so as to tally with a fashionable philosophy.

Pedagogical moral. If one wants to show how a theory has actually or could have been built then he should try to spot every piece in the original scaffolding. But if one wishes to teach the theory itself then he had better forget its historical beginnings and expound it in a coherent way.

Finally, the moral of this set of morals is: If you want to live happily, learn the tricks of the trade, put out plenty of solid uncontroversial (if possible dull) stuff, and avoid the qualms and pangs that accompany the search for depth — because this search is unending, for physics is bottomless: its foundations are temporary.

Bibliography

AHARONOV, Y., P. G. BERGMANN, and J. L. LEBOWITZ: Time symmetry in the quantum process of measurement. Phys. Rev. **134**, B 1410 (1964).

—, and D. BOHM: Significance of electromagnetic potentials in the quantum theory. Phys. Rev. **115**, 485 (1959).

ALEMBERT, J. D': Traité de dynamique (1743). Paris: Gauthier-Villars 1921. 2 vols.

AMPÈRE, A. M.: Essai sur la philosophie des sciences. Paris: Bachelier 1834, 1843. 2 vols.

ARZELIÈS, H.: Transformation relativiste de la température et de quelques autres grandeurs thermodynamiques. Nuovo cimento **35**, 792 (1965).

BANACH, S.: Mechanics. Warszawa-Wrocław: Monografie matematyczne 1951.

BARGMANN, V., L. MICHEL, and V. L. TELEGDI: Precession of the polarization of particles moving in a homogeneous electromagnetic field. Phys. Rev. Letters **2**, 435 (1959).

BECK, G., and H. M. NUSSENZVEIG: Uncertainty relation and diffraction by a slit. Nuovo cimento **9**, 1068 (1958).

BERGMANN, P. G.: Non-linear field theories. Phys. Rev. **75**, 680 (1949).

— Conservation laws in general relativity as the generators of coordinate transformations. Phys. Rev. **112**, 287 (1958).

— Observables in general relativity. Revs. Modern Phys. **33**, 510 (1961).

— The general theory of relativity. In: S. FLÜGGE (ed.), Handbuch der Physik, Bd. IV. Berlin-Göttingen-Heidelberg: Springer 1962.

—, and J. BRUNINGS: Non-linear field theories. II. Revs. Modern Phys. **21**, 480 (1949).

BETH, E. W.: The foundations of mathematics. Amsterdam: North-Holland 1959.

BIRKHOFF, G.: Dynamical systems. New York: Amer. Math. Soc. 1927.

—, and J. v. NEUMANN: The logic of quantum mechanics. Ann. Math. **37**, 823 (1936).

BLOCHINZEW, D. I.: Grundlagen der Quantenmechanik. Berlin: Deutscher Verlag der Wissenschaften 1953.

BOHM, D.: Quantum theory. Englewood Cliffs (N. J.): Prentice-Hall 1951.

— A suggested interpretation of the quantum theory in terms of 'hidden variables'. Phys. Rev. **85**, 166, 180 (1952).

— Causality and chance in modern physics. London: Routledge & Kegan Paul 1957.

—, and J. BUB: A proposed solution of the measurement problem in quantum mechanics by a hidden variable theory. Revs Modern Phys. **38**, 453 (1966).

BOHR, N.: La théorie atomique et la description des phénomènes. Paris: Gauthier-Villars 1932.

— Kausalität und Komplementarität. Erkenntnis **6**, 293 (1936).

— Atomic physics and human knowledge. New York: John Wiley & Sons 1958.

BOLTZMANN, L.: Populäre Schriften. Leipzig: Johann Ambrosius Barth 1905.

BONDI, H.: Negative mass in general relativity. Revs. Modern Phys. **29**, 423 (1957).

BOPP, F. (Hrsg.): Werner Heisenberg und die Physik unserer Zeit. Braunschweig: F. Vieweg & Sohn 1961.

BORN, M.: Natural philosophy of cause and chance. Oxford: Clarendon Press 1949.

— Physics in my generation. London and New York: Pergamon Press 1956.

— Vorhersagbarkeit in der klassischen Mechanik. Z. Physik **153**, 372 (1958).

BORN, M.: Bemerkungen zur statistischen Deutung der Quantenmechanik. In: Bopp 1961.

BOUWKAMP, C. J.: Diffraction theory. Repts. Progr. in Phys. **17**, 35 (1954).

BRAITHWAITE, R. B.: Scientific explanation. Cambridge: Cambridge University Press 1953.

— Axiomatizing a scientific system by axioms in the form of identifications. In: L. HENKIN, P. SUPPES, and A. TARSKI (eds.), The axiomatic method. Amsterdam: North-Holland Publ. Co. 1959.

BRIDGMAN, P. W.: The logic of modern physics. New York: MacMillan & Co. 1927.

— Einstein's theories and the operational point of view. In: SCHILPP 1949.

— Reflections of a physicist. New York: Philosophical Library 1955.

BROGLIE, L. DE: The current interpretation of wave mechanics: A critical study. Amsterdam-London-New York: Elsevier 1964.

BUNGE, M.: A picture of the electron. Nuovo cimento **13**, 977 (1955a).

— Strife about complementarity. Brit. J. Phil. Sci. **6**, 1, 141 (1955b).

— The philosophy of the space-time approach to the quantum theory. Methodos **7**, 295 (1955c).

— A critique of the frequentist theory of probability. Congresso Internacional de Filosofía, vol. III. São Paulo: Sociedade Brasileira de Filosofía 1956a.

— Survey of the interpretations of quantum mechanics. Am. J. Phys. **24**, 272 (1956b).

— Lagrangian formulation and mechanical interpretation. Am. J. Phys. **25**, 211 (1957).

— Causality: The place of the causal principle in modern science. Cambridge (Mass.): Harvard University Press 1959a; 2nd ed. with Appendix: Cleveland and New York: Meridian Books 1963.

— Metascientific queries. Springfield (Ill.): Ch. C. Thomas 1959b.

— Cinemática del electrón relativista. Tucumán: Universidad Nacional de Tucumán 1960.

— The weight of simplicity in the construction and assaying of scientific theories. Phil. Sci. **28**, 120 (1961a). Repr. in: M. H. FOSTER and M. L. MARTIN (eds.), Probability, confirmation, and simplicity. Readings in the philosophy of inductive logic. New York: Odyssey Press 1966.

— Laws of physical laws. Am. J. Phys. **29**, 518 (1961b).

— Causality, chance, and law. Am. Scientist **49**, 432 (1961c).

— Intuition and science. Englewood Cliffs (N. J.): Prentice-Hall 1962a.

— Causality: A rejoinder. Phil. Sci. **29**, 306 (1962b).

— Cosmology and magic. Monist **47**, 116 (1962c).

— The myth of simplicity. Englewood Cliffs (N. J.): Prentice-Hall 1963a.

— A general black box theory. Phil. Sci. **30**, 346 (1963b).

— Phenomenological theories. In: BUNGE (ed.), The critical approach. In Honor of Karl Popper. New York: The Free Press 1964.

— Physics and reality. Dialectica **19**, 195 (1965).

— Mach's critique of Newtonian mechanics. Am. J. Phys. **34**, 585 (1966).

— (ed.): Studies in the foundations methodology and philosophy of science, vol. 1 Delaware seminar in the foundations of physics. Berlin-Heidelberg-New York: Springer 1967a.

— Studies in the foundations methodology and philosophy of science, vol. 3/I Scientific research I. The search for system and vol. 3/II Scientific research II. The search for truth. Berlin-Heidelberg-New York: Springer 1967b.

BUNGE, M.: The maturation of science. In: I. LAKATOS and A. MUSGRAVE (eds.), Proceedings of the International Colloquium in the Philosophy of Science, London 1965, vol. III. Amsterdam: North-Holland 1967c.
— (ed.:) Quantum theory and reality. Berlin-Heidelberg-New York: Springer 1967d.
CARATHÉODORY, C.: Zur Axiomatik der speziellen Relativitätstheorie. Sitz.ber. preuss. Akad. Wiss. Physik.-math. Kl. **1924**, 12.
CARNAP, R.: Physikalische Begriffsbildung. Karlsruhe: Braun 1926.
— Foundations of logic and mathematics. Chicago: University of Chicago Press 1939.
— Philosophical foundations of physics (M. GARDNER, ed.). New York: Basic Books 1966.
CARTAN, E.: Sur les équations de la gravitation d'Einstein. J. Math. pur. et appl. **1**, 141 (1922).
— Sur les varietés à connexion affine et la théorie de la relativité généralisée. Ann. École Norm. Sup. **40**, 325 (1923); **41**, 1 (1924).
CHEW, G.: The dubious role of the space-time continuum in microscopic physics. Sci. Progr. **51**, 529 (1963).
CHIU, H.-Y., and W. F. HOFFMANN (eds.): Gravitation and relativity. New York: Benjamin 1964.
COLEMAN, B. C.: Thermodynamics of materials with memory. Arch. Rational Mech. Anal. **17**, 1 (1964).
—, H. MARKOVITZ, and W. NOLL: Viscometric flows of non-newtonian fluids. Berlin-Heidelberg, New York: Springer 1966.
CORSON, E. M.: Introduction to tensors, spinors and relativistic wave equations. Cambridge (Mass.): Harvard University Press 1949.
DE GROOT, S. R., and J. VLIEGER: Derivation of Maxwell's equations. Physica **31**, 254 (1965).
DEHNEN, H.: Zur allgemein-relativistischen Dynamik. Ann. Phys. (Lpz.) **13**, 101 (1964).
DE-SHALIT, A., H. FESHBACH, and L. VAN HOVE (eds.): Preludes in theoretical physics. In Honor of V. F. Weisskopf. Amsterdam: North-Holland 1966.
DESTOUCHES-FÉVRIER, P.: La structure des théories physiques. Paris: Presses Universitaires de France 1951.
DIRAC, P. A. M.: Classical theory of radiating electrons. Proc. Roy. Soc. (London) A **167**, 148 (1938).
— The physical interpretation of quantum mechanics. Proc. Roy. Soc. (London) A **180**, 1 (1942).
— The theory of magnetic poles. Phys. Rev. **74**, 817 (1948).
— Generalized Hamiltonian dynamics. Can. J. Math. **2**, 129 (1950).
— The Hamiltonian form of field dynamics. Can. J. Math. **3**, 1 (1951).
— The principles of quantum mechanics, 4th ed. Oxford: Clarendon Press 1958.
DUFFIN, R. J.: Pseudo-Hamiltonian mechanics. Arch. Rational Mech. Anal. **9**, 309 (1962).
DUHEM, P.: La theórie physique, 2nd ed. Paris: Rivière 1914. Engl. transl.: The aim and structure of physical theory. New York: Atheneum 1962.
EDDINGTON, A. S.: The mathematical theory of relativity, 2nd ed. Cambridge: Cambridge University Press 1924.
— The philosophy of physical science. Cambridge: Cambridge University Press 1939.
EDELEN, D. G. B.: The structure of field space. Berkeley and Los Angeles: University of California Press 1962.

EHLERS, J., and W. KUNDT: Exact solutions of the gravitational field equations. In: WITTEN, L. (ed.), Gravitation: An introduction to current research. London and New York: John Wiley & Sons.

EINSTEIN, A.: Einfluß der Schwerkraft auf die Ausbreitung des Lichtes. Ann. Phys. 35, 898 (1911).

— Grundlagen der allgemeinen Relativitätstheorie. Ann. Phys. 49, 769 (1916).

— Prinzipielles zur allgemeinen Relativitätstheorie. Ann. Phys. 55, 241 (1918).

— Äther und Relativitätstheorie. Berlin: Springer 1920.

— Autobiographical notes. In: SCHILPP 1949.

— Out of my later years. New York: Philosophical Library 1950.

EXNER, F.: Vorlesungen über die physikalischen Grundlagen der Naturwissenschaften. Leipzig und Wien: Franz Deuticke 1922.

FERMI, E.: Notes on quantum mechanics. Chicago: The University of Chicago Press 1961.

FOCK, V.: The theory of space, time and gravitation, 2nd ed. Oxford: Pergamon Press 1964.

FORWARD, R. L.: General relativity for the experimentalist. Proc. I.R.E. 49, 892 (1961).

FRANK, P.: Foundations of physics. Chicago: The University of Chicago Press 1946.

GEORGE, A. (ed.): Louis de Broglie, physicien et penseur. Paris: Albin Michel 1953.

GIORGI, G., and A. CABRAS: Questioni relativistiche sulle prove della rotazione terrestre. Rendic. Accad. Naz. Lincei 9, 513 (1919).

GOOD, J. (ed.): The scientist speculates. London: Heinemann 1962.

GRAD, H.: Levels of description in statistical mechanics and thermodynamics. In: BUNGE (ed.) 1967a.

GUENIN, J. DE: Optimum distribution of effort. Operations Research 9, 1 (1961).

HAVAS, P.: The range of application of the lagrange formalism. Nuovo cimento, Suppl. 5, 363 (1957).

— Four-dimensional formulations of Newtonian mechanics and their relation to the special and general theory of relativity. Revs. Modern Phys. 36, 938 (1964).

— Relativity and causality: In: Y. BAR-HILLED (ed.), Logic, methodology and philosophy of science. Amsterdam: North-Holland 1965.

— Foundations problems in general relativity. In: BUNGE (ed.) 1967a.

—, and J. N. GOLDBERG: Lorentz-invariant equations of motion of point masses in the general theory of relativity. Phys. Rev. 128, 398 (1962).

HEISENBERG, W.: Über den anschaulichen Inhalt der quantentheoretischen Kinematik und Mechanik. Z. Physik 43, 172 (1927).

— The physical principles of the quantum theory. Chicago: University of Chicago Press 1930.

— Die beobachtbaren Größen in der Theorie der Elementarteilchen. Z. Physik 120, 513, 673 (1943).

— Physics and philosophy. New York: Harper & Brothers 1958.

— Die Rolle der phänomenologischen Theorien im System der theoretischen Physik. In: DE-SHALIT et al. 1966.

HEITLER, W.: Der Mensch und die naturwissenschaftliche Erkenntnis, 3. Aufl. Braunschweig: F. Vieweg & Sohn 1964.

HELLINGER, E.: Die allgemeinen Ansätze der Mechanik der Kontinua. Enz. math. Wiss. Bd. 4, S. 602. Leipzig: Teubner 1914.

HEMPEL, C. G.: Fundamentals of concept formation in empirical science. Chicago: University of Chicago Press 1952.

— Aspects of scientific explanation. New York: Free Press 1965.

HENKIN, L., P. SUPPES, and A. TARSKI (eds.): The axiomatic method. Amsterdam:
North-Holland 1959.

HERMES, H.: Einführung in die mathematische Logik. Stuttgart: B. G. Teubner
1963.

HERTZ, H.: The principles of mechanics (1894). New York: Dover 1956.

HILBERT, D.: Mathematische Probleme. Arch. Math. Phys. 1, 44, 213 (1901).

— Die Grundlagen der Physik. Nachr. kgl. Ges. Wiss. Göttingen 1915, 395; 1917,
53.

— Axiomatisches Denken. Math. Ann. 78, 405 (1918).

—, u. P. BERNAYS: Grundlagen der Mathematik, 2 Bände. Berlin: Springer 1934
u. 1939.

HOLTON, G.: Sources of EINSTEIN's early work in relativity theory. Proceedings
of the XIth International Congress of the History of Science.

IKEDA, M., and Y. MIYACHI: Coordinate systems in general relativity and geometrical representation for physical quantities. Progr. Theoret. Phys. (Kyoto),
Suppl. 9, 45 (1959).

INFELD, L. (ed.): Proceedings on theory of gravitation. Warszawa: PWN; Paris:
Gauthier-Villars 1964.

—, and A. E. SCHEIDEGGER: Radiation and gravitational equations of motion.
Can. J. Math. 3, 195 (1951).

JÁNOSSY, L.: Theory and practice of the evaluation of measurements. Oxford:
Clarendon Press 1965.

KANT, I.: Kritik der reinen Vernunft (1781, 1787). Hamburg: Meiner 1952.

KEMBLE, E. C.: The fundamental principles of quantum mechanics (1937). New
York: Dover 1958.

KENNEDY, F. J., and E. H. KERNER: Note on the inequivalence of classical and
quantum Hamiltonians. Am. J. Phys. 33, 463 (1965).

KERNER, E. H.: Gibbs ensemble and biological ensemble. Ann. N.Y. Acad. Sci. 96,
975 (1962).

— Dynamical aspects of kinematics. Bull. Math. Biophys. 26, 333 (1964).

KIRCHHOFF, G.: Vorlesungen über mathematische Physik. I. Mechanik. Leipzig:
B. G. Teubner 1883.

KLEIN, F.: Über die geometrische Grundlagen der Lorentzgruppe. Jber. Deutsch.
Mathem. Vereinig. 19, 281 (1910).

KNEALE, W., and M.: The development of logic. Oxford: Clarendon Press 1962.

KOLMOGOROFF, A.: Foundations of the theory of Probability (1933). New York:
Chelsea 1950.

KOOPMAN, B. O.: Hamiltonian systems and transformations in Hilbert space. Proc.
Nat. Acad. Sci. U.S. 17, 315 (1931).

KRETSCHMANN, E.: Über den physikalischen Sinn der Relativitätspostulate. Ann.
Phys. 53, 575 (1917).

LANCZOS, C.: The variational principles of mechanics. Toronto: University of
Toronto Press 1949.

LANDÉ, A.: Zur Quantentheorie der Messung. Z. Physik 153, 389 (1959).

— New foundations of quantum mechanics. Cambridge: Cambridge University
Press 1965.

LAWVERE, F. W.: An elementary theory of the category of sets. Proc. Nat. Acad.
Sci. U.S. 52, 1506 (1964).

LEJEWSKI, C.: A contribution to LEŚNIEWSKI's mereology. Polskie Towarzystwo
Naukowe na Obczyżnie. Yearbook. London 1954—1955.

— Studies in the axiomatic foundations of Boolean algebra. Notre Dame J. of
Formal Logic 1, 23, 91 (1960); 2, 79 (1961).

LEVI-CIVITA, T.: The absolute differential calculus. London and Glasgow: Blackie & Sons 1929.

—, e U. AMALDI: Lezioni di meccanica razionale, vol. II. Bologna: Zanichelli 1927.

LICHNEROWICZ, A.: Théories relativistes de la gravitation et de l'electromagnétisme. Paris: Masson & Cie. 1955.

LIEBES, S.: Gravitational lenses. Phys. Rev. **133**, 835 (1964).

LIPKIN, D. M.: Existence of a new conservation law in electromagnetic theory. J. Math. Phys. **5**, 698 (1964).

LONDON, F., and E. BAUER: La théorie de l'observation en mécanique quantique. Paris: Hermann 1939.

LORENTZ, H. A.: The theory of the electron. Leipzig: Teubner 1909.

LUDWIG, G.: Die Grundlagen der Quantenmechanik. Berlin-Göttingen-Heidelberg: Springer 1954.

— Gelöste und ungelöste Probleme des Meßprozesses in der Quantenmechanik. In: BOPP 1961.

— Versuch einer axiomatischen Grundlegung der Quantenmechanik und allgemeinerer physikalischer Theorien. Z. Physik **181**, 233 (1964).

MACH, E.: The science of mechanics, 4th ed. (1883). La Salle (Ill.) and London: Open Court 1942.

MACKEY, G. W.: Mathematical foundations of quantum mechanics. New York and Amsterdam: Benjamin 1963.

MARGENAU, H.: Quantum-mechanical description. Phys. Rev. **49**, 240 (1936).

— The nature of physical reality. New York: McGraw-Hill Book Co. 1950.

— Measurements in quantum mechanics. Ann. Phys. (N.Y.) **23**, 469 (1963).

— Measurements and quantum states. Phil. Sci. **30**, 1, 138 (1963).

—, and J. PARK: Objectivity in quantum mechanics. In: BUNGE (ed.) 1967a.

MARTIN, J. L.: Generalized dynamics and the "classical analogue" of a Fermi oscillator. Proc. Roy. Soc. (London) A **251**, 536 (1959).

MARTIN, R. M.: A homogeneous system for formal logic. J. Symb. Log. **8**, 1 (1943).

McKINSEY, J. C. C., A. C. SUGAR, and P. SUPPES: Axiomatic foundations of classical particle mechanics. J. Rational Mech. Anal. **2**, 253 (1953).

—, and P. SUPPES: On the notion of invariance in classical mechanics. Brit. J. Phil. Sci. **5**, 290 (1955).

MERZBACHER, E.: Quantum mechanics. New York and London: John Wiley & Sons 1961.

MESSIAH, A.: Quantum mechanics. Amsterdam: North-Holland 1961, 2 vols.

MORGAN, T. A.: Two classes of new conservation laws for the electromagnetic field and for other massless fields. J. Math. Phys. **5**, 1659 (1964).

MOYAL, J. E.: Quantum mechanics as a statistical theory. Proc. Cambridge Phil. Soc. **45**, 99 (1949).

NEUMANN, J. v.: Mathematical foundations of quantum mechanics (1931). Princeton: Princeton University Press 1955.

— Zur Operatorenmethode in der klassischen Mechanik. Ann. Math. **33**, 587 (1932).

NOETHER, E.: Invariante Variationsprobleme. Nachr. kgl. Ges. Wiss. Göttingen **1918**, 235.

NISHIYAMA, Y.: The relation between classical mechanics and wave mechanics. Progr. Theoret. Phys. (Kyoto) **30**, 657 (1963).

NOLL, W.: A mathematical theory of the mechanical behavior of continuous media. Arch. Rational Mech. Anal. **2**, 197 (1958).

— The foundations of classical mechanics. In: HENKIN et al. (eds.) 1959.

NOLL, W.: La mécanique classique, basée sur un axiome d'objectivité. In: J.-L. DESTOUCHES and F. AESCHLIMANN (eds.): La méthode axiomatique dans les mécaniques classiques et nouvelles. Paris: Gauthier-Villars 1963.
— Space-time structures in classical mechanics. In: BUNGE (ed.) 1967a.
NYBORG, P.: On classical theories of spinning particles. Nuovo cimento 23, 1057 (1962).
— Macroscopic motion of classical spinning particles. Nuovo cimento 26, 821 (1962).
OSTWALD, W.: Vorlesungen über Naturphilosophie. Leipzig: Veit & Co. 1902.
OTT, H.: Lorentz-Transformation der Wärme und der Temperatur. Z. Physik 175, 70 (1963).
PAULI, W.: Aufsätze und Vorträge über Physik und Erkenntnistheorie. Braunschweig: F. Vieweg & Sohn 1961.
PETROW, A. S.: Einstein-Räume. Berlin: Akademie-Verlag 1964.
PHILIPS, T. O.: Lorentz invariant localized states. Phys. Rev. 136, B 893 (1964).
PHILLIPS, M.: Classical electrodynamics. In: S. FLÜGGE (ed.): Handbuch der Physik, Bd. IV. Berlin-Göttingen-Heidelberg: Springer 1962.
PLANCK, M.: Die Einheit des physikalischen Weltbildes. Physik. Z. 10, 62 (1909).
— Theory of electricity and magnetism. London: McMillan & Co. 1932.
— Where is science going? London: Allen & Unwin 1933.
POLYA, G.: Mathematics and plausible reasoning. Princeton: Princeton University Press 1954, 2 vols.
POPPER, K. R.: The logic of scientific discovery (1935), 2nd ed. London: Hutchinson; New York: Basic Books 1959.
— Probability magic or knowledge out of ignorance. Dialectica 11, 354 (1957a).
— The propensity interpretation of the calculus of probability and the quantum theory. In: S. KÖRNER (ed.): Observation and interpretation. London: Butterworths Sci. Publ. 1957b.
— The propensity interpretation of probability. Brit. J. Phil. Sci. 10, 25 (1959).
— Conjectures and refutations: The growth of scientific knowledge. London: Routledge & Kegan Paul; New York: Basic Books 1963a.
— Creative and non-creative definitions in the calculus of probability. Synthèse 15, 167 (1963b).
POST, E. J.: Formal structure of electromagnetics. Amsterdam: North-Holland 1962.
— On the physical necessity for general covariance in electromagnetic theory. In: BUNGE (ed.) 1967a.
PRZIBRAM, K. (ed.): Briefe zur Wellenmechanik. Wien: Springer 1963.
PROCA, A.: Sur un nouveau type d'électron. Portugaliae Phys. 1, 59 (1943).
QUINE, W. V.: Mathematical logic (1951). New York: Harpers 1962.
RAFANELLI, K., and R. SCHILLER: Classical motions of spin-$\frac{1}{2}$ particles. Phys. Rev. 135, B 279 (1964).
RANKIN, B.: Quantum mechanical time. J. Math. Phys. 6, 1057 (1965).
REICHENBACH, H.: Axiomatik der Raum-Zeit-Lehre. Braunschweig: F. Vieweg & Sohn 1924.
— The philosophy of space and time (1927). New York: Dover 1957.
— Philosophical foundations of quantum mechanics. Berkeley and Los Angeles: University of California Press 1944.
— The theory of probability. Berkeley and Los Angeles: University of California Press 1949.
— The rise of scientific philosophy. Berkeley and Los Angeles: University of California Press 1951.
RENNINGER, M.: Messung ohne Störung des Meßobjekts. Z. Physik 158, 417 (1960).

ROBINSON, A.: Introduction to model theory and to the metamathematics of algebra. Amsterdam: North-Holland 1963.
ROHRLICH, F.: The principle of equivalence. Ann. Phys. (N.Y.) **22**, 169 (1963).
— Classical charged particles. Readings (Mass.): Addison-Wesley 1965.
ROLL, P. G., R. KROTKOV, and R. H. DICKE: The equivalence of inertial and gravitational mass. Ann. Phys. (N.Y.) **26**, 442 (1964).
ROSENFELD, L.: Strife about complementarity. Sci. Progr. No. 163, 393 (1953).
— Foundations of quantum theory and complementarity. Nature **190**, 384 (1961).
RUSSELL, B.: Our knowledge of the external world (1914). London: Allen & Unwin 1952.
SADEH, D.: Experimental evidence for the constancy of the velocity of gamma rays, using annihilation in flight. Phys. Rev. Letters **10**, 271 (1963).
SCHEIBE, E.: Die kontingenten Aussagen in der Physik. Frankfurt a.M.: Athenäum 1964.
SCHILLER, R.: Relations of quantum to classical physics. In: BUNGE (ed.) 1967a.
SCHILPP, P. A. (ed.): Albert Einstein philosopher-scientist. Evanston (Ill.): The Library of Living Philosophers 1949.
SCHLICK, M.: Sur le fondement de la connaissance. Paris: Hermann 1935.
SCHÖNBERG, M.: Classical theory of the point electron. Phys. Rev. **69**, 211 (1946).
SCHRÖDINGER, E.: Space-time structure. Cambridge: Cambridge University Press 1954.
SEGAL, I. E.: Postulates for general quantum mechanics. Ann. Math. **48**, 930 (1947).
SHANKLAND, R. S.: Conversations with Einstein. Am. J. Phys. **31**, 47 (1963).
— Michelson-Moreley experiment. Am. J. Phys. **32**, 16 (1964).
SHAPIRO, I. I.: Fourth test of general relativity. Phys. Rev. Letters **13**, 789 (1964).
SHELKUNOFF, S. A.: Theory of antennas of arbitrary size and shape. Proc. I.R.E. **29**, 493 (1941).
SMOLUCHOWSKI, M. v.: Über den Begriff des Zufalls und den Ursprung der Wahrscheinlichkeitsgesetze in der Physik. Naturwissenschaften **6**, 253 (1918).
SOBOCINSKI, B.: Studies in LEŚNIEWSKI's mereology. Polskie Towarzystwo Naukowe na Obczyźnie. Yearbook. London 1954—1955.
STOLL, R. R.: Introduction to set theory and logic. San Francisco and London: W. H. Freeman 1961.
STROCCHI, F.: Complex coordinates and quantum mechanics. Revs. Modern Phys. **38**, 36 (1966).
SUPPES, P.: Introduction to logic. Princeton (N. J.): Van Nostrand 1957.
— Probability concepts in quantum mechanics. Phil. Sci. **28**, 378 (1961).
— The role of probability in quantum mechanics. In: B. BAUMRIN (ed.): Philosophy of science. The Delaware Seminar, vol. 2. New York: John Wiley & Sons 1963.
SUTCLIFFE, W. G.: Lorentz transformations of thermodynamic quantities. Nuovo cimento **39**, 683 (1965).
SYNGE, J. L.: Relativity: The general theory. Amsterdam: North-Holland 1960.
TARSKI, A.: Introduction to logic. New York: Oxford University Press 1941.
— The semantic conception of truth and the foundations of semantics. Phil. and Phenomenol. Res. **4**, 341 (1944).
— Logic, semantics, metamathematics. Oxford: Clarendon Press 1956.
— What is elementary geometry? In: HENKIN et al. (eds.) 1959.
—, A. MOSTOWSKI, and R. M. ROBINSON: Undecidable theories. Amsterdam: North-Holland 1953.
THOMSON, G.: Research in theory and practice. Contemp. Phys. **5**, 103 (1963).

TOLMAN, R. C.: Relativity, thermodynamics and cosmology. Oxford: Clarendon Press 1934.

TRAUTMAN, A.: Conservation laws in general relativity. In: WITTEN (ed.) 1962.

— Foundations and current problems of general relativity. In: A. TRAUTMAN, F. A. E. PIRANI, and H. BONDI: Lectures on general relativity. Brandeis Summer Institute in Theoretical Physics, vol. I. Englewood Cliffs (N. J.): Prentice-Hall 1965.

TRUESDELL, C.: The mechanical foundations of elasticity and fluid dynamics. J. Rational Mech. Anal. **1**, 125 (1952); **2**, 593 (1953).

— Principles of continuum mechanics. Dallas: Socony Mobil Oil Co. 1961.

— Six lectures on modern natural philosophy. New York: Springer 1966.

—, and W. NOLL: The non-linear field theories of mechanics. In: S. FLÜGGE (ed.): Handbuch der Physik, Bd. III/3. Berlin-Heidelberg-New York: Springer 1965. and R. TOUPIN: The classical field theories. In: S. FLÜGGE (ed.): Handbuch der Physik, Bd. III/1. Berlin-Göttingen-Heidelberg: Springer 1960.

URBANIK, K.: Joint probability distributions of observables in quantum mechanics. Studia Math. **21**, 117 (1961).

WAKITA, H.: Measurement in quantum mechanics: II. Progr. Theor. Phys. (Kyoto) **27**, 139 (1962).

WEBER, J.: General relativity and gravitational waves. New York and London: Interscience Publ. 1961.

WHEELER, J. A.: Geometrodynamics. New York and London: Academic Press 1962.

—, and R. P. FEYNMAN: Classical electrodynamics in terms of direct interparticle action. Revs. Modern Phys. **21**, 425 (1949).

WHITEHEAD, A. N.: An enquiry concerning the principles of natural knowledge. Cambridge University Press 1919.

WIGNER, E. P.: Die Messung quantenmechanischer Operatoren. Z. Physik **133**, 101 (1952).

— Relativistic invariance and quantum phenomena. Revs. Modern Phys. **29**, 255 (1957).

— Remarks on the mind-body question. In: GOOD (ed.) 1962.

— The problem of measurement. Am. J. Phys. **31**, 6 (1963).

WITTEN, L. (ed.): Gravitation: An introduction to current research. London and New York: John Wiley & Sons 1962.

WOODGER, J.: The axiomatic method in biology. Cambridge: Cambridge University Press 1937.

ZANSTRA, H.: A study of relative motion in connection with classical mechanics. Phys. Rev. **23**, 528 (1924).

ZEEMAN, E. C.: Causality implies the Lorentz group. J. Math. Phys. **5**, 490 (1964).

Subject Index

SPRINGER-VERLAG
BERLIN·HEIDELBERG·NEW YORK

Springer Tracts in Natural Philosophy